Liquids and Liquid Mixtures

Butterworths Monographs in Chemistry

Butterworths Monographs in Chemistry is a series of occasional texts by internationally acknowledged specialists, providing authoritative treatment of topics of current significance in chemistry and chemical engineering

Published titles

Kinetics and Dynamics of Elementary Gas Reactions
Prostaglandins and Thromboxanes
Silicon in Organic Synthesis
Solvent Extraction in Flame Spectroscopic Analysis

Forthcoming titles

Compleximetric Titration
Coordination Catalysis in Organic Chemistry
Strategy in Organic Synthesis

Butterworths Monographs in Chemistry

Liquids and Liquid Mixtures

J S Rowlinson
Dr Lee's Professor of Chemistry
University of Oxford

F L Swinton
Professor of Chemistry
New University of Ulster

Third edition

Butterworth Scientific
London · Boston · Sydney · Wellington · Durban · Toronto

First published 1982

© Butterworth & Co (Publishers) Ltd 1982

British Library Cataloguing in Publication Data

Rowlinson, J. S.
    Liquids and liquid mixtures.—3rd ed.—
(Butterworths monographs in chemistry
and chemical engineering)
1. Liquids
I. Title   II. Swinton, F. L.
530.4'2   QC145.2

ISBN 0–408–24192–6
0–408–24193–4 Pbk

Filmset by Northumberland Press Ltd, Gateshead, Tyne and Wear
Printed and bound in Great Britain
by Page Bros (Norwich) Ltd

Preface to the Third Edition

The aim of this book remains unchanged: to give an account of the equilibrium properties of liquids and liquid mixtures, and to relate these to the properties of the constituent molecules by the methods of statistical thermodynamics.

Much has happened in the twelve years since the publication of the second edition. The properties of pure and mixed fluids near their critical points have now been thoroughly studied and methods of describing such states thermodynamically have been worked out. Tricritical points have received much attention. Measurements of the excess properties of mixtures have grown beyond number, and, in some cases, are now of an accuracy that was not even attempted in the 1960s. The study of liquid mixtures at high pressures has not shown the same quantitative increase, although the accuracy of a higher proportion of the work than hitherto now approaches that of the leading workers.

The interpretation of this material has been put on a much better footing by the use of the methods of computer simulation and by the developments of perturbation theories of liquids and mixtures. Both techniques were in their infancy in 1969. The use of simulated data for testing theories has eliminated the dangerous step of guessing at the intermolecular forces. These are now known accurately for the inert gases and some of their mixtures, but it is our ignorance of their form and strength that looks like being the main bar to progress in the interpretation of the properties of less simple systems. If we knew the forces more accurately, it now seems that there would be no real difficulties in the way of a full statistical interpretation of the thermodynamic properties.

We thank many friends and colleagues for useful comments and for sending us work before publication. We are particularly indebted to S. Angus, J. C. G. Calado, M. L. McGlashan, I. A. McLure, R. L. Scott, L. A. K. Staveley, W. B. Streett, D. J. Tildesley and C. J. Wormald. A grant from the Leverhulme Trust greatly assisted the co-operation of two authors who normally work on different islands. Finally, we thank Miss H. McCollum and Miss W. E. Nelson for typing the manuscript.

J.S.R.
F.L.S.

June 1981

Contents

Chapter 1

Introduction

1.1 The liquid state

Everyone can recognize a liquid. It is popularly defined as the state of matter which, when placed in a closed vessel, conforms to the shape of the vessel but does not necessarily fill the whole of its volume. The first property distinguishes it from a solid and the second from a gas. Although this simple definition is adequate for many purposes, it does not go very deeply into the relationship between the three principal states of matter, solid, liquid and gas.

When a crystalline solid is heated, it changes to either a liquid or a gas at a temperature which is a function of the applied pressure. At this melting, or sublimation, temperature one of the fluid states and the solid state can exist in equilibrium. Only at a fixed temperature and pressure, the triple point, can the three phases of a pure substance, solid, liquid and gas, remain in mutual equilibrium (*Figure 1.1*). This restriction is a consequence of Gibbs's phase rule, which requires that the sum of the number of degrees of freedom and the number of phases shall be three for a pure substance. Thus, in a two-dimensional phase diagram of pressure as a function of temperature (*Figure 1.1*(b)), a single phase is represented by an area, two coexistent phases by a line and three by the intersection of three lines at a point.

Figure 1.1(a) represents the volume of a given amount of a substance, say one mole, as a function of pressure and temperature. If this amount is placed in a vessel of the pressure, volume and temperature represented by the point x, then the vessel will contain liquid characterized by the point y and gas (or vapour) characterized by the point z. That is, the liquid and gas have, respectively, the pressure, temperature and volume per mole of these points. Since the overall molar volume is that of the vessel, x, it follows that the ratio of the masses of the two phases (m_y/m_z) is the ratio of the lengths (xz/xy).

A line that joins two coexistent phases in such a way that the position of a point on the line is a measure of the relative amounts of the phases is called a *tie-line*. Thus all points on the line B_2C are liquid states for they are states of the fluid in equilibrium with infinitesimal amounts of gas, and all points on B_3C represent vapour in equilibrium with infinitesimal amounts

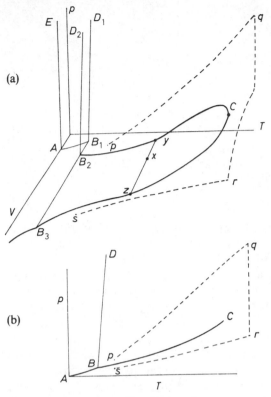

Figure 1.1 (a) *The* (p, V, T) *surface for a given mass of a simple substance, and* (b) *the* (p, T) *projection of this surface*
A volume of the solid at absolute zero
AB_1 *vapour-pressure curve of the solid*
B_1, B_2 *and* B_3 *volumes of the solid, liquid and gas at the triple point*
BD and BC melting curve of the solid and vapour-pressure curve of the liquid; the letter ends at the critical point C

of liquid. B_2C and B_3C are thus the locus of *bubble points* and *dew points* respectively. The line B_2C is also often called the *orthobaric liquid curve* or the *saturation curve* and this latter term can be conveniently applied both to B_2C (a saturated liquid) and to B_3C (a saturated gas). All points on the curved surface to the right of B_2D_2 represent the fluid state, which cannot be divided into liquid and gas in any but an arbitrary way except for points along the continuous curve B_2CB_3. Thus a fluid at point p, which might be said to be 'obviously' a liquid, can be changed to point s, 'obviously' a gas, first by heating at constant volume to q, expanding at constant temperature to r, and then cooling at constant volume to s. At no point in this three-stage transformation has a change of phase occurred, and no dividing meniscus, such as that which separates y and z, would have been observed.

There is, in fact, no qualitative physical test that could distinguish fluid at point p from that at point s without making a change of pressure, volume or temperature that brings the fluid into the two-phase region of B_2CB_3. The two fluids differ only in degree. At higher temperatures and pressures and particularly in the vicinity of C, the *gas–liquid critical point*, any attempt to divide the fluid state must become even more artificial. It would, however, be pedantic to describe all states except B_2C and B_3C as fluids every time they are mentioned, and so in this book the word liquid is used freely for fluids at or near the line B_2C and the word gas for fluids of low density, in contexts where the meaning is clear.

Over the past few decades much evidence has accumulated to show that the fluid states, gas and liquid, possess many structural similarities and that both are quite distinct from the crystalline solid state of the same substance. Some of this evidence has resulted from the computer simulation of assemblies of particles that interact with particularly simple intermolecular potentials, but most of it is a consequence of the direct experimental investigation into the structure of liquids and highly-compressed gases using x-ray and neutron diffraction[1].

The diffraction pattern of a liquid is very similar to that of the same substance in the gaseous form when it is above the critical temperature and compressed to a comparable density, and both are totally different to that of the crystalline solid. Diffraction studies on compressed gases are difficult to carry out and the majority of such measurements have been performed only over the past two decades. Previously, because dilute gases possess no x-ray diffraction patterns apart from those due to internal molecular structure, diffraction studies were used as evidence for the contrary postulate, the similarity of the solid and liquid states. During the same period this misconception was compounded by the application to the liquid state of statistical theories based on the cell model[2] in which the individual molecules are imagined to be confined, and to move, within cells composed of their nearest neighbours. Such a model is solid-like in nature and imposes too great a degree of structure on the fluid. When applied to more appropriate systems such as clathrate compounds that actually do possess a cage-like configuration, such theories are extremely successful[3], but their extensive use in the field of liquids and particularly liquid mixtures has been singularly unhelpful.

In the neighbourhood of the triple point the change from the solid state to the fluid or from the liquid to the gas is a sharp one, at which the characteristic equilibrium properties of the substance change discontinuously at a so-called first-order transition. The changes for the molar entropy, heat capacity and volume of argon are shown in *Figure 1.2* and other properties such as the coefficients of thermal expansion and of compressibility, refractive index and dielectric constant show similar behaviour. The relative values of the three thermodynamic properties shown in this figure are typical of almost all substances, but a few, of which water and the elements antimony, bismuth, gallium and germanium are the most familiar, are atypical and contract on melting so that $V(s) > V(l)$.

For all substances in the region of the triple point, the change from solid

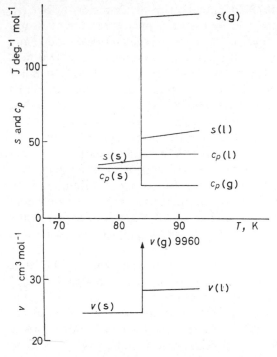

Figure 1.2 Changes of molar entropy (s), heat capacity (c_p) and volume (v) at the triple point of argon. The entropy is measured from an arbitrary zero

to liquid is less drastic than that from liquid to gas. However, at high pressures the situation is very different, for whereas the solid–fluid discontinuities are but little affected by pressure, as is shown by the near-vertical slope of *BD* in *Figure 1.1*, the size of the liquid–gas discontinuities decreases, slowly at first, but then with increasing rapidity as the two states become indistinguishable at the critical point *C*. A fascinating account of the early experimental attempts to establish the unity, or continuity, of the fluid states has been given recently by Levelt Sengers[4].

The continuity of the two fluid states contrasts sharply with the apparent lack of continuity between the fluid and the solid state. It is, of course, impossible to prove that the lines B_1D_1 and B_2D_2 of *Figure 1.1*(a) never meet at a critical point but the solid–fluid transition has been followed experimentally to, for example, 7400 bar and 50 K for helium and to 12 000 bar and 330 K for argon without changing its character in any way[5]. Computer simulation of model systems under extreme conditions also fails to give evidence for a solid–fluid critical point.

The temperatures of the triple points of substances range from 14 K for hydrogen to temperatures too high for accurate measurement for substances such as diamond[6]. Triple-point pressures are never very high, that of carbon dioxide being one of the highest known at 5.2 bar. Higher pressures have been reported only for the elements carbon, phosphorus and arsenic, for which the triple points are not well established. Most pressures are of the order of 10^{-3} bar and a few, such as that of n-pentane, are as low as 10^{-7} bar. The triple point differs slightly from the normal melting point in the presence of air at atmospheric pressure, both because of the finite slope of B_2D_2 in *Figure 1.1*(a), which represents the change of the melting point with pressure, and also because of the solubility of air in the liquid. The former effect is usually the larger and for water lowers the melting temperature by 0.0075 K whereas the air solubility depresses the melting temperature by a further 0.0025 K, a total of 0.010 K. The triple point is more reproducible than the normal melting point and that of water, 273.1600 K, is now used as the single thermometric fixed point that defines the kelvin[7].

Gas liquid critical temperatures also extend over an enormous range[8], starting with 5.2 K for helium and rising to 2079 K for caesium, the highest experimentally-determined value and to an estimated 23 000 K for tungsten[9]. Critical pressures are more uniform, with most substances possessing values around 50 bar but with those of helium and hydrogen being unusually low at 2.3 bar and 13 bar, that of water being unusually high for a non-metallic substance at 221 bar, and the experimentally-determined value for mercury extremely high at 1510 bar.

The *normal liquid range* of a substance is sometimes defined as the temperature interval between the normal melting point and the normal boiling point, the latter being the temperature at which the vapour pressure is one atmosphere. This is an artificial definition which has little to recommend it. As may be seen from the large range of triple-point and critical pressures quoted above, a pressure of one atmosphere has no fundamental significance and refers to quite different relative conditions for different substances. In this book no undue emphasis is placed on that part of the liquid state which happens to lie below one atmosphere pressure.

The triple or melting point of a substance is much more susceptible to small variations in the symmetry of the intermolecular potential than is the critical point. For example, the isomeric hydrocarbons n-octane and 2,2,3,3-tetramethylbutane have the very comparable critical temperatures of, respectively, 568.8 and 567.8 K but their melting points of 216.4 and 373.9 K are quite different. Often highly-symmetric, pseudo-spherical molecules such as adamantane (melting point 541.2 K) have unusually high melting temperatures and, therefore, small liquid ranges, and many such substances (e.g., hexachloroethane, melting point 460.0 K) sublime directly at atmospheric pressure.

Such materials undergo a first-order transition in the solid state from a normal crystalline solid to a so-called *plastic crystal* or *rotator-phase solid* stable between the transition temperature and the melting point[10]. In the plastic crystalline phase the molecules possess a considerable degree of rotational motion although full three-dimensional translational order is maintained under normal pressures.

The wide variation in the properties of liquids makes it useful to attempt a qualitative classification of liquids before discussing any of their properties in detail. A satisfactory classification can be made only in terms of the intermolecular forces, for it is these forces that are the sole determinants of all the physical properties discussed in this book. At low temperatures, the strength and symmetry of these forces determine the properties of the crystal. In the fluid states at higher temperatures, the symmetry becomes less important.

The equilibrium properties of a substance, whether it is solid or fluid, are the result of a balance between the cohesive or potential energy on the one hand, and the kinetic energy of the thermal motions on the other. The translational kinetic energy is the same for all molecules at a given temperature—the classical principle of the equipartition of energy—and so it is solely the differences in the strength and types of the intermolecular energies that cause the properties of one substance to differ from those of another at this fixed temperature. If liquids are compared, not at the same temperature, but at the same fraction of, say, their critical temperatures, then it is found that it is differences in type and symmetry of the intermolecular energies that are responsible for the differences. The following classes of liquids may be distinguished, in increasing order of complexity:

(1) *The noble gases*   The molecules are monatomic and so, in isolation, spherically symmetrical. The force between a pair of molecules is entirely central; that is, the force acts through the centres of gravity and neither molecule exerts a torque on the other.

(2) *Homonuclear diatomic molecules such as hydrogen, nitrogen, oxygen and the halogens, and heteronuclear diatomic molecules with negligible dipole moments, such as carbon monoxide*   The internuclear distances are here much smaller than the mean separation of the molecules in the liquid, and so the intermolecular forces do not depart greatly from spherical symmetry.

(3) *The lower hydrocarbons and other simple non-polar substances such as carbon tetrachloride*   The lower hydrocarbons are comparable in symmetry with diatomic molecules, but as the number of carbon atoms increases, the maximum internuclear distance within one molecule ceases to be small when compared with the mean intermolecular separation. That is, the forces depart greatly from spherical symmetry.

(4) *Simple polar substances such as sulphur dioxide and the hydrogen and methyl halides*   These molecules have a direct electrostatic interaction between permanent dipole moments superimposed on a force field which is otherwise reasonably symmetrical. Molecules with large quadrupole moments, such as carbon dioxide and acetylene (ethyne), are few in number but must be included in this class.

(5) *Molecules of great polarity or electrical asymmetry*   Water, hydrogen fluoride and ammonia are in this class. The polarity is often localized in one part of the molecule, as in organic alcohols, amines, ketones and nitriles.

These divisions are rather arbitrary—carbon disulphide, for example, possesses properties that could place it in either class 3 or 4—but do place in a rough order of increasing complexity many of the substances discussed in this book. As far as is possible, examples will be chosen from the earlier substances in this list, since these illustrate more clearly many of the essential properties of the liquid state without the complications introduced by the polarity and shape of the larger molecules. These complications are discussed explicitly in some sections as, for example, in the case of the mixtures of substances from class 5 above which are discussed in Chapter 5, but where the property under discussion is exhibited by the simple as well as by the complex molecules, then the former have been chosen as examples.

There are several important omissions from the above list of liquids. Because this book is concerned solely with non-electrolytes, molten salts and solutions of ionic species are excluded entirely as are liquid metals. Certain types of non-electrolyte are also omitted. The first is helium, which must be excluded from the first class because of the peculiarity of its liquid structure; the second is that class of substances, many of high molecular weight, which form glasses on cooling; and the third is the large class of organic molecules with highly-anisotropic intermolecular potentials that form an ordered fluid phase on cooling, so-called *liquid crystals*. None is discussed here in detail but it may be useful to indicate briefly how these three classes of liquid depart from normality.

The phase diagram of helium is shown in *Figure 1.3*. There is no triple point at which gas, liquid and solid are in equilibrium, since the solid is stable only at pressures above 24 bar. There is, however, a second liquid phase, of most unusual dynamic properties, which is formed on cooling the normal liquid in equilibrium with the vapour[11].

This behaviour is partly a consequence of the fact that ^{4}He obeys Bose–Einstein statistics, but is mainly due to the light mass and weak intermolecular forces. The mass (and moments of inertia) of a molecule are not determinants of the equilibrium properties of fluids that obey the laws of classical mechanics—a consequence of the equipartition of energy—but are relevant to the quantal description of the system. The importance of the mass is measured by the size of the dimensionless quantity $(h^2/m\sigma^2\varepsilon)^{1/2}$, called the de Broglie wavelength, where h is Planck's constant, m is the molecular mass, and σ and ε are a length and an energy proportional to the size of the molecule and to the greatest energy of interaction between a pair of molecules[12].

The departure from classical behaviour is large for helium since m, σ and ε are all small and the value of the de Broglie wavelength is 2.67. Hydrogen, deuterium and neon show much smaller departures, usually termed quantum corrections, and have de Broglie wavelengths of 1.73, 1.22 and 0.59, respectively. All other liquids are adequately described by classical theory.

A simple explanation of this quantum effect is that, in the condensed state, the low mass leads to a high frequency of vibration and so to a large

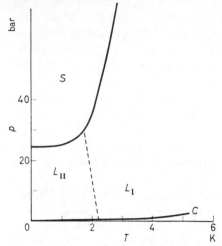

Figure 1.3 The phase diagram of normal helium. The melting point of the solid is at 24 bar or above. The liquid state is divided into two liquids L_I and L_{II} by a second-order transition which is shown by the dashed line. C is the critical point. Compare this diagram with Figure 1.1(b)

zero-point energy. At zero pressure and temperature, the cohesive forces of helium are not strong enough to restrain the molecules within a solid lattice. Only when an additional restraint of 24 bar external pressure is applied does the liquid crystallize to a normal solid.

The *glassy state*[13] is rarely met in simple substances but is common with polymers and with substances such as polyalcohols and sugars. Their liquids can be readily cooled below the normal or true melting temperature, their viscosity increases dramatically and, eventually, a hard glass is formed. The volume and heat content of a substance in the solid, liquid and glassy states are shown in *Figure 1.4*. Such behaviour is typical of a material

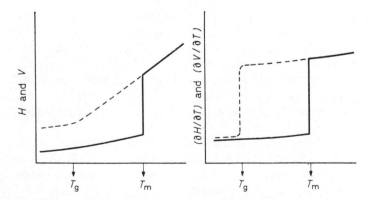

Figure 1.4 The glassy state: H and V, and their derivatives, change on passing from the liquid state (above T_m) to the solid state (full line), to the supercooled liquid (dashed line) and, below T_g, to the glassy state (dashed line)

undergoing a second-order transition where the heat capacity and thermal expansion coefficients (second-order differentials of the Gibbs function with respect to temperature and to pressure and temperature) show discontinuities at the glass temperature.

The glassy state is metastable with respect to the crystalline but it is usually sufficiently reproducible for the meaningful measurement of thermodynamic properties. The actual value of the glass temperature is dependent on the time-scale of the experiment used in its determination, the shorter the time-scale, the higher the transition temperature.

A glass has all the properties of a liquid except that of obvious mobility. It is isotropic, has no ordered crystalline structure, no cleavage planes, flows slowly under stress, and has no sharp melting point but softens gradually to a normal liquid without the absorption of an enthalpy of fusion. Many polymers, such as polytetrafluoroethane, exhibit more than one glass transition. These occur as different portions of the polymer chain become frozen into disordered, amorphous regions at different temperatures as the temperature is reduced.

Molecules that form *liquid crystals*, often called *mesophases* to emphasize the fact that the state is intermediate between that of the crystal and the liquid, are usually either rod-like or plate-like in shape and frequently possess large permanent dipole moments that assist in the ordering process[14]. In a normal crystal there is complete translational order in all three dimensions, whereas in a liquid crystal there is either two or one-dimensional order. The transitions between crystal, liquid crystal and isotropic liquid and between two different liquid crystalline states of the same substance are usually first order, although the individual enthalpies and volumes of transition are much smaller than the comparable values for the direct crystal-to-liquid fusion process.

There are three distinct types of liquid crystal. *Nematic* (thread-like) liquid crystals are ordered in only one dimension with rod-like molecules pointing in the same direction. Unlike a normal crystal this semi-ordered state does not extend over long distances but *swarms* of about 10^5 molecules exist, within which there is near-perfect alignment, but there is usually little correlation between the relative orientations of neighbouring swarms.

Smectic (soap-like) liquid crystals possess two-dimensional order with a layered structure but with no correlation in orientation between adjacent layers. Frequently a substance can exist in more than one smectic phase, the phases differing in the alignment of the individual molecules within each layer. Often, on heating, a smectic mesophase will change to a nematic mesophase before melting completely to the isotropic liquid at a yet higher temperature.

The third type of liquid crystal, the *cholesteric* mesophase, is formed from plate-like molecules that are invariably derivatives of cholesterol. These also form a layered structure with the molecules aligned in one dimension only in each layer and the direction of alignment changing regularly in a helical manner within a stack of adjacent layers. The individual molecules of substances that form cholesteric mesophases are always optically active and the regular helical arrangement within the cholesteric mesophase enhances the optical activity by many orders of magnitude. Another characteristic is the range of bright colours exhibited when a cholesteric liquid crystal is viewed by reflected light.

References

1 PINGS, C. J., in *The Physics of Simple Liquids*, Eds. Temperley, H. N. V., Rowlinson, J. S., and Rushbrooke, G. S., Chap. 10, North-Holland, Amsterdam (1968); KARNICKY, J. F., and PINGS, C. J., *Adv. chem. Phys.*, **34**, 157 (1976); EGELSTAFF, P. A., LITCHINSKY, D., McPHERSON, R., and HAHN, L., *Molec. Phys.*, **36**, 445 (1978); SULLIVAN, J. D., and EGELSTAFF, P. A., *Molec. Phys.*, **39**, 329 (1980)
2 PRIGOGINE, I., *The Molecular Theory of Solutions*, North-Holland, Amsterdam (1957)
3 VAN DER WAALS, J. H., *Trans. Faraday Soc.*, **52**, 184 (1956); CHILD, W. C., *Q. Rev. chem. Soc.*, **18**, 321 (1964)
4 LEVELT SENGERS, J. M. H., *Physica*, **98A**, 363 (1979)
5 HOLLAND, F. A., HUGGILL, J. A. W., and JONES, G. O., *Proc. R. Soc.*, **A207**, 268 (1951); ROBINSON, D. W., *Proc. R. Soc.*, **A225**, 393 (1954); BABB, S. E., *Rev. mod. Phys.*, **35**, 400 (1963)
6 ZERNIKE, J., *Chemical Phase Theory*, 23, Kluwer, Antwerp (1956); BUNDY, F. P., *J. chem. Phys.*, **38**, 631 (1963)
7 PAGE, C. H., and VIGOUREUX, P. (Eds.), *The International System of Units*, HMSO, London (1977)
8 AMBROSE, D., *Vapour–Liquid Critical Properties, NPL Report*, Chem 107, National Physical Laboratory, Teddington (1980)
9 FRANCK, E. U., *Rev. mod. Phys.*, **35**, 400 (1963); **40**, 697 (1968)
10 KLEIN, M. L., and VENABLES, J. A. (Eds.), *Rare Gas Solids*, Academic Press, London (1976); SHERWOOD, J. N. (Ed.), *The Plastically Crystalline State*, Wiley, Chichester (1979)
11 MENDELSSOHN, K., in *Handbuch der Physik*, Ed. Flügge, S., Vol. 15, 370–416, Springer, Berlin (1956)
12 HIRSCHFELDER, J. O., CURTISS, C. F., and BIRD, R. B., *Molecular Theory of Gases and Liquids*, Wiley, New York (1954)
13 MACKENZIE, J. D., (Ed.), *Modern Aspects of the Vitreous State*, Vol. 1–3, Butterworths, London (1961–64); *The Vitreous State, Discuss. Faraday Soc.*, **50**, (1970)
14 GRAY, G. W., *Molecular Structure and the Properties of Liquid Crystals*, Academic Press, London (1962); JOHNSON, J. F., and PORTER, R. S. (Eds.), *Liquid Crystals and Ordered Fluids*, Plenum, New York (1970); DE GENNES, P. G., *The Physics of Liquid Crystals*, University Press, Oxford (1975)

Chapter 2
The thermodynamic properties

2.1 Summary of thermodynamic relations

The more important of the thermodynamic equations used in this book are set out in this section but they are not in the order in which they occur in formal textbooks of thermodynamics, as this section is in no way intended as a substitute for such books[a]. Many of these equations are well known, but those describing the change of thermodynamic functions along the saturation curve of a liquid are derived and discussed in greater detail than usual, since these are the derivatives which are most readily measured for a liquid. The equations in this section are applicable to pure substances and to mixtures of fixed composition. The change of thermodynamic functions with composition is discussed in Chapters 4 and 5.

Small changes in the energy U, the heat-content (or enthalpy) H, the Helmholtz free energy A, and the Gibbs free energy G are given by the following four fundamental equations for systems of constant composition and total mass:

$$dU = T\,dS - p\,dV \tag{2.1}$$

$$dH = T\,dS + V\,dp \tag{2.2}$$

$$dA = -S\,dT - p\,dV \tag{2.3}$$

$$dG = -S\,dT + V\,dp \tag{2.4}$$

where S is the entropy. It follows that

$$(\partial U/\partial V)_S = (\partial A/\partial V)_T = -p \tag{2.5}$$

$$(\partial H/\partial p)_S = (\partial G/\partial p)_T = V \tag{2.6}$$

$$(\partial A/\partial T)_V = (\partial G/\partial T)_p = -S \tag{2.7}$$

[a] Three of the most suitable for the subject of this book are cited in ref. 1.

The changes of U, H, S, A and G with temperature are related to the heat capacities at constant volume and at constant pressure:

$$C_V = T(\partial S/\partial T)_V = (\partial U/\partial T)_V = -T(\partial^2 A/\partial T^2)_V \tag{2.8}$$

$$C_p = T(\partial S/\partial T)_p = (\partial H/\partial T)_p = -T(\partial^2 G/\partial T^2)_p \tag{2.9}$$

$$U = A - T(\partial A/\partial T)_V \tag{2.10}$$

$$H = G - T(\partial G/\partial T)_p \tag{2.11}$$

The isothermal changes of U, H, S, A and G with volume or pressure can be expressed solely in terms of pressure, volume, temperature and their mutual derivatives, as, for example, in equations (2.5) and (2.6). Similar equations for the entropy (Maxwell's equations) can be obtained by the differentiation of equations (2.5)–(2.7). Thus

$$-(\partial^2 A/\partial V \, \partial T) = (\partial S/\partial V)_T = (\partial p/\partial T)_V \tag{2.12}$$

$$-(\partial^2 G/\partial p \, \partial T) = (\partial S/\partial p)_T = -(\partial V/\partial T)_p \tag{2.13}$$

Similar equations for U and H are obtained by differentiation of the Gibbs–Helmholtz equations, (2.10) and (2.11):

$$(\partial U/\partial V)_T = -p + T(\partial p/\partial T)_V \tag{2.14}$$

$$(\partial H/\partial p)_T = V - T(\partial V/\partial T)_p \tag{2.15}$$

whence

$$(\partial C_V/\partial V)_T = T(\partial^2 p/\partial T^2)_V \tag{2.16}$$

$$(\partial C_p/\partial p)_T = -T(\partial^2 V/\partial T^2)_p \tag{2.17}$$

The pair of equations (2.14) and (2.15) are known as the *thermodynamic equations of state*.

It is usually unprofitable to express U, A and C_V as functions of the pressure or H, G and C_p as functions of the volume, but such equations can readily be obtained if they are needed. However, one 'cross-relation' of this kind is useful—that for the difference between C_p and C_V,

$$\left(\frac{\partial H}{\partial T}\right)_p = \left(\frac{\partial (U + pV)}{\partial T}\right)_p$$

$$= \left(\frac{\partial U}{\partial T}\right)_V + \left(\frac{\partial U}{\partial V}\right)_T \left(\frac{\partial V}{\partial T}\right)_p + p \left(\frac{\partial V}{\partial T}\right)_p \tag{2.18}$$

whence, from equation (2.14),

$$C_p - C_V = T \left(\frac{\partial p}{\partial T}\right)_V \left(\frac{\partial V}{\partial T}\right)_p \tag{2.19}$$

This difference is equal to the 'work of expansion' only if $(\partial U/\partial V)_T$ is zero, as, for example, in a perfect gas. However, for liquids the second term on the right-hand side of equation (2.18) is far from negligible and is generally larger than the third. Equation (2.19) may be put into two other forms by using an identity connecting the three mutual derivatives of p, V and T. The specification of any two of these determines the third in a system of one phase and of fixed composition. Hence

$$dp = (\partial p/\partial V)_T \, dV + (\partial p/\partial T)_V \, dT \tag{2.20}$$

$$\left(\frac{\partial V}{\partial p}\right)_T \left(\frac{\partial T}{\partial V}\right)_p \left(\frac{\partial p}{\partial T}\right)_V = -1 \tag{2.21}$$

The alternative forms of equation (2.19) are therefore

$$C_p = C_V - T\left(\frac{\partial V}{\partial T}\right)_p^2 \bigg/ \left(\frac{\partial V}{\partial p}\right)_T \tag{2.22}$$

$$C_p = C_V - T\left(\frac{\partial p}{\partial T}\right)_V^2 \bigg/ \left(\frac{\partial p}{\partial V}\right)_T \tag{2.23}$$

Equation (2.21) may be compared with the identity connecting the three mutual derivatives along a fixed line on the p, V, T surface as, for example, along a line of constant entropy, or along the saturation (or orthobaric) curve which is denoted here by the subscript σ. For such derivatives

$$\left(\frac{\partial V}{\partial p}\right)_\sigma \left(\frac{\partial T}{\partial V}\right)_\sigma \left(\frac{\partial p}{\partial T}\right)_\sigma = +1 \tag{2.24}$$

The three mutual derivatives of p, V and T are named as follows:

$(1/V)(\partial V/\partial T)_p$ is the coefficient of thermal expansion and is denoted here by α_p. However, if a saturated liquid is heated, its vapour pressure increases, and so the volume increase of a pure substance in contact with its own vapour is not that at constant pressure. The measured derivative is that along the saturation curve and is denoted here by α_σ. This is equal to $(1/V)(\partial V/\partial T)_\sigma$. The difference between the two coefficients of thermal expansion is shown in *Figure 2.1*. It is usually negligible below the normal boiling point but is appreciable at higher vapour pressures. It follows directly from the rules of partial differentiation (or from this figure) that

$$\left(\frac{\partial V}{\partial T}\right)_p = \left(\frac{\partial V}{\partial T}\right)_\sigma - \left(\frac{\partial V}{\partial p}\right)_T \left(\frac{\partial p}{\partial T}\right)_\sigma \tag{2.25}$$

where $(\partial p/\partial T)_\sigma$ is the rate of increase of vapour pressure with temperature. This equation, like most of those involving derivatives with a subscript σ, is formally applicable to any fixed line of the p, V, T surface, including that of the saturated vapour, if the meaning of σ is suitably extended.

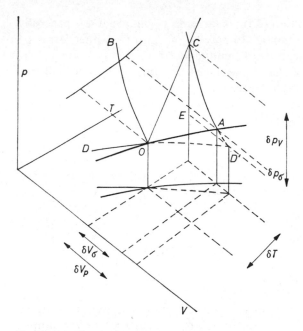

Figure 2.1 The change of the orthobaric volume of a liquid with p, V and T is shown by OA. The slope of its projections on the (V, T) and (p, T) planes are $(\partial V/\partial T)_\sigma$ and $(\partial p/\partial T)_\sigma$. OB and AC represent the isothermal compression of the liquid, and OC represents the heating at constant volume, with a slope of $(\partial p/\partial T)_V$. DOD' is an isobar whose slope at O is $(\partial V/\partial T)_p$. Equation (2.25) is readily derived from this figure

$(-1/V)(\partial V/\partial p)_T$ is the isothermal coefficient of bulk compressibility and is denoted here by β_T. The coefficient β_σ is of little importance. It is usually negative and much larger than β_T. Such a coefficient can be useful in the discussion of a solid–fluid boundary.

$(\partial p/\partial T)_V$ is the thermal pressure coefficient and is denoted here by γ_V. It follows that the slope of a vapour-pressure curve, $(\partial p/\partial T)_\sigma$, may be written γ_σ—a notation which will be used whenever it is necessary to contrast this derivative with γ_V. At the critical point γ_V and γ_σ become equal. This fact was known to van der Waals and has been 'rediscovered' several times since then[2].

The change of α_p with pressure is the complement of the change of β_T with temperature:

$$\left(\frac{\partial}{\partial p}\right)_T\left(\frac{\partial V}{\partial T}\right)_p = \left(\frac{\partial}{\partial T}\right)_p\left(\frac{\partial V}{\partial p}\right)_T \qquad \left(\frac{\partial \alpha_p}{\partial p}\right)_T = -\left(\frac{\partial \beta_T}{\partial T}\right)_p \qquad (2.26)$$

Experimental results are often expressed in terms of these coefficients, and it is therefore useful to rewrite some of the equations above in the following forms:

$$\alpha_p = \beta_T \gamma_V \tag{2.27}$$

$$\alpha_\sigma = \alpha_p - \beta_T \gamma_\sigma = \alpha_p \left(1 - \frac{\gamma_\sigma}{\gamma_V} \right) \tag{2.28}$$

$$\gamma_V - \gamma_\sigma = \alpha_\sigma / \beta_T \tag{2.29}$$

$$(\partial S / \partial V)_T = \alpha_p / \beta_T = \gamma_V \tag{2.30}$$

$$(\partial S / \partial p)_T = - V \alpha_p \tag{2.31}$$

$$(\partial U / \partial V)_T = -p + T\alpha_p / \beta_T = -p + T\gamma_V \tag{2.32}$$

$$(\partial H / \partial p)_T = V(1 - T\alpha_p) \tag{2.33}$$

$$C_p = C_V + TV\alpha_p \gamma_V \tag{2.34}$$

$$= C_V + TV\alpha_p^2 / \beta_T \tag{2.35}$$

$$= C_V + TV\beta_T \gamma_V^2 \tag{2.36}$$

In practice, C_p (like α_p) is not usually measured directly for a liquid on its saturation curve, as an increase of temperature at constant pressure causes complete evaporation. A heat capacity more closely related to experiment is C_σ, that of a liquid which is maintained at all temperatures in equilibrium with an infinitesimal amount of vapour. This heat capacity is the amount of energy supplied per unit rise of temperature to heat a liquid along its saturation curve, and therefore is given by

$$T \left(\frac{\partial S}{\partial T} \right)_\sigma \equiv C_\sigma = T \left(\frac{\partial S}{\partial T} \right)_V + T \left(\frac{\partial S}{\partial V} \right)_T \left(\frac{\partial V}{\partial T} \right)_\sigma \tag{2.37}$$

$$= C_V + T \left(\frac{\partial p}{\partial T} \right)_V \left(\frac{\partial V}{\partial T} \right)_\sigma \tag{2.38}$$

This equation may be combined with equations (2.34)–(2.36) to give several equations, of which the most useful are

$$C_\sigma = C_V + TV\alpha_\sigma \gamma_V = C_V + TV\alpha_p (\gamma_V - \gamma_\sigma) \tag{2.39}$$

$$= C_p - TV\alpha_p \gamma_\sigma = C_p - TV(\alpha_p - \alpha_\sigma)\gamma_V \tag{2.40}$$

The second pair of equations shows that the difference between C_p and C_σ is often negligible at low vapour pressures. Neither C_p nor C_σ is, in general, equal to the change of heat content with temperature along the saturation curve. This is given by

$$\left(\frac{\partial H}{\partial T} \right)_\sigma = \left(\frac{\partial H}{\partial T} \right)_p + \left(\frac{\partial H}{\partial p} \right)_T \left(\frac{\partial p}{\partial T} \right)_\sigma \tag{2.41}$$

$$= C_p + V(1 - T\alpha_p)\gamma_\sigma = C_\sigma + V\gamma_\sigma \tag{2.42}$$

Similarly

$$\left(\frac{\partial U}{\partial T}\right)_{\sigma} = \left(\frac{\partial H}{\partial T}\right)_{\sigma} - \left(\frac{\partial(pV)}{\partial T}\right)_{\sigma} = C_{\sigma} - pV\alpha_{\sigma} \tag{2.43}$$

If the product $(T\alpha_p)$ is positive but less than unity, as it generally is near the normal boiling point, then

$$\left(\frac{\partial H}{\partial T}\right)_{\sigma} > C_p > C_{\sigma} > \left(\frac{\partial U}{\partial T}\right)_{\sigma} > C_V \tag{2.44}$$

The differences between the first four of these quantities are generally much less than that between the fourth and fifth.

These equations apply also to the saturated-gas boundary, but here the difference between C_p and C_{σ} may never be ignored, as $(V\alpha_p)$ is always large and, indeed, may be large enough to make C_{σ} negative. In practice, this heat capacity of the saturated gas is positive for large molecules, for which that of the perfect gas is above about $80\,\mathrm{J\,K^{-1}\,mol^{-1}}$, and negative for the saturated vapours of substances such as argon, carbon dioxide, ammonia and steam.

The adiabatic coefficient of bulk compressibility is defined by

$$\beta_S = -\frac{1}{V}\left(\frac{\partial V}{\partial p}\right)_S \tag{2.45}$$

It is related to the speed of sound in a fluid by

$$W^2 = v/\mathcal{M}\beta_S \tag{2.46}$$

where W is the speed, v the molar volume and $\mathcal{M}$ the molar mass. The speed *defined* by this equation is, of course, a purely thermodynamic quantity. The experimental speed is equal to this speed over a wide range of frequencies and amplitudes for most fluids, and so may be regarded as an equilibrium property. However, it is sometimes larger than the speed defined by equation (2.46) at very high frequencies where the thermal properties of a fluid depend on the rate of heating. This anomalous speed is discussed in Section 2.5.

The ratio of β_T to β_S may be derived as follows:

$$dS = (\partial S/\partial V)_p \, dV + (\partial S/\partial p)_V \, dp \tag{2.47}$$

whence

$$\left(\frac{\partial V}{\partial p}\right)_S = -\frac{(\partial S/\partial p)_V}{(\partial S/\partial V)_p} = -\frac{(\partial S/\partial T)_V(\partial T/\partial p)_V}{(\partial S/\partial T)_p(\partial T/\partial V)_p} = \frac{C_V}{C_p}\left(\frac{\partial V}{\partial p}\right)_T \tag{2.48}$$

or

$$\beta_T/\beta_S = C_p/C_V \tag{2.49}$$

The equilibrium speed of sound is, therefore, given by

$$W^2 = (C_p/C_V)(v/\mathcal{M}\beta_T) \tag{2.50}$$

An expression for the difference between β_T and β_S follows from equation (2.49) and the equations for the difference between C_p and C_V:

$$\beta_T - \beta_S = TV\alpha_p^2/C_p \tag{2.51}$$

or

$$(1/\beta_S) - (1/\beta_T) = TV\gamma_V^2/C_V \tag{2.52}$$

The adiabatic thermal pressure coefficient, γ_S, is given by

$$\gamma_S \equiv \left(\frac{\partial p}{\partial T}\right)_S = -\frac{(\partial S/\partial T)_p}{(\partial S/\partial p)_T} = \frac{C_p}{T(\partial V/\partial T)_p} \tag{2.53}$$

Alternative forms for this last equation are

$$\gamma_S = \frac{C_p}{TV\alpha_p} = \frac{\alpha_p}{\beta_T - \beta_S} = \gamma_V + \frac{C_V}{TV\alpha_p} = \gamma_\sigma + \frac{C_\sigma}{TV\alpha_p} \tag{2.54}$$

The third adiabatic coefficient, α_S, is of little importance. It is generally negative and is given by

$$\alpha_S \equiv \frac{1}{V}\left(\frac{\partial V}{\partial T}\right)_S = -\frac{C_V}{TV\gamma_V} = -\beta_S\gamma_S \tag{2.55}$$

The chemical potential is an intensive property which may be defined by the equations

$$\mu = (\partial G/\partial n)_{p,T} = g \tag{2.56}$$

(where n is the amount of substance) for a one-component system maintained at constant pressure and temperature. The condition for equilibrium between two or three phases is that the three quantities, pressure, temperature and chemical potential, shall be equal in each phase.

The requirements of thermal and mechanical stability impose limitations on the signs and sizes of many of these thermodynamic functions. Consider a system of energy U, entropy S and volume V, which may be represented as a point on a surface in a three dimensional (U, S, V)-space. The surface itself represents all possible states of the system, and for a small displacement on such a surface,

$$\delta U = \left(\frac{\partial U}{\partial S}\right)_V (\delta S) + \left(\frac{\partial U}{\partial V}\right)_S (\delta V) + \tfrac{1}{2}\left(\frac{\partial^2 U}{\partial S^2}\right)_V (\delta S)^2$$

$$+ \left(\frac{\partial^2 U}{\partial S\,\partial V}\right)(\delta S)(\delta V) + \tfrac{1}{2}\left(\frac{\partial^2 U}{\partial V^2}\right)_S (\delta V)^2 + \cdots \tag{2.57}$$

Now, for equilibrium at a point on this surface it is necessary[3] that the surface lies above the tangent plane at that point. The surface is therefore essentially convex and

$$\left(\frac{\partial^2 U}{\partial S^2}\right)_V > 0 \qquad \left(\frac{\partial^2 U}{\partial V^2}\right)_S > 0 \tag{2.58}$$

$$\left(\frac{\partial^2 U}{\partial S^2}\right)_V \left(\frac{\partial^2 U}{\partial V^2}\right)_S - \left(\frac{\partial^2 U}{\partial S \partial V}\right)^2 > 0 \tag{2.59}$$

From equation (2.1)

$$(\partial U/\partial S)_V = T \qquad (\partial U/\partial V)_S = -p \tag{2.60}$$

and so (2.58) becomes

$$(\partial T/\partial S)_V > 0 \qquad \text{or} \qquad (T/C_V) > 0 \tag{2.61}$$

and

$$-(\partial p/\partial V)_S > 0 \qquad \text{or} \qquad (1/V\beta_S) > 0 \tag{2.62}$$

The first of these conditions, (2.61), is the condition of thermal stability, namely that $1/C_V$ is positive. The second can be called the condition of mechanical stability, namely that $1/\beta_S$ is positive, but this expression is used also for the stronger condition that follows from (2.59). This may be written

$$-\left(\frac{T}{C_V}\right)\left(\frac{\partial p}{\partial V}\right)_S - \left(\frac{\partial T}{\partial V}\right)_S^2 > 0 \tag{2.63}$$

By substitution from equations (2.52) and (2.55)

$$-\left(\frac{C_V}{T}\right)\left(\frac{\partial p}{\partial V}\right)_T > 0 \tag{2.64}$$

that is,

$$-(\partial p/\partial V)_T > 0 \qquad \text{or} \qquad (1/\beta_T) > 0 \tag{2.65}$$

The condition of (isothermal) mechanical stability can be derived more simply by considering the (A, V, T) surface, rather than the more primitive surface of (U, S, V). If local fluctuations of volume are to lead to an increase of A in an isothermal system, then

$$(\partial^2 A/\partial V^2)_T > 0 \tag{2.66}$$

which leads at once to (2.65). However, this surface, which has one intensive variable and only two extensive variables, is not so powerful an instrument as the (U, S, V) surface. It is, for example, impossible to derive the condition of thermal stability from the (A, V, T) surface, since the only states that may be represented on such a surface are those of uniform temperature.

These conditions of stability are necessary but not sufficient in a multi-component system. In these there is a further condition, that of material stability, which is discussed in Chapter 4.

There are no restrictions on the signs of α_p or γ_V except the obvious consequence of (2.65) that both coefficients have the same sign. They are generally positive. The coefficient γ_σ is always positive for liquid–gas equilibrium and usually positive for solid–fluid equilibrium. It follows from equations (2.49) and (2.52) that

$$C_p > C_V \quad \text{and} \quad \beta_T > \beta_S \tag{2.67}$$

These conditions are equalities only if α_p, and therefore γ_V, are zero. The coefficient γ_S has the same sign as γ_V and its modulus is larger.

2.2 Vapour pressure

The vapour pressure of a pure liquid is a function only of temperature, since a system of one component and two phases has only one degree of freedom. If a gas is compressed isothermally, at a temperature below the critical, the pressure rises until the vapour pressure is reached and the first drop of liquid is formed. This pressure and volume define the *dew point* for that temperature (*see* Chapter 1). The pressure remains constant at p_σ until the last trace of gas disappears at the *bubble point*, and then rises rapidly in the liquid phase. The volumes of the gas and liquid at the dew and bubble points are the orthobaric volumes. The absence of a change of pressure between these points is often a sensitive test of purity and is used to demonstrate the absence of dissolved air from a liquid.

The process of condensation described above is that for a fluid which is always at equilibrium. It is sometimes possible to compress a gas to a pressure greater than p_σ without producing immediate condensation, as the liquid phase is formed readily only in the presence of suitable nuclei about which the liquid drops can grow. Such nuclei are generally foreign particles, such as dust, and so the cleaner the gas the more readily is this non-equilibrium state produced. The state is usually described as *supercooled*, since it may also be produced by cooling at constant pressure a gas which is near to its saturation curve.

It is similarly possible, although less easy, to over-expand or *superheat* a liquid. It is even possible, by confining the liquid in a capillary, to carry this expansion into the region of negative pressures without evaporating the liquid. This is best done by applying tension to the liquid by spinning the capillary in a centrifuge. In this way, negative pressures of -270 bar have been reached in water, and -310 bar in chloroform[4]. Other methods of stretching the liquid lead to smaller and less reproducible negative pressures. In the complementary process of direct superheating, a temperature of 270 °C has been achieved[5] for liquid water at 1 bar.

The variation with temperature of the vapour pressures of some liquids is shown in *Figure 2.2*. Such graphs form a remarkably regular family of curves and it is rare for two of them to cross. Such a crossing (a *Bancroft point*) occurs only when the molecules in the two liquids are dissimilar in chemical type or in shape. In *Figure 2.2* the logarithm of the pressure is plotted against the temperature. Such a graph is satisfactory for direct

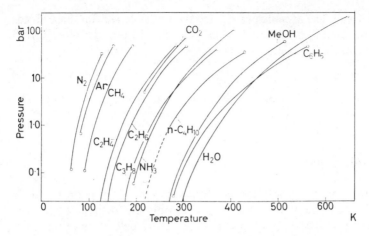

Figure 2.2 The logarithm of the vapour pressures of 12 liquids as functions of the absolute temperature. The circles mark the triple and critical points

reading. The logarithm of the pressure is, however, much closer to a linear function of the reciprocal of the absolute temperature than to this itself, and so can be represented by the well-known equation

$$\log_{10} p_\sigma = a - b/T \tag{2.68}$$

This representation is shown in *Figure 2.3* for benzene.

Although it is widely used, and is generally quite satisfactory, this equation cannot fit measurements of the highest accuracy[a] even over small ranges of temperature. The accuracy may be improved by adding further parameters, as, for example, in the equation of Antoine[7]

$$\log_{10} p_\sigma = a - b/(T + c) \tag{2.69}$$

where c is a temperature which is small compared to T. Forziati, Norris and Rossini[8] have used this equation to represent their accurate measurements of the vapour pressures of the hydrocarbons. Many other variants on equation (2.68) have been proposed, for example, the addition of terms in $1/T^2$ and/or in $\ln(T/K)$; these are fully reviewed by Ambrose[9]. One complication, with which only the most recent equations deal adequately, is that $(d^2 p_\sigma/dT^2)$ is almost certainly weakly singular at the critical point (see Chapter 3) so that p_σ cannot be an analytic function of T.

The slope of the vapour-pressure curve is the important coefficient γ_σ, and is proportional to the enthalpy of evaporation. It is readily found if an equation such as (2.69), or one of its more modern variants has been fitted to the experimental points and supplemented, if necessary, with a deviation graph. The relation of this slope to the enthalpy may be derived by the following argument. Let the pressure and temperature of a two-phase system be changed from p and T to $p + \delta p$ and $T + \delta T$ in such a way that

---

[a] The measurement of vapour pressure is described fully in a review by Ambrose[6].

neither phase, liquid or gas, disappears. The condition for the preservation of this equilibrium is

$$\mu^l(p, T) = \mu^g(p, T) \tag{2.70}$$

$$\mu^l(p + \delta p, T + \delta T) = \mu^g(p + \delta p, T + \delta T) \tag{2.71}$$

or

$$\left[\left(\frac{\partial \mu^l}{\partial p}\right)_T - \left(\frac{\partial \mu^g}{\partial p}\right)_T\right]\delta p + \left[\left(\frac{\partial \mu^l}{\partial T}\right)_p - \left(\frac{\partial \mu^g}{\partial T}\right)_p\right]\delta T = 0 \tag{2.72}$$

The chemical potential is equal to the molar Gibbs free energy in a system of one component, and so, from equations (2.6) and (2.7)

$$\left(\frac{\partial p}{\partial T}\right)_\sigma = \gamma_\sigma = \frac{S^g - S^l}{V^g - V^l} \tag{2.73}$$

The difference $(S^g - S^l)$ is denoted ΔS and is equal to $\Delta H/T$, so giving Clapeyron's equation

$$(\partial p/\partial T)_\sigma = \Delta S/\Delta V = \Delta H/T\Delta V \tag{2.74}$$

This equation is exact and is applicable to liquid–gas, to solid–fluid and to solid–solid phase boundaries with, of course, the appropriate extension of meaning of the suffix σ. However, for the liquid–gas equilibrium, the volume difference takes a particularly simple form at low pressures. In the limit, V^g differs from its value in a perfect gas by an amount that is small compared

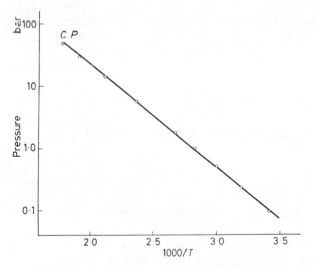

Figure 2.3 *The logarithm of the vapour pressure of benzene as a function of the reciprocal of the absolute temperature. The points are those of Table 2.3, the highest (CP) being the critical point. The straight line has the equation*

$$\log_{10}(p/\text{bar}) = 4.6757 - 1660 \text{ K}/T$$

to V^g, and V^l approaches a constant value that is similarly small. Hence

$$(\partial p/\partial T)_\sigma = (p\Delta h)/(RT^2) \tag{2.75}$$

or

$$\left(\frac{\partial \ln p}{\partial(1/T)}\right)_\sigma = -\frac{\Delta h}{R} \tag{2.76}$$

This approximation, the Clapeyron–Clausius equation, has been widely used for calculating latent heats from the slopes of vapour-pressure curves, but such use is dangerous without considering carefully the validity of the two assumptions necessary for its deduction from the exact equation (2.74).

Consider, for example, the vapour-pressure curve of benzene. At the normal boiling point, the observed slope[8,10] is 23.41 mmHg K^{-1}. The unit of pressure (mmHg) is defined[11] as 133.322 N m^{-2}, and therefore

$$(\partial p/\partial T)_\sigma = 1.102_6 \, \text{J cm}^{-3} = \Delta h/\Delta v$$

Hence, from the approximate equation (2.76), and with a gas constant of 8.314 J K^{-1} mol^{-1},

$$\Delta h = 31\,960 \, \text{J mol}^{-1}$$

However, a better estimate of Δh may be obtained by correcting the gas volume by means of the virial equation of state which, to the second term, may be written

$$v^g = RT/p + B \tag{2.77}$$

The second virial coefficient of benzene[12] at 80 °C is approximately $-960 \, \text{cm}^3 \, \text{mol}^{-1}$ and the molar volume of the liquid[13] is 96 cm^3 mol^{-1}. The difference between corrected gas and liquid volumes, calculated from equation (2.77), gives

$$\Delta h = 30\,800 \, \text{J mol}^{-1}$$

The difference between the two calculated enthalpies is due almost entirely to the neglect of the departure of the vapour from the perfect-gas laws in the derivation of equation (2.75). The measured enthalpy[14] is

$$\Delta h = 30\,780 \, \text{J mol}^{-1}$$

with which that calculated from equation (2.74) agrees, within the experimental errors of $(\partial p/\partial T)_\sigma$ and B. The departure of a saturated vapour from the perfect-gas laws increases with increasing temperature and vapour pressure.

Equation (2.76) can be integrated to give the empirical vapour-pressure equation (2.68) if it is assumed that the enthalpy of evaporation is independent of temperature—an assumption which is never exactly correct. Thus a derivation of equation (2.68) can be made only after making three assumptions that are reasonable near the triple point but quite incorrect at high vapour pressures. The fact that this simple equation holds quite well over the whole of the liquid range means that both Δh and Δv change with temperature in such a way that, fortuitously, the slope $[\partial \ln p / \partial(1/T)]_\sigma$ remains almost constant. Both Δh and Δv vanish at the critical point but their ratio remains finite. The change of Δh with temperature is given by

$$\left(\frac{\partial(\Delta h)}{\partial T}\right)_\sigma = \left(\frac{\partial(T \Delta s)}{\partial T}\right)_\sigma = \gamma_\sigma \left[\Delta v + T\left(\frac{\partial \Delta v}{\partial T}\right)_\sigma\right] + T \Delta v \left(\frac{\partial^2 p}{\partial T^2}\right)_\sigma \qquad (2.78)$$

$$- \Delta c_\sigma + \Delta s = \Delta c_\sigma + \gamma_\sigma \Delta v \qquad (2.79)$$

$$= \Delta c_p + \gamma_\sigma \Delta[(\partial h/\partial p)_T] = \Delta c_p + \gamma_\sigma \Delta[v - T(\partial v/\partial T)_p] \qquad (2.80)$$

The first pair of equations, (2.79), is of little use, as c_σ of the gas is not experimentally accessible, but the second pair, (2.80), reduces to the simple equation

$$[\partial(\Delta h)/\partial T]_\sigma = \Delta c_p \qquad (2.81)$$

at low temperatures where γ_σ is negligible. The coefficient of γ_σ in these equations is non-zero down to the lowest temperatures, as it is the difference between the isothermal Joule–Thomson coefficients, $(\partial h/\partial p)_T$, of the gas and liquid phases. This coefficient is non-zero and negative in a real gas in the limit of zero density, and non-zero and positive in a liquid at low temperatures. It is zero for a liquid at the temperature at which the inversion curve cuts the vapour-pressure curve. This temperature, at which α_p is equal to $(1/T)$, is near the critical

The normal boiling point of a liquid is the temperature at which the vapour pressure is 1 atm. It does not differ measurably from the temperature at which the liquid boils freely in dry air at a pressure of 1 atm. The process of boiling expels the dissolved air from the system, so that equilibrium is set up between pure liquid and pure vapour. In this respect, the boiling point differs from the melting point which is affected by air dissolved in the liquid phase.

The molar entropy of evaporation of a liquid at its normal boiling point usually lies between $9R$ and $14R$, where R is the gas constant—a much smaller range than is covered by the latent heats and the temperatures separately. The range of entropies is even less for those common organic solvents whose boiling points lie between 30 and 150 °C. This rough constancy is embodied in *Trouton's rule*[15] which is now usually stated that the molar entropy of evaporation of a liquid at its normal boiling point (the *Trouton constant*) is about 10–11R.

Table 2.1 shows the Trouton constants of some common liquids. Some simple substances have been excluded from this list because of known abnormalities. Hydrogen and helium are not comparable with the other simple liquids because of the large size of the 'quantum corrections'. Those substances that exist as strongly bound dimers (or higher groups) in either gaseous or liquid states have been excluded. Substances that are strongly dimerized in the liquid have unusually large Trouton's constants, as, for example, nitric oxide[16] with a value of $13.7R$. If the gas is also dimerized,

Table 2.1 Entropies of evaporation at the normal boiling
point (Trouton's constant) divided by the
gas constant

Ar	9.0	CH_4	8.8	H_2S	10.6
Kr	9.1	C_2H_6	9.6	SO_2	11.4
Xe	9.2	$n-C_4H_{10}$	9.9	CH_3Cl	10.4
Rn	9.4	$i-C_4H_{10}$	9.8	$CHCl_3$	10.6
N_2	8.7	$n-C_7H_{16}$	10.3	$(CH_3)_2O$	10.4
O_2	9.1	$c-C_6H_{12}$	10.2	NH_3	11.7
Cl_2	10.3	C_6H_6	10.5	H_2O	13.1
CO	8.9	HCl	10.3	CH_3OH	12.6
N_2O	10.8	HBr	10.3	C_2H_5OH	13.2

then the constant is low, as with acetic acid (7.5) and hydrogen fluoride
(3.1).

The variation shown by the more normal substances in *Table 2.1* are not
immediately explicable. The most useful generalization that can be made
from these figures is that Trouton's constant increases with the increasing
complexity of the intermolecular forces. Superimposed on this increase is
the tendency noted by Barclay and Butler[15] for the Trouton's constants of
chemically similar substances to increase with increasing boiling point.
These generalizations cover a much wider range of substances than can be
cited here, but even for this limited sample it is seen that the constants of
non-polar liquids are distributed fairly evenly from 8.5 to $11R$ and do not
all lie in the usually quoted range of $10-11R$. The latter is that covered by
the non-polar common laboratory solvents.

2.3 Mechanical properties

It is convenient to discuss together the coefficients of thermal expansion, α_p,
of isothermal compressibility, β_T, and of thermal pressure, γ_V, which are
interrelated by equation (2.27). These coefficients and those along the
saturation curve, α_σ and γ_σ, may be called the *mechanical coefficients* of a
liquid, and may be contrasted with the *adiabatic coefficients*, β_S, γ_S and W
(the speed of sound), to which they are related through the *thermal
coefficients*, C_V, C_p and C_σ.

The name *mechanical* is not entirely appropriate, since the adiabatic
coefficients can also be measured by purely mechanical means; that is, by
experiments in which there are no measurements of quantities of heat. It is
only the equations which relate the two classes that contain the thermal
coefficients. The most useful distinction is to be found at the molecular level
where it will be shown that the mechanical coefficients are determined, to a
high degree of accuracy, solely by the intermolecular forces, whilst the
adiabatic and the thermal coefficients depend also on the internal properties
of the molecules.

In practice, knowing the adiabatic coefficients is of little use without a knowledge of either the thermal or the mechanical coefficients. The measurement of both the adiabatic and the mechanical coefficients is a way of measuring heat capacities without measuring a quantity of heat, as, for example, in the method of Lummer and Pringsheim (the measurement of γ_S) and in the determination of heat capacities from the speed of sound.

In this section we discuss the mechanical coefficients, the methods of measuring them and the accuracy with which they are known. For most liquids it is necessary to consider together both the mechanical and adiabatic coefficients in order to determine the 'best' set of coefficients. Discussion is, therefore, deferred until Section 2.5.

The coefficient that is most easily measured is α_σ which may be determined from measurements of the orthobaric liquid density over the whole liquid range. Measurements of the density of a liquid in contact with air at 1 atm probably give a coefficient which is nearer to α_p than to α_σ, but they are only possible at low vapour pressure where the difference between these coefficients is negligible. If the orthobaric density of the pure liquid is known to 1 part in 10^5 at intervals of 5 K, then it is not difficult to obtain α_σ to about $\frac{1}{2}$ per cent by some suitable method of differentiation.

Such tables of density are common in works of reference but particular attention may be drawn to the measurements of Young, since these are reliable and generally extend from room temperature to the critical points. Their accuracy at room temperature is often a little less than that of the best modern results, but the latter rarely cover such wide ranges of temperature. Most of Young's work is quoted by Timmermans[17]. Several empirical equations can be used to represent the change of α_σ with temperature, the most common at low temperatures being a linear or quadratic equation. The coefficient becomes infinite at the critical point.

The isothermal compressibility is not so readily measured. It is usually about 10^{-4} bar^{-1} for a liquid near its triple point but decreases rapidly with increasing pressure (at constant temperature), often falling to half its initial value at 500 bar.

Measurements of β_T have been made by two methods. The first, and more reliable, is to observe the changes of volume when the pressure is increased in steps of about 100 bar to pressures of 1000 bar or more. The value of β_T at the vapour pressure is obtained by an extrapolation of the mean values over these large changes of pressure. Such an extrapolation needs very great care if it is to give accurate values of β_T at low pressures.

With such measurements, there is little difficulty in measuring the changes of volume, which are comparatively large, nor is there much trouble in performing the compression isothermally. If a liquid is suddenly compressed to, say, 200 bar, then its temperature rises by about 5 K. (The amount of the increase is given by the adiabatic coefficient, γ_S.) This rise of temperature is soon dissipated if the liquid is enclosed in a metal vessel in a thermostat, and so the final (isothermal) change of volume is soon observable.

In the second method, an attempt is made to measure directly the limiting value of β_T at zero pressure by observing the small change of volume following a pressure change of about 1 bar. It is now very difficult to conduct the experiment isothermally, for the initial (adiabatic) rise of temperature is only about 10^{-2} K and such an increase will not be dissipated smoothly unless the thermostat is maintained constant to 10^{-3} or 10^{-4} K. This is technically very difficult, and it is rare for truly isothermal changes to have been measured by this method. In a successful measurement of an isothermal compressibility it is clear that a necessary condition is

$$\Delta p \cdot \beta_T \gg \delta T \cdot \alpha_p \tag{2.82}$$

where Δp is the experimental change of pressure and δT the size of the temperature fluctuations of the thermostat. It is interesting to notice that most of the results published by Quincke[18] in 1883 as isothermal compressibilities are within 2 per cent of the now accepted adiabatic coefficients but are up to 30 per cent below the true isothermal compressibilities. His pressure change was rarely more than $\frac{1}{2}$ bar.

Direct measurement of β_T is, therefore, best based on measurements of volume as a function of pressure up to several hundred bars. The calculation of β_T from such observations is made by fitting the results to an empirical equation[19], and the two most commonly used are those of Tait and of Hudleston.

Tait[20] sought an equation to represent the compressibility of sea-water, the density of which had been measured from the oceanographic research vessel HMS Challenger, and proposed

$$\frac{V_0 - V}{V_0 p} - \frac{A}{B + p} \tag{2.83}$$

where V_0 is the volume at zero pressure and A and B are positive parameters. This equation was later interpreted, apparently first by Tammann[21], as a differential equation which could be written

$$\left(\frac{\partial V}{\partial p}\right)_T = -\frac{A}{B + p} \tag{2.84}$$

and integrated to give

$$V - V_0 = A \ln[B/(B + p)] \tag{2.85}$$

which for practical purposes is usually written

$$\frac{V - V_0}{V_0} = C \log_{10}\left[\frac{B}{B + p}\right] \tag{2.86}$$

This substitution was detected first by Kell[22] and independently by Hayward[23]. Both observe that Tait's original equation (2.83) is generally more accurate and more convenient than the modified equation (2.86) to which his name is now most commonly attached. *Table 2.2* shows Tait's original fitting of equation (2.83) to Amagat's measurements of the volume of water.

Hudleston[24] proposed the equation

$$\ln\left[\frac{pV^{2/3}}{V_0^{1/3} - V^{1/3}}\right] = A + B(V_0^{1/3} - V^{1/3}) \tag{2.87}$$

and this has been widely used to represent the best modern measurements of the compression of hydrocarbons. For these it is superior to Tait's equation, although it is less accurate for water. It is also less convenient to use than either form of Tait's equation, but this is of little moment if the fitting is done by computer.

Table 2.2 The relative volumes of water at 0 °C

p/atm	1	501	1001	1501	2001	2501	3001
V (obs. Amagat)	1.000 00	0.976 68	0.956 45	0.939 24	0.923 93	0.910 65	0.898 69
V (calc. Tait)	1.000 00	0.976 57	0.956 52	0.939 16	0.923 99	0.910 62	0.898 75

The earliest measurements of β_T that are of more than historical interest are those of Amagat[25] in 1893. He measured volume as a function of pressure and temperature for 12 common liquids to 3000 bar and, in some cases, to 200 °C. The accuracy of his measurements is good, even by modern standards, and their range has rarely been surpassed. He calculated α_p and β_T from his results—generally mean values over 20 °C and 50 bar, respectively—and, from their ratio, found the thermal pressure coefficient, γ_V.

He was followed by Bridgman[26] who, in 1913, studied almost the same 12 liquids up to 12 kbar but not to such high temperatures. He again measured volume as a function of pressure and temperature. Twenty years later he published three papers[27] in which he reported similar measurements, up to 95 °C, on a very wide range of organic liquids—hydrocarbons, halides, alcohols, glycols and esters—and in 1942 he extended his pressure range to 50 kbar in a search for a solid–fluid critical point[28]. This series of papers and the book[28] in which they are summarized are still an important source of knowledge of the volumetric behaviour of fluids at very high pressures.

Other substantial sets of measurements on liquids are those on organic liquids of Gibson[29] (who used the pseudo-Tait equation), those on hydrocarbons of Doolittle[30] (who used Hudleston's equation) and of several other groups[31], and those on water of Kell and Whalley[32]; the last of whom has reviewed the field[33].

The third mechanical coefficient, γ_V, is more easily measured than β_T, but such measurements were until recently almost entirely confined to the work of Hildebrand and his colleagues[34,35]. Their apparatus is essentially a constant-volume thermometer. The liquid is confined to a glass bulb with a capillary neck in which its level may be maintained at a fixed point by an electrical contact to a mercury interface. The coefficient γ_V is found directly as the slope of a graph of pressure against temperature for a liquid in such an apparatus.

This graph is always close to a straight line, even at the critical point, and so its slope is easily measured. For example, Amagat's measurements of α_p and β_T for ethyl ether at 20 °C give the following figures for the slope and curvature of the isochores

$$\left(\frac{\partial p}{\partial T}\right)_V = 9.0 \, \text{bar K}^{-1} \qquad \left(\frac{\partial^2 p}{\partial T^2}\right)_V = -0.009 \, \text{bar K}^{-2}$$

Thus, in a typical experiment to measure γ_V directly, in which the temperature is raised by 5 K, the pressure would increase by 45 bar. The change in γ_V over this range would be only 0.05 bar K^{-1}, about 0.6 per cent, and so the curvature of a plot of p against T would be almost undetectable.

The sign of the second derivative is important since it determines the change of C_V with pressure or density (2.16). Unfortunately, measurements of this derivative are rare, but those of β_T by Gibson and co-workers[29] confirm that it is small and negative for organic liquids at temperatures near their normal boiling point. A negative derivative implies an increase of C_V with pressure. Such an increase has been observed directly for ethanol by Bryant and Jones[36], but such calorimetric evidence of the sign of the derivative is probably harder to obtain than the mechanical evidence.

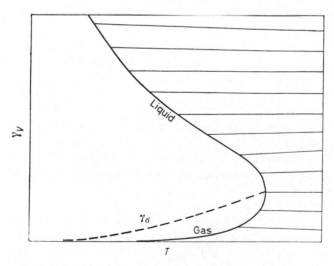

Figure 2.4 The variation of the thermal pressure coefficient with temperature for the saturated gas and liquid, and for the homogeneous fluid at temperatures above saturation

The thermal-pressure coefficient changes smoothly with temperature on passing along the coexistence curve from liquid to gas through the critical point. At this point it is equal to γ_σ, the limiting slope of the vapour-pressure curve.

The relation between the two coefficients is shown in *Figure 2.4*. The full line passing through the critical point represents γ_V along the coexistence curve. The shape of the (p, V, T) surface near this point makes it clear that here the change of γ_V with saturation temperature is infinitely rapid. The full lines to the right of this curve show γ_V along a set of isochores, for each of which the volume is the orthobaric volume at the temperature at which the line cuts the saturation curve.

The slope of these lines are always small, as indicated by the figure. The slope of the vapour-pressure curve, γ_σ, is shown by the dashed line which crosses the two-phase region and cuts the coexistence curve at the critical point. It is seen that γ_V is a much more convenient function for interpolation or extrapolation than α_p or β_T, as it remains finite up to the critical point where its value is easily determined from vapour-pressure measurements.

2.4 Heat capacities

None of the three heat capacities, C_V, C_p or C_σ, can conveniently be measured directly for a liquid on its saturation curve.

A direct measurement of C_V is difficult at low temperatures where γ_V is large, since it is impossible to make a vessel strong enough to confine the liquid to constant volume as the temperature is raised. However, with good design the correction for the expansion of the vessel need not be large, and Jones and his colleagues attempted such measurements with some success[36]. Their accuracy at low temperatures (high γ_V) is less than that of the best measurements of C_σ. Direct measurements of C_V are more practicable near the critical point where γ_V is much smaller.

The heat capacity at constant pressure, C_p, may be measured directly for a fluid at a pressure a little greater than the saturation vapour pressure by using a flow calorimeter, but such measurements are now rarely made for liquids of appreciable vapour pressure. The direct measurement of C_p of a saturated liquid is clearly impossible, since the application of heat at constant pressure simply evaporates the liquid.

The third heat capacity, C_σ, is more closely related to experiment and could, in principle, be measured directly in a calorimeter whose volume at each temperature was adjusted to be that of the saturated liquid. This would, however, be difficult to realize experimentally.

In practice, it is customary not to attempt the direct measurement of any of these but to measure the heat capacity at constant *total* volume of a liquid in equilibrium with a small amount of its vapour. In a two-phase system, this is closely related to C_σ of the liquid and less closely to C_p. At zero vapour pressure, it is clearly equal to both.

The differences at high vapour pressure were first discussed as problems of practical importance when heat capacities were required for liquid ammonia, carbon dioxide and methyl chloride, all of which are substances used in the refrigeration industry. Osborne and van Dusen[37] and Babcock[38] made full analyses of the heat capacity of the two-phase system in order to derive C_σ and C_p for liquid ammonia. However, their methods are lengthy and their equations are not always in the simplest form. A later treatment by Hoge[39] is clearer and more concise, but he derives only C_σ.

It is simple to obtain both C_σ and C_p from the heat capacity of the two-phase system at constant total volume by using the equations already derived for α_σ and C_σ. Let there be n^l moles of liquid (molar volume, v^l) and n^g moles of vapour (molar volume, v^g) in a closed vessel of total volume V. Let δn be the number of moles that evaporate when the temperature is raised δT. The total volume is unchanged, and so

$$n^l v^l + n^g v^g = (n^l - \delta n)v^l(1 + \alpha_\sigma^l \delta T) + (n^g + \delta n)v^g(1 + \alpha_\sigma^g \delta T) \tag{2.88}$$

or

$$\frac{dn^g}{dT} = -\frac{n^l v^l \alpha_\sigma^l + n^g v^g \alpha_\sigma^g}{\Delta v} \tag{2.89}$$

where Δv is $(v^g - v^l)$.

The coefficient α_σ^g is large and negative, and so (dn^g/dT) may have either sign, according to the size of the ratio (n^l/n^g). If this ratio is large, as is generally the case in a calorimeter, then (dn^g/dT) is negative—that is, vapour condenses on raising the temperature and saturation pressure.

The heat capacity of the two-phase system is composed of three parts— the saturation heat capacity of the liquid, that of the gas, and the heat required to evaporate (δn) moles of liquid. That is,

$$C = n^l c_\sigma^l + n^g c_\sigma^g - T\gamma_\sigma(n^l v^l \alpha_\sigma^l + n^g v^g \alpha_\sigma^g) \tag{2.90}$$

where $(T\gamma_\sigma)$ has been substituted for $(\Delta h/\Delta v)$ by using Clapeyron's equation. This equation is formally symmetrical in the properties of the liquid and gas and may be used as it stands. It is more convenient, however, to eliminate c_σ^g in favour of $(\partial^2 p/\partial T^2)_\sigma$ by using equations (2.78) and (2.79).

This gives

$$C = n(c_\sigma^l - T\gamma_\sigma v^l \alpha_\sigma^l) + n^g T\Delta v(\partial^2 p/\partial T^2)_\sigma \tag{2.91}$$

$$= n[c_p^l - T\gamma_\sigma v^l(2\alpha_p^l - \beta_T^l \gamma_\sigma)] + n^g T\Delta v(\partial^2 p/\partial T^2)_\sigma \tag{2.92}$$

where n is the total amount of substance.

The first of these equations is a combination of two of Hoge, with the minor difference that he replaces $n^g \Delta v$ by the more easily observable difference $(V - nv^l)$. It is seen from this equation that C does not approach (nc_σ^l) in the limit of zero vapour space ($n^g \to 0$). The heat capacity changes discontinuously at the phase boundary from the value given by equation

(2.91) with n^g equal to zero, to the heat capacity at constant volume of the homogeneous liquid. The change at the boundary is equal to $[Tv\alpha_\sigma(\gamma_V - \gamma_\sigma)]$ and is negative at both bubble and dew points. This is illustrated by plotting entropy as a function of $\ln T$ in *Figure 2.5*. The slopes of an isochore on such a graph are C_V, of an isobar, C_p and of the saturation curve, C_σ. This figure shows that $C_V < C_\sigma < C_p$ for the saturated liquid, that C_V of the two-phase system is a little less than C_σ, and that C_p of the two-phase system is infinite. In this figure, C_σ is small but negative on the gas boundary.

The relative sizes of the terms in equation (2.91) may be illustrated by considering a calorimeter of $10\,cm^3$ internal volume containing $9.5\,cm^3$ of liquid benzene and $0.5\,cm^3$ of benzene vapour at the normal boiling point. The three terms on the right-hand side of this equation are here 14.87, -0.01 and $+0.01\,J\,K^{-1}$, respectively. That is, the 'vapour correction' to be applied to the measured C to obtain (nc_σ^l) is negligible for this filling. The heat capacity (nc_p^l) is larger than (nc_σ^l) by 0.1 per cent, equation (2.92). At higher vapour pressures these corrections rapidly become more important. For example, the difference between C_p and C_σ for ammonia[37] at 30 °C, where the vapour pressure is 11 bar, is 1 per cent of C_σ. The correction for nitrogen at this vapour pressure is 5 per cent (see below).

It is seen from these two equations that C_σ is more easily obtained from the two-phase heat capacity than is C_p. The correction to obtain C_σ requires a knowledge of α_σ, but that to obtain C_p requires also a knowledge of β_T which, as shown in the previous section, is not so readily available.

Many of the published heat capacities are at such low vapour pressures that the distinction between C_σ and C_p is negligible. Some authors distinguish between the two and tabulate both, but others omit to do so even

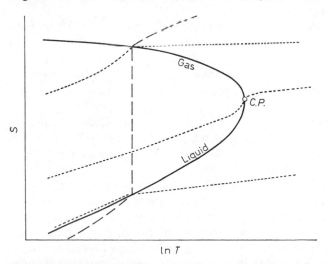

Figure 2.5 The change of entropy with $\ln T$ *along the saturation curve (full line), along the isobar (dashed line), and along three isochores (dotted lines), the second of which passes through the critical point*

where the difference is not entirely negligible. It is reasonable to assume that papers in which there is no mention of the correction to obtain C_p (that is, of β_T or equivalent information) are, in fact, reporting measurements of C_σ. If the heat capacities are to be used for computing entropies of the saturated liquid, then C_σ is the correct one to use. Most of the best measurements have been made for this purpose, and it is therefore rare to find good measurements above 25 °C, the standard temperature at which entropies are usually computed.

The measurement of the heat capacity of the two-phase system is made by supplying electrically a small amount of heat to a closed vessel nearly full of liquid and observing the rise in temperature. The thermal capacity of the calorimeter may be calculated from the known heat capacity of the material used in its construction or, more usually, is determined from a second experiment with a liquid of known heat capacity in the calorimeter. Comparatively simple equipment will give the heat capacity to about 1 per cent, but if a higher accuracy is needed, then the equipment can become quite elaborate.

The best practice is to use an 'adiabatic' calorimeter in which the heat loss to the surroundings is eliminated by ensuring that a shield around the vessel containing the liquid is maintained, at all times, at the same temperature as the liquid. The experimental arrangements are described in recent books[40]. Timmermans[17] tabulates C_p for most common organic liquids.

2.5 Adiabatic properties

The most important of the adiabatic coefficients is β_S, the bulk compressibility. It can be measured directly or can be calculated from the speed of sound by using equation (2.46).

The direct measurements are usually made by compressing the liquid in a glass bulb to a pressure about 1 bar above atmospheric. The isentropic volume change on suddenly releasing the excess pressure is about 10^{-4} of the total volume and may easily be observed if the vessel has a capillary neck. A small correction must be made to the apparent value of β_S for the compressibility of the glass vessel.

Such measurements were first made for common organic liquids by Tyrer[41] in 1913, but his work was not followed up until 1952. Since then, Staveley and his colleagues[42] have measured β_S for a considerable number of liquids and binary mixtures up to 70 °C, and similar measurements have been made by Harrison and Moelwyn-Hughes[43]. These results and those of Tyrer agree well and are apparently accurate to about 1 per cent.

Much more common and more accurate are values of β_S derived from the speed of sound. The accurate measurement of this speed is very inconvenient for sound of audible frequency owing to its long wavelength and the consequent large size of the apparatus. However, measurements at frequencies above the audible (ultrasonic frequencies) are readily made on samples of liquid of about 100 cm^3 or less. Such sound waves are generated by applying an alternating electric field of suitable frequency to a crystal of quartz which is thereby set into resonant longitudinal oscillations (the piezoelectric effect). Ultrasonic waves are generated from the free surface of such an oscillating crystal, and their wavelengths can be measured by setting up standing waves in the liquid between the crystal surface and a parallel reflector. By this and similar methods using the oscillating quartz, the speeds and sometimes the coefficients of absorption have been measured for several hundred liquids.

Most of the measurements have been at or near room temperature, but fortunately the simple liquids such as argon, nitrogen and oxygen have been adequately studied. The frequency used by Freyer, Hubbard and Andrews[44], the pioneers in this field, was 0.4 MHz (that is $0.4 \times 10^6\,s^{-1}$) and most of the more recent work has been at this, or even higher frequencies. A full description of the apparatus and methods, and many of the results, are given in reviews[45].

At these frequencies it is necessary to consider carefully whether the observed speed of sound is the equilibrium speed. At low frequencies, W is independent of the frequency v, and the small coefficient of absorption per unit length, α, increases as v^2. This absorption is due to the 'classical' effects of the shear and bulk viscosity and the thermal conductivity of the fluid. These values of W are correctly related to the static compressibility, β_S, by equation (2.46).

However, many fluids show dispersion (that is, change of W with v) at high frequencies due to relaxation effects[45,46]. W increases slowly with v, often over several decades, and α contains a 'non-classical' component with a peak at a frequency of $(1/2\pi\tau)$, where τ is the time of relaxation of the physical process that is responsible for the dispersion. When v is large compared with τ^{-1}, then W resumes a constant but higher value that is not directly related to the static compressibility. These relaxation effects are due to processes of equipartition of energy or change of internal structure in the fluid whose times of relaxation are generally less than $10^{-6}\,s$. Such processes have, in the past, been described formally as a bulk viscosity, and this is, indeed, a valid mathematical description if a frequency-dependent coefficient of viscosity is admitted. However, the name is not a helpful one and disguises the nature of the processes responsible. There was once some doubt about the existency of true (frequency-independent) bulk viscosity in a monatomic fluid, but good evidence for it in liquid argon has been obtained by Naugle[47].

It is usually not difficult to be sure that measurements are made at a frequency below the relaxation region, and so to obtain the static β_S. One very powerful test has been developed recently by comparing W at a frequency of, say, $10^6\,s^{-1}$ with that of hypersonic waves at frequencies of upwards of $10^{10}\,s^{-1}$. If these speeds are the same, then it is unlikely that any relaxation process is occurring.

These hypersonic speeds are found indirectly from the Brillouin scattering of light from a laser. The fluctuations of density that are inevitably present in a liquid can be regarded, after Fourier analysis, as a set of random hypersonic waves propagated at frequencies of the order of $10^{10}\,s^{-1}$, which are so highly damped that each wave moves only a distance of a few molecular diameters. Nevertheless, such a wave can act as a diffraction grating for a monochromatic beam of light and can produce a line at the appropriate Bragg angle. Since the 'grating' is moving, the spectrum consists of a closely spaced doublet on each side of the Rayleigh line. The separation of the doublet is a result of the Doppler difference between waves moving in opposite directions but otherwise identical, and so is a measure of the speed of (hyper)sound. Since the speed of light is large compared with that of sound, the Doppler spacing is small (about $0.2\,cm^{-1}$) and can only be observed accurately with the highly monochromatic light from a laser (line width, about $0.04\,cm^{-1}$).

This type of scattering was first predicted by Brillouin[48] and has been much used in the last few years for studying the dynamical properties of fluids[45,49]. Thus the separation of the doublet gives the speed of sound, the width of the Brillouin lines is a measure of the length of the 'grating' and so of the coefficient of absorption, α, and the ratio of the intensity of the Brillouin and Rayleigh lines is a function of β_T/β_S.

For the purpose of this chapter, however, the valuable feature of these results is that they can show convincingly that dispersion is absent in many liquids below $10^{10}\,s^{-1}$. Thus Comley[50] has shown that there is no relaxation in aliphatic hydrocarbons, but that it is present in benzene. Chiao and Fleury[49] have demonstrated its presence also for carbon disulphide, toluene and acetic acid.

Conventional ultrasonic methods cover the range 10^5 to $10^8\,s^{-1}$ which is often sufficient to show if there is a relaxation process with a time longer than $10^{-9}\,s$. Thus the results on argon, discussed below, cover a frequency range of 10^2, and the essentially constant speed found is good evidence for the absence of relaxation. Indeed, it would be hard to imagine a mechanism that could produce it.

The measurements of Heasell and Lamb[51] on 94 liquids from 100 to 200 MHz have shown that absorption, and therefore dispersion, is negligible up to these frequencies for most alcohols, amines, ethers and alkyl halides. Heavy absorption was found in esters, benzene, ethylene dichloride and in amines[51], and has been reported by others at much lower frequencies for some branched paraffins[52], for cyclohexane derivatives[53] (0.2 MHz) and for toluene[54] (0.1 MHz).

It is clear that as soon as one leaves argon and comes to much more complicated liquids, it is difficult to predict the frequency range in which any given liquid will show dispersion. The two cases that have been most thoroughly investigated are those of carbon disulphide[49,55] and acetic acid[49,56]. Their study has led to some elucidation of the processes responsible for dispersion in liquids.

In the case of carbon disulphide, it is clear that it is the vibrational energy of the molecules which is responsible for the dispersion. Such energy can here only be acquired from translational energy after a lag of about 10^{-6} s, and it is this lag which is responsible for the failure of the vibrational energy to follow the heating and cooling of the adiabatic acoustical cycle. Such lags are known to be responsible for the dispersion of sound in gases—a phenomenon which is much better understood than dispersion in liquids.

Ethylene dichloride is probably similar to carbon disulphide but is unusual in showing two regions of thermal dispersion in both gas and liquid states[45,46,57].

With acetic acid, it is probable that there are two independent mechanisms: a thermal relaxation below 7 MHz which is similar to that in carbon disulphide, and a lag in the equilibrium of monomers and dimers above 15 MHz. This interpretation is not yet beyond doubt, although it is certain that structural as well as thermal relaxation can occur in hydrogen-bonded liquids.

A third process of which the relaxation can occasionally be observed is that of chemical reaction. The classic example amongst gases is the dimerization of nitrogen dioxide. In liquids, the association of acetic acid can be regarded as a chemical reaction, but a clearer example is the acid–base equilibrium in an aqueous solution of an amine[58].

However, even when the mechanism of relaxation is obscure, there is usually no mistaking its presence, and no difficulty in choosing values of the speed of sound which are unaffected by it. Such values lead to accurate values of β_S and are our principal source of knowledge of this coefficient.

The principal use of values of β_S is to calculate β_T, the isothermal coefficient. This may be done directly by equation (2.51), if α_p and C_p are known. However, in the absence of any independent knowledge of β_T it is only α_σ and C_σ that are generally available. It is therefore necessary to combine equations (2.28), (2.39), (2.40), (2.49) and (2.51) to give the following equation for β_T

$$\beta_T = \frac{\beta_S C_\sigma + TV\alpha_\sigma(\alpha_\sigma + \beta_S\gamma_\sigma)}{C_\sigma - TV\gamma_\sigma(\alpha_\sigma + \beta_S\gamma_\sigma)} \tag{2.93}$$

This, fortunately explicit, equation for β_T is the most direct way of calculating the isothermal coefficient from the adiabatic coefficient and the three experimentally accessible coefficients C_σ, α_σ and γ_σ. The values of β_T so derived are those for the saturated liquid, as the pressures used in measuring β_S, either directly or from the speed of sound, are always low. Once β_T is known, all the other coefficients α_p, γ_V, C_p and C_V may readily be derived.

This calculation is illustrated for eight liquids, argon, nitrogen, hydrogen chloride, methane, carbon tetrachloride, benzene, n-heptane, and water. The speed of sound has been accurately measured in all these liquids. In all but benzene it is clear that there was no dispersion under the conditions used;

for benzene it appears, from the concordance of different routes to β_S and β_T that the results quoted below are the equilibrium speed of sound.

Argon

Measurements on this liquid are unusually extensive, since it is the most commonly chosen 'model' substance used to test theories. They have been carefully reviewed by the IUPAC Group under Angus[59], and by Rabinovich[60]. Their values of the vapour pressure and orthobaric volumes are the same and are used here. From them we calculate α_σ and γ_σ. The latter was checked against Rabinovich's values of $(\Delta h/T\Delta v)$. The best value of the triple point appears to be that of Pavese[61]. There are concordant measurements of the speed of sound[62] from which β_S and hence β_T can be obtained. This leads to the values of γ_V, shown by the smooth curve in *Figure 2.6*. The heat capacities C_p recommended by Rabinovich agree well with those measured by Jones and Walker[36], and are consistent with their directly observed values of c_V, *Figure 2.7*.

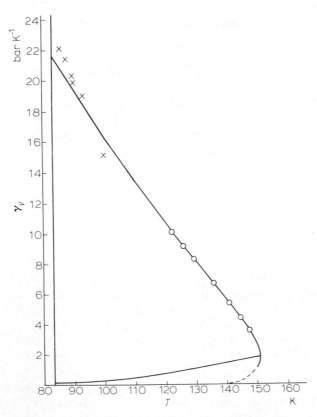

Figure 2.6 Thermal pressure coefficient of argon. The circles are values measured by Michels et al.[63], and the crosses values calculated from measurements of β_T by van Itterbeek et al.[64]

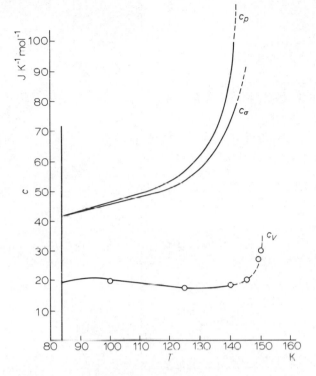

Figure 2.7 The heat capacities of liquid argon. The circles are values of c_V measured directly by Jones and Walker[36]

The direct measurement of γ_V by Michels[63] agrees well with the values obtained indirectly from W, but the values calculated from the measurements of β_T by van Itterbeek[64] are probably somewhat in error (*Figure 2.6*). The smoothed, and consistent, thermodynamic functions are shown in *Table 2.3*.

Nitrogen

We follow the *IUPAC Tables*[65] for the vapour pressures and orthobaric densities, and go to the original papers for C_σ and W. The values for the former[66] agree well with $(\partial h/\partial T)_\sigma$ as calculated by means of equation (2.42) from the *IUPAC Tables*. The values of W measured by van Itterbeek and his colleagues[62], and confirmed by other measurements[62,67], do not agree with those in the *IUPAC Tables*, and are used in preference. (The results of Pine are for hypersonic frequencies, from Brillouin scattering.) The results are shown in *Table 2.4*.

The properties of oxygen are similar to those of nitrogen, although the triple point is lower and the critical temperature higher, giving a much longer liquid range. An *IUPAC Table* is expected shortly and so no summary of its properties is attempted here.

Table 2.3 Argon

T	p_σ	v	$10^3\alpha_\sigma$	$10^3\alpha_p$	$10^4\beta_S$	$10^4\beta_T$	γ_σ	γ_V	c_σ	c_p	c_V	W
K	bar	$cm^3\,mol^{-1}$	K^{-1}		bar^{-1}		$bar\,K^{-1}$		$J\,K^{-1}\,mol^{-1}$			$m\,s^{-1}$
83.80^t	0.689	28.24	4.37	4.39	0.949	2.037	0.0803	21.55	41.8	41.9	19.4	863
85	0.790	28.39	4.40	4.42	0.972	2.088	0.0894	21.17	42.1	42.2	19.7	855
87.28^b	1.013	28.69	4.45	4.47	1.023	2.189	0.1082	20.42	42.8	42.9	20.0	838
90	1.338	29.04	4.54	4.57	1.078	2.330	0.134	19.61	43.4	43.6	20.2	821
95	2.137	29.72	4.77	4.82	1.213	2.674	0.190	18.03	44.6	44.9	20.4	783
100	3.247	30.47	5.13	5.21	1.367	3.154	0.259	16.5	45.9	46.3	20.1	747
105	4.735	31.29	5.55	5.68	1.563	3.79	0.341	15.0	47.0	47.6	19.6	708
110	6.665	32.21	6.02	6.22	1.81	4.60	0.437	13.5	48.1	49.1	19.3	667
120	12.13	34.43	7.41	7.91	2.50	7.34	0.672	10.8	51	53	18	587
130	20.23	37.48	9.9	11.3	3.99	14.0	0.971	8.1	57	62	18	485
140	31.67	42.40	16.0	21	7.59	38	1.35	5.6	73	90	18	374
150.86^c	48.98	74.6	∞	∞	∞	∞	1.8	1.8	∞	∞	∞	0

t triple point
b boiling point
c critical point

Table 2.4 Nitrogen

T	p_σ	v	$10^3 x_\sigma$	$10^3 \alpha_p$	$10^4 \beta_S$	$10^4 \beta_T$	γ_σ	γ_w	c_σ	c_p	c_V	W
K	bar	cm³ mol⁻¹	K⁻¹		bar⁻¹		bar K⁻¹		J K⁻¹ mol⁻¹			m s⁻¹
63.148ᵗ	0.1252	32.28	4.31	4.31	1.164	1.84	0.0230	23.5	56.2	56.2	35.6	995
65	0.1742	32.54	4.47	4.48	1.222	1.97	0.0300	22.7	56.4	56.5	34.9	975
70	0.3858	33.32	4.95	4.96	1.390	2.40	0.0565	20.7	56.8	56.9	33.0	925
75	0.7612	34.20	5.44	5.47	1.594	2.92	0.0957	13.7	57.3	57.4	31.3	875
77.347ᵇ	1.013	34.64	5.66	5.70	1.707	3.22	0.1191	17.7	57.5	57.7	30.6	851
80	1.370	35.18	5.94	5.99	1.85	3.60	0.1498	15.6	57.7	58.0	29.9	823
85	2.290	36.29	6.50	6.60	2.17	4.46	0.220	14.8	58.2	58.6	28.6	773
90	3.607	37.55	7.14	7.31	2.59	5.59	0.308	13.1	59.3	60.1	27.8	720
95	5.409	39.00	7.95	8.25	3.18	7.26	0.413	11.4	60.5	61.8	27.1	662
100	7.789	40.68	9.00	9.53	4.03	9.74	0.538	9.8	62.6	64.7	26.8	600
105	10.84	42.69	10.4	11.3	5.23	13.6	0.682	3.4	65.6	69.1	26.7	540
110	14.67	45.2	12.4	14.1	7.09	20.2	0.848	7.0	69.8	75.7	26.6	477
115	19.39	48.5	15.9	19.5	10.7	35	1.039	5.6	77	88	27.2	403
120	25.13	53.3	24.0	32	19.3	—	1.266		—		—	314
126.20ᶜ	34.00	89.2	∞	∞	∞	∞	1.6	.6	∞	∞	∞	0

t triple point
b boiling point
c critical point

Hydrogen chloride

Hydrogen chloride is one of the simplest dipolar substances with a dipole moment of $1.08\,D$ ($3.60 \times 10^{-30}\,Cm$) and, in addition, a sizeable quadrupole moment of $3.8 \times 10^{-26}\,esu\,cm^2$ ($12.7 \times 10^{-40}\,C\,m^2$)[68]. It therefore possesses a long liquid range extending over $166\,K$ and there is some recent evidence of weak hydrogen bonding in both the crystalline and liquid states[69].

Thermodynamic data on this important compound are rather fragmentary and only the orthobaric density and the vapour pressure have been determined over the complete temperature range from the triple point temperature to the critical temperature. Both these functions have been measured by Nunes da Ponte and Staveley[70] between 160 and 240 K and by Thomas[71] from 273 K up to the critical temperature. The densities listed in the *TRC Tables*[72] are in reasonable agreement with Thomas's measurements but show considerable deviation at low temperatures from the results of Nunes da Ponte and Staveley, and are superseded by the latter.

Chihara and Inaba[73] have measured C_p at two temperatures just above the triple point temperature and their results are in good agreement with the earlier measurements of Giauque and Wiebe[74] that extend up to the normal boiling temperature. The early measurements of Clausius[75] and of Karwat and Eucken[75] can now be discounted. There do not appear to be any measurements of C_p in the liquid state at pressures above 1 bar although there are extensive results for the supercritical gas at pressures up to $2000\,bar$[76]. These results are shown in *Table 2.5* which is much less complete than those for argon and nitrogen.

Table 2.5 Hydrogen chloride

T	p_σ	v	$10^3\alpha_\sigma$	$10^3\alpha_p$	$10^4\beta_T$	γ_σ	γ_V	c_σ, c_p	c_V
K	bar	$cm^3\,mol^{-1}$	K^{-1}			bar^{-1}	$bar\,K^{-1}$		$J\,K^{-1}\,mol^{-1}$
159.05^t	0.138	29.05	1.89			0.011		57.97	
160	0.149	29.10	1.90			0.012		57.99	
180	0.623	30.27	2.04	2.05	2.88	0.039	7.1	58.66	50.7
188.18^b	1.013	30.81	2.10	2.12	3.00	0.057	7.0	58.93	50.4
200	1.896	31.62	2.22	2.25	3.21	0.094	6.9		
220	4.630	33.15	2.51	2.58	3.68	0.187	6.8		
240	9.641	35.31	3.05			0.322			
260	18.22	37.79	3.72			0.515			
280	31.11	41.02	4.63			0.765			
300	49.65	45.9	7.01			1.105			
320	75.7	57.6	23.5			1.511			
324.65^c	83.1	81.0	∞	∞		1.628		∞	∞

t   triple point
b   boiling point
c   critical point

Methane

We follow the *IUPAC Tables*[77] for p_σ, v and C_p. The speed of sound they recommend agrees quite well with the measurements of van Itterbeek and his colleagues[78] and of Blagoi *et al.*[79], but we have used the experimental results directly. The results are shown in *Table 2.6*.

Carbon tetrachloride

For the vapour pressure we follow the recommendations of Timmermans[17] (based on Young's measurements) and of Boublik, Fried and Hala[80]. The density has been measured many times; we rely principally on Young[17] and Campbell and Chatterjee[81]. The heat capacity at constant pressure[42,82] and the speed of sound [44,83] are well-established. The results are shown in *Table 2.7*.

Benzene

The vapour pressure has been measured accurately by Ambrose and his colleagues[10,84], the density by Young[17], Campbell and Chatterjee[81] and Diaz Peña and Tardajos[31]. The thermal pressure coefficient has been measured directly[34,35], and the heat capacity is well-established[82,86]. The speed of sound is known to show dispersion under some conditions[50,51] but there are now several sets of results, reviewed by Bobik[87], which agree with each other and are consistent with the equilibrium thermodynamic properties. All these results have been used in compiling *Table 2.8*.

n-Heptane

The density and vapour pressure are well-established, principally by Young's measurements[17,80] at high temperatures. The compressibility β_T has been measured several times[30,31] and the thermal pressure coefficient at least twice (Westwater *et al.*[34], and Bagley *et al.*[35]). The heat capacity C_p is particularly accurately known at room temperatures and below[86,88], since n-heptane has been chosen as a calorimetric sub-standard[89]. The results of Amirkhanov and his colleagues[90] appear, however, to be a little low. The dispersion-free speed of sound is reviewed by Zotov and his colleagues[91]. *Table 2.9* summarizes these results.

Water

The density of water is at a maximum at 4 °C, and at this temperature

$$\alpha_\sigma = 0 \qquad \alpha_p \simeq 0 \qquad \gamma_V = \gamma_\sigma \tag{2.94}$$

$$\beta_T \simeq \beta_S \tag{2.95}$$

$$C_\sigma = C_V \simeq C_p \tag{2.96}$$

Table 2.6 Methane

T	p_σ	v	$10^3\alpha_\sigma$	$10^3\alpha_p$	$10^4\beta_S$	$10^4\beta_T$	γ_σ	γ_V	c_σ	c_p	c_V	W
K	bar	cm³ mol⁻¹	K⁻¹		bar⁻¹		bar K⁻¹		J K⁻¹ mol⁻¹			m s⁻¹
90.68^t	0.1172	35.55	2.92	2.93	0.947	1.471	0.0151	19.88	52.7	52.7	34.0	1530
100	0.3448	36.55	3.10	3.11	1.094	1.745	0.0359	17.81	54.3	54.3	34.1	1443
110	0.8839	37.76	3.36	3.38	1.293	2.145	0.0751	15.74	55.5	55.6	33.5	1349
111.631^b	1.0132	37.97	3.41	3.43	1.332	2.226	0.0835	15.41	55.7	55.8	33.4	1333
120	1.919	39.13	3.70	3.74	1.561	2.71	0.1358	13.78	56.7	56.9	32.7	1250
130	3.680	40.70	4.16	4.24	1.922	3.55	0.221	11.96	58.1	58.6	31.8	1149
140	6.419	42.54	4.77	4.93	2.46	4.83	0.332	10.21	60.0	61.0	31.0	1039
150	10.40	44.78	5.64	5.97	3.30	6.98	0.470	8.55	63.1	65.0	30.7	919
160	15.92	47.66	6.98	7.68	4.77	11.0	0.638	6.98	68	72	31	789
170	23.27	51.63	9.35	11.04	7.64	20.2	0.839	5.46	77	85	32	649
180	32.83	58.1	15.1	—	14.8	—	1.081	—	—	—	—	494
190.555^c	45.95	98.9	∞	∞	∞	∞	1.42	1.42	∞	∞	∞	0

t triple point
b boiling point
c critical point

Table 2.7 Carbon tetrachloride

T	p_σ	v	$10^3\alpha_\sigma$	$10^5\alpha_p$	$10^4\beta_S$	$10^4\beta_T$	γ_σ	γ_V	c_σ	c_p	c_V	W
°C	bar	cm³ mol⁻¹	K⁻¹		bar⁻¹		bar K⁻¹		JK⁻¹ mol⁻¹ $c_\sigma - c_p$			m s⁻¹
-22.9^t	0.0109	91.7	1.14	1.14	0.512	0.741	0.0006	15.4	130	130	90	1079
-20	0.0134	92.1	1.14	1.14	0.524	0.757	0.0008	15.1	130	130	90	1069
-10	0.0253	93.1	1.16	1.15	0.564	0.816	0.0015	14.2	131	131	90	1036
0	0.0447	94.20	1.177	1.177	0.609	0.881	0.0024	13.36	131.0	131.0	90.5	1003
10	0.0753	95.33	1.198	1.198	0.659	0.954	0.0037	12.56	131.3	131.3	90.7	970
20	0.1215	96.49	1.219	1.220	0.715	1.032	0.0056	11.82	131.7	131.7	91.0	938
30	0.1890	97.68	1.240	1.241	0.772	1.117	0.0080	11.11	132.0	132.0	91.2	907
40	0.2845	98.91	1.265	1.266	0.838	1.213	0.0112	10.44	132.3	132.3	91.4	876
50	0.416	100.18	1.294	1.296	0.912	1.322	0.0152	9.80	132.6	132.7	91.6	845
60	0.591	101.50	1.324	1.327	0.991	1.437	0.0201	9.20	133	133	92	816
70	0.822	102.87	1.345	1.349	1.082	1.563	0.0262	8.63	133	133	92	786
76.72^b	1.013	103.8	1.38	1.39	1.16	1.67	0.0308	8.3	134	134	92	766
80	1.118	104.3	1.39	1.40	1.19	1.72	0.0353	8.1	134	134	93	756
100	1.945	107.3	1.49	1.50	1.44	2.10	0.0511	7.1	135	135	93	697
200	14.57	129.4	—	—	—	—	0.228	—	—	—	—	552
283.2^c	45.6	276	∞	∞	∞	∞	0.58	0.58	∞	∞	∞	0

t triple point
b boiling point
c critical point

Table 2.8 Benzene

T	p_σ	v	$10^3\alpha_\sigma$	$10^3\alpha_p$	$10^4\beta_S$	$10^4\beta_T$	γ_σ	γ_V	c_σ	c_p	c_V	W
°C	bar	cm³ mol⁻¹	K⁻¹		bar⁻¹		bar K⁻¹		J K⁻¹ mol⁻¹			m s⁻¹
5.524ᵗ	0.0480	87.34	1.20	1.20	0.573	0.828	0.0025	14.5	132	132	91	1397
20	0.1000	88.86	1.22	1.22	0.648	0.935	0.0048	13.0	134.9	134.9	93.5	1325
40	0.2437	91.08	1.25	1.25	0.771	1.091	0.0102	11.5	139.2	139.2	98.4	1230
60	0.5219	93.46	1.29	1.29	0.921	1.281	0.0187	10.1	144.2	144.3	103.7	1140
80.10ᵇ	1.0103	95.90	1.35	1.35	1.107	1.52	0.0305	8.9	149	149	109	1053
100	1.800	98.53	1.44	1.45	1.343	1.85	0.0486	7.8	153	153	112	969
150	5.822	106.9	1.79	1.83	2.38	3.27	0.1186	5.6	170	171	125	758
200	14.35	118.3	2.43	2.60	4.99	—	0.230	3.6	—	—	—	551
250	29.85	140	—	—	—	—	0.399	2.2	—	—	—	—
288.94ᶜ	48.98	255.3	∞	∞	∞	∞	0.62	0.62	∞	∞	∞	0

t triple point
b boiling point
c critical point

Table 2.9 n-Heptane

T	p_σ	v	$10^3\alpha_\sigma$	$10^3\beta_p$	$10^4\beta_s$	$10^4\beta_T$	γ_σ	γ_v	c_σ	c_p	c_v	W
°C	bar	$cm^3\,mol^{-1}$	K^{-1}		bar^{-1}		$bar\,K^{-1}$		$J\,K^{-1}\,mol^{-1}$			$m\,s^{-1}$
-90.60^t	—	129.5	—	—	—	—	—	—	203.1	203.1	—	—
-80	—	130.9	—	—	—	—	—	—	201.7	201.7	—	—
-60	—	133.6	—	—	—	—	—	—	202.0	202.0	—	—
-40	—	136.6	—	—	—	—	—	—	205.1	205.1	—	—
-20	—	139.7	—	—	—	1.21	—	—	209.9	209.9	—	—
0	0.0152	143.0	1.211	1.211	0.94	1.390	—	10.0	216.0	216.0	169	—
20	0.0473	146.6	1.234	1.234	1.097	1.653	—	8.88	223.0	223.0	176.0	1155
40	0.1230	150.3	1.29	1.29	1.313	1.986	—	7.80	230.6	230.6	183.2	1069
60	0.2780	154.4	1.36	1.36	1.588	2.419	—	6.85	238.8	238.9	190.8	985
80	0.5790	158.8	1.45	1.45	1.943	3.01	—	5.99	247.4	247.6	198.7	903
100	1.060	163.6	1.57	1.58	2.42	5.81	0.030	5.25	256.8	257.1	206.8	821
150	3.71	179.0	2.03	2.08	4.65	14.8	0.079	3.58	282.5	283.7	227.0	620
200	9.68	202.3	3.03	3.28	11.6	120	0.162	2.22	313	318	249	417
250	21.31	259	9.1	12	76	∞	0.301	1.0	400	450	287	184
267.1^c	27.3	430	∞	∞	∞		0.35	0.35	∞	∞	∞	0

Normal boiling point 98.43 °C

t triple point
c critical point

Table 2.10 Water

T	p_σ	v	$10^3\alpha_\sigma$	$10^3\alpha_p$	$10^4\beta_S$	$10^4\beta_T$	γ_σ	γ_V	c_σ	c_p	c_V	W
°C	bar	cm³ mol⁻¹	K⁻¹		bar⁻¹		bar K⁻¹		J K⁻¹ mol⁻¹			m s⁻¹
0.01ᵗ	0.006111	18.0191	−0.0685	−0.0685	0.5085	0.5089	0.000444	−1.35	75.98	75.98	75.92	1402.5
10	0.012276	18.0216	+0.0878	+0.0878	0.4775	0.4781	0.000822	+1.84	75.53	75.53	75.44	1447.4
20	0.023384	18.0485	0.2066	0.2067	0.4558	0.4589	0.001447	4.51	75.34	75.34	74.83	1482.5
40	0.073812	18.1574	0.3851	0.3853	0.4311	0.4424	0.003931	8.71	75.27	75.27	73.35	1529.0
60	0.19933	18.3238	0.5228	0.5232	0.4228	0.4450	0.009220	11.76	75.38	75.38	71.62	1551.1
80	0.47375	18.5386	0.6403	0.6412	0.4258	0.4614	0.01918	13.90	75.58	75.59	69.76	1554.6
100.00ᵇ	1.01325	18.7980	0.7485	0.7503	0.4382	0.4902	0.03617	15.31	75.93	75.95	67.89	1543.1
150	4.757	19.645	1.021	1.029	0.506	0.622	0.1276	16.54	77.7	77.8	63.1	—
200	15.55	20.833	1.35	1.38	0.64	0.87	0.3254	15.7	80.6	81.0	60	—
250	39.78	22.55	1.85	1.95	0.95	1.47	0.672	13.3	86.1	87.6	57	—
300	85.9	25.29	2.90	3.3	1.8	3.3	1.211	10.0	98	104	56	—
350	165.3	31.37	6.8	9.8	5	15	2.03	6.5	143	182	57	—
373.85ᶜ	220.3	55.83	∞	∞	∞	∞	2.64	2.64	∞	∞	∞	—

t triple point
b boiling point
c critical point

It is clear that at the temperature at which α_σ is zero, α_p and γ_V are not exactly zero, since neither β_T nor γ_σ is zero. However, the temperature at which α_p and γ_V are zero lies very close to that at which α_σ is zero and, for all practical purposes, the relations (2.94)–(2.96) may all be taken to be equalities.

The thermodynamic properties have been measured with exceptional care, by Kell, Ambrose and others. Here we follow the review of Kell[92] for p_σ, v, α_σ at 1 atm, and β_T at 1 atm. His values of p_σ are those of Ambrose and Lawrenson[93]. We follow Kell for C_p at 1 atm (based on earlier work of de Haas[94]), and Del Grosso and Mader[95] and Chen and Millero[95] for W. The best values of the critical constants are discussed by Balfour, Sengers and Levelt Sengers[96]. The results are set out in *Table 2.10*.

2.6 Residual and configurational properties

Changes in energy can be measured experimentally, but no absolute value can be given to such a quantity, for it can only be reckoned from an arbitrary zero. A natural choice of the zero is the state of the fluid when each molecule has no kinetic energy, is in its ground level of rotational, vibrational and electronic energy, and is separated from all other molecules by an infinite distance. This is the state of a gas at the absolute zero of temperature and at zero density.

Another possible choice is that of the crystal at absolute zero. This differs from the former state by the lattice energy of the crystal—an energy that is the sum of the intermolecular energies of the molecules and the zero-point energies of oscillation.

Both these conventional zeros are widely used and both have their advantages, but neither is the most convenient reference state if one is studying the properties of a fluid as a manifestation of its intermolecular forces. The energy difference of a given fluid from that of the perfect gas at absolute zero is made up of two parts. These may be related to the energy changes of the two steps of the most natural path leading from the reference state of zero temperature and infinite volume to the state (T, V) of the given fluid.

First, the perfect gas may be heated from 0 to T whilst the volume remains infinite. The energy needed for this change is a purely molecular energy; that is, it is that needed to increase the internal energies and the translational kinetic energies of the molecules to those appropriate to the temperature T. If the fluid is now compressed to the volume V at constant temperature T, there will be a further change in its energy which is due solely to the effects of the intermolecular forces.

It is this second energy change which is called the *residual energy*, as it is the energy of the fluid, less that of the same substance as a perfect gas at the same temperature. The name *internal energy* is sometimes chosen for this quantity, but this usage can lead to confusion, as the name is also used for the total energy, U, above that of the gas at absolute zero. More formally, the residual value of any function X may be denoted $X^*(V, T)$ and defined[97,98]

$$X^*(V, T) = \int_{\infty}^{V} \left[\left(\frac{\partial X}{\partial V} \right)_T - \left(\frac{\partial X}{\partial V} \right)_T^{\text{perfect gas}} \right] dV \qquad (2.97)$$

The measurement of a residual property of a liquid is made in two steps. First, the integral of (2.97) is evaluated from infinite volume to V^g, the orthobaric vapour volume. Secondly, there is added to this quantity the change in X on isothermal condensation, less the change (if any) in this property on compressing a perfect gas from V^g to V^l. The most informative of the residual functions are U^*, H^*, C_V^* and C_p^*. The energies are calculated by the two-step process just described, but the heat capacities may be found more simply.

These are absolute quantities, and so the residual heat capacities are obtained by subtracting the absolute values for the liquid and for the perfect gas. The latter are most conveniently calculated for simple molecules from a knowledge of the frequencies of their normal modes of vibration[99,100]. Direct calorimetric values of C_V or C_p must be used for more complex molecules for which no certain assignment of frequencies can be made. These must be corrected for departures from the perfect-gas laws[98]. Such departures are usually far from negligible even at the normal boiling point.

The first of the two terms in the residual heat content, H^*, is that of the saturated gas. At low vapour pressure, this is given by

$$H_{\text{gas}}^* = p_\sigma [B - T(dB/dT)] \qquad (2.98)$$

where p_σ is the vapour pressure and B the second virial coefficient. The calculation of H^* for the gas at higher vapour pressures, and particularly near the critical point, can only be made if the equation of state has been well established or, more rarely, from direct measurements of the isothermal Joule–Thomson coefficient, $(\partial H/\partial p)_T$. If the volume of the gas is known as a function of temperature and pressure, then

$$H_{\text{gas}}^* = \int_{\infty}^{V} \left[T \left(\frac{\partial p}{\partial T} \right)_V + V \left(\frac{\partial p}{\partial V} \right)_T \right] dV = \int_{0}^{p} \left[V - T \left(\frac{\partial V}{\partial T} \right)_p \right] dp \qquad (2.99)$$

of which equation (2.98) is a special case. The second term in H^* of the orthobaric liquid is the enthalpy of evaporation, ΔH, which, as shown above, may be obtained either by direct measurement or from the slope of the vapour-pressure curve. This second term is usually much the larger part of the residual heat content. A typical graph of the variation of H^* with temperature is shown in *Figure 2.8*.

The molar residual energy, u^*, and the molar residual heat capacities are related to h^* by

$$h^* = u^* + pv - RT \qquad (2.100)$$

$$(\partial h^*/\partial T)_p = c_p^* \qquad (\partial u^*/\partial T)_v = c_v^* \qquad (2.101)$$

These equations follow from equation (2.97). The slope $(\partial h^*/\partial T)_\sigma$ is not equal to c_p^* but approaches it closely at low vapour pressures.

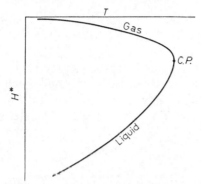

Figure 2.8 Sketch of the variation with temperature of the residual heat content of orthobaric gas and liquid. The difference between the two curves is the enthalpy of evaporation

The residual properties are, perhaps, the most direct measures of the effects of the intermolecular forces, but in the statistical theory of fluids it is more convenient to work with slightly different functions—the *configurational properties* (*see* Chapter 7). The differences between the two sets of functions are trivial. They are the configurational properties of the perfect gas, and it will be shown that these vanish for u and c_v. If the configurational properties are denoted h', u', etc., then

$$u' = u^* \qquad c'_v = c_v^* \tag{2.102}$$

$$h' = h^* + RT \qquad c'_p = c_p^* + R \tag{2.103}$$

Argon (Table 2.11)

The enthalpies of evaporation are those recommended by Rabinovich[60] and are consistent with the values of γ_σ in *Table 2.3*. The values of c'_p and c'_v are obtained directly from those in that table by subtracting the molar heat capacity of the perfect gas which is $\frac{3}{2}R$ for argon.

Nitrogen (Table 2.12)

The enthalpies of evaporation and configurational properties are those of the *IUPAC Tables*[65], and the values of c_p and c_v are those of *Table 2.4*, less the perfect gas contribution which is $\frac{5}{2}R$ at all relevant temperatures.

Hydrogen chloride

The results in *Table 2.5* are not sufficiently extensive to justify a table of configurational properties. The values of c'_p and c'_v at the normal boiling point are 38.1 and 29.6 J K^{-1} mol^{-1}, respectively.

Methane (Table 2.13)

The results shown in this table are calculated directly from the figures in the *IUPAC Tables*[77].

Water (Table 2.14)

The values shown are derived from *Table 2.10*, the *NEL Tables*[101] and spectroscopic values of C_V for the perfect gas[102].

For most organic molecules it is difficult to compile tables with the range or precision of *Tables 2.11–2.14* since the equation of state and heat capacities of the liquid above the normal boiling point are not known accurately. For carbon dioxide, the *IUPAC Table*[103] lists Δh and (after

Table 2.11 Configurational properties of liquid argon

T	Δh	h'	u'	c'_p	c'_v
K	J mol^{-1}			J K^{-1} mol^{-1}	
83.80^t	6520	− 5790	− 5790	29.4	6.9
85	6490	− 5700	− 5700	29.7	7.2
87.28^b	6420	− 5630	− 5640	30.4	7.5
90	6350	− 5550	− 5560	31.1	7.7
95	6180	− 5390	− 5400	32.4	7.9
100	6010	− 5220	− 5230	33.8	7.6
105	5810	− 5050	− 5070	35.1	7.1
110	5590	− 4870	− 4890	36.6	6.8
120	5060	− 4480	− 4530	41	6
130	4380	− 4040	− 4120	50	6
140	3400	− 3500	− 3630	78	6
150.86^c	0	− 1860	− 2230	∞	∞

t   triple point
b   boiling point
c   critical point

Table 2.12 Configurational properties of liquid nitrogen

T	Δh	h'	u'	c'_p	c'_v
K	J mol^{-1}			J K^{-1} mol^{-1}	
63.148^t	6030	− 5520	− 5520	35.4	14.8
65	5980	− 5460	− 5460	35.7	14.1
70	5820	− 5280	− 5280	36.1	12.2
75	5650	− 5090	− 5100	36.6	10.5
77.347^b	5570	− 5000	− 5010	36.9	9.8
80	5470	− 4900	− 4910	37.2	9.1
85	5260	− 4720	− 4730	37.8	7.8
90	5040	− 4520	− 4540	39.3	7.0
95	4790	− 4320	− 4340	41.0	6.3
100	4500	− 4110	− 4140	43.9	6.0
105	4160	− 3880	− 3920	48.3	5.9
110	3760	− 3630	− 3690	54.9	5.8
115	3270	− 3340	− 3440	67	6.4
120	2610	− 2990	− 3130	—	—
126.20^c	0	− 1760	− 2060	∞	∞

Table 2.13 Configurational properties of liquid methane

T	Δh	h'	u'	c_p'	c_v'
K	J mol^{-1}			J K^{-1} mol^{-1}	
90.68^t	8720	−7890	−7890	27.7	9.1
100	8500	−7720	−7720	29.3	9.2
110	8240	−7420	−7420	30.6	8.6
111.631^b	8190	−7370	−7370	30.8	8.5
120	7930	−7100	−7110	31.9	7.8
130	7560	−6770	−6790	33.6	6.9
140	7130	−6420	−6450	36.0	6.1
150	6610	−6040	−6080	40.0	5.8
160	5960	−5610	−5690	47	6
170	5120	−5110	−5230	54	7.
180	3950	−4480	−4670	67	—
190.555^c	0	−2620	−3070	∞	∞

Table 2.14 Configurational properties of liquid water

t	Δh	h'	u'	c_p'	c_v'
°C	J mol^{-1}			J K^{-1} mol^{-1}	
0.01^t	45 053	−42 780	−42 780	50.81	50.75
10	44 628	−42 290	−42 290	50.32	50.23
20	44 204	−41 790	−41 790	50.09	49.58
40	43 348	−40 820	−40 820	49.93	48.01
60	42 478	−39 800	−39 800	49.93	46.17
80	41 585	−38 780	−38 780	50.01	44.18
100.00^b	40 655	−37 790	−37 790	50.23	42.17
150	38 090	−35 250	−35 260	51.7	37.0
200	34 960	−32 610	−32 640	54.4	33
250	30 900	−29 740	−29 830	61	30
300	25 300	−26 430	−26 650	75	27
350	16 090	−21 970	−22 490	150	27
373.85^c	0	−15 220	−16 460	∞	∞

some simple arithmetic) h' and u' over the whole of the liquid range, but not c_p' or c_v'. For ethylene, the *IUPAC Table*[104] list only Δh outside the critical region. Ethane and propane are covered in the older tables edited by Din[105].

2.7 van der Waals's equation

The empirical equation of van der Waals[106,107] played an important part in the early development of theories of the liquid state and of solutions. It is still not without interest, since it is the simplest form of equation that gives a qualitatively adequate account of the process of condensation (including the metastable states) and of the properties of the liquid. Its merits are that it is easy to manipulate and that it never predicts physically absurd results. It may be used in theoretical work for a quick qualitative examination of a new problem. Some of the properties of a 'van der Waals fluid' are summarized here, others are discussed in the next chapter.

The equation is

$$(p + a/v^2)(v - b) = RT \tag{2.104}$$

where v is the molar volume, R is the gas constant, and a and b are parameters which change from one substance to another. They may be expressed in terms of two of the critical constants p^c, v^c and T^c, as is shown in any elementary text. The equation obtained by eliminating a and b in favour of the critical constants is in *reduced form*; that is, it is a function only of the dimensionless ratios p/p^c, v/v^c and T/T^c. The vapour pressure and orthobaric volumes may be found by equating the temperature, pressure and chemical potential of the two coexisting phases[a]. The coefficients α_p, β_T and γ_V are readily derived by differentiation. The last is particularly simple, as it is a function only of density:

$$\gamma_V = \frac{R}{v - b} \qquad \frac{\gamma_V T^c}{p^c} = \frac{8(v^c/v)}{3 - (v^c/v)} \tag{2.105}$$

Figure 2.9 shows that, qualitatively, the vapour pressure, orthobaric densities and γ_V behave correctly. The prediction that γ_V is independent of T (that is, that the isochores are linear) is a good approximation, but is not exactly true, as shown above and in the next chapter. At very low temperatures, the equation becomes less realistic, since it does not predict the occurrence of a solid phase. The limiting values for the orthobaric liquid at zero temperature, of the three dimensionless coefficients $(\alpha_p T^c)$, $(\beta_T p^c)$ and $(\gamma_V T^c/p^c)$ are, respectively, $(8/27)$, 0 and ∞.

There are two other consequences of this equation of state that are fulfilled approximately by many real liquids. These are not entirely independent but are best considered separately. The first is that the function $v^2(\partial u/\partial v)_T$ should be a constant[42,109]. In practice, it is not exactly constant but does not change very much over the whole of the fluid range. It is necessarily finite both at the critical point and in the limit of zero gas density, where it is equal to $RT^2(\mathrm{d}B/\mathrm{d}T)$. It is therefore not surprising that it is a well-behaved function along both branches of the orthobaric curve. It may be related to γ_V by the thermodynamic equation of state (2.14). Its calculation is sensitive to small errors in γ_V and, near the critical point, in V. Some typical values are shown in *Figure 2.10*. The variations shown cannot be dismissed as experimental error.

The second simple consequence[109] of van der Waals's equation is that, at low temperatures,

$$(\partial u/\partial v)_T = \Delta u/v^l \tag{2.106}$$

---

[a] See ref. 98. The method of solution used there is not original but was devised (though not published) by Gibbs[108].

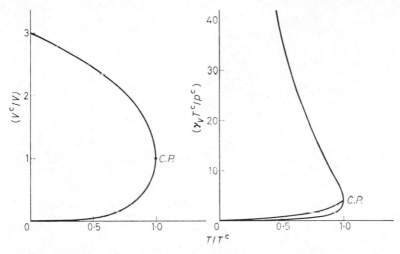

Figure 2.9 The reduced orthobaric densities, slope of the vapour-pressure curve, and thermal pressure coefficient for van der Waals's equation

where Δu is the energy of evaporation. This equation follows directly from the facts that van der Waals's equation requires the residual energy to be a linear function of density, and that the orthobaric vapour density may be neglected at low temperatures. The energy of evaporation at low temperatures is given by

$$\Delta u = \Delta h - p_\sigma \Delta v \qquad (2.107)$$

Figure 2.11 shows the closeness with which equation (2.106) is obeyed by some of the liquids discussed above. The figures for carbon tetrachloride are

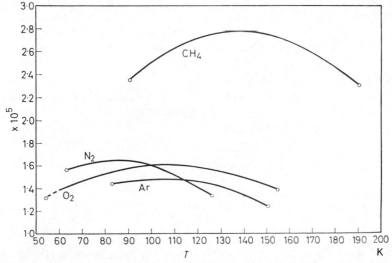

Figure 2.10 The function $v^2(\partial u/\partial v)_T$, in J cm³ mole⁻²; the circles mark the triple and critical points

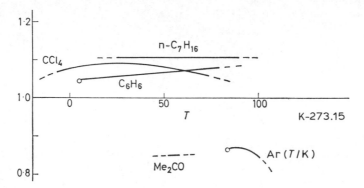

Figure 2.11 The ratio $[v(\partial u/\partial v)_T/\Delta u]$ at low temperatures. The circles mark the triple points

those of Benninga and Scott[34]. The agreement is poor for argon and nitrogen and for all polar liquids. Water is, as often, anomalous, as the left-hand side of equation (2.106) vanishes at 4 °C.

A discussion of all empirical equations proposed since that of van der Waals would be out of place here. However, one of these deserves mention because of its wide use in chemical engineering. This is the equation of Redlich and Kwong[110],

$$p = \frac{RT}{v - b} - \frac{a}{T^{1/2}v(v + b)} \tag{2.108}$$

It is clearly based on that of van der Waals and has much of its simplicity. However, its greater numerical accuracy, particularly when modified[110] to allow for departures from the principle of corresponding states, makes it of great practical value.

Van der Waals's equation is, however, more correct in its (a/v^2) term than in its term in $RT/(v - b)$, as is shown by the results in *Figures 2.10* and *2.11*. It is, therefore, generally more useful to modify the term $(v - b)^{-1}$ to obtain better agreement with experiment.

Equations (2.104) and (2.108) and these other modifications are readily extended to mixtures by treating a as a quadratic function and b as a linear function of mole fraction (*see* Chapter 8).

References

1 PRIGOGINE, I., and DEFAY, R., *Chemical Thermodynamics*, Vol. 1, Transl. Everett, D. H., Longmans, London (1954); GUGGENHEIM, E. A., *Thermodynamics*, 5th Edn, North-Holland, Amsterdam (1967); McGLASHAN, M. L., *Chemical Thermodynamics*, Academic Press, London (1979)
2 KEESOM, W. H., *Communs phys. Lab. Univ. Leiden*, No. 75 (1901)
3 GIBBS, J. W., *Collected Works*, Vol. 1, 42, Yale University Press, New Haven (1928); *see also* ref. 1
4 BRIGGS, L. J., *J. appl. Phys.*, **21**, 721 (1950); *J. chem. Phys.*, **19**, 970 (1951); DONOGHUE, J. J., VOLLRATH, R. E., and GERJUOY, E., *ibid.* 55; TEMPERLEY, H. N. V., and TREVENA, D. H., *Liquids and their Properties*, Chap. 10, Horwood, Chichester (1978)

5 KENRICK, F. B., GILBERT, C. S., and WISMER, K. L., *J. phys. Chem., Ithaca*, **28**, 1297 (1924)

6 AMBROSE, D., in *Experimental Thermodynamics*, Eds. Le Neindre, B., and Vodar, B., Vol. 2, Chap. 13, Butterworths, London (1975)

7 ANTOINE, C., *C. r. hebd. Séanc. Acad. Sci., Paris*, **107**, 681, 778, 836, 1143 (1888)

8 FORZIATI, A. F., NORRIS, W. R., and ROSSINI, F. D., *J. Res. natn. Bur. Stand.*, **43**, 555 (1949); WILLINGHAM, C. B., TAYLOR, W. J., PIGNOCCO, J. M., and ROSSINI, F. D., *J. Res. natn. Bur. Stand.*, **35**, 219 (1945)

9 AMBROSE, D., *NPL Report*, Chem 19 (1972) and Chem 114 (1980), National Physical Laboratory, Teddington

10 AMBROSE, D., BRODERICK, B. E., and TOWNSEND, R., *J. chem. Soc.*, A, 633 (1967)

11 *British Standards 2520, 3763; Misc. Publs natn. Bur. Stand.*, No. 253 (1963)

12 LAMBERT, J. D., ROBERTS, G. A. H., ROWLINSON, J. S., and WILKINSON, V. J., *Proc. R. Soc.*, **A196**, 113 (1949)

13 *Table 2.8*

14 FIOCK, E. F., GINNINGS, D. C., and HOLTON, W. B., *J. Res. natn. Bur. Stand.*, **6**, 881 (1931)

15 TROUTON, F., *Phil. Mag.*, **18**, 54 (1884); BARCLAY, I. M., and BUTLER, J. A. V., *Trans. Faraday Soc.*, **34**, 1445 (1938)

16 RICE, O. K., *J. chem. Phys.*, **4**, 367 (1936); SMITH, A. L., KELLER, W. E., and JOHNSTON, H. L., *J. chem. Phys.*, **19**, 189 (1951); GUGGENHEIM, E. A., *Molec. Phys.*, **10**, 401 (1966); **11**, 403 (1966); SCOTT, R. L., *Molec. Phys.*, **11**, 399, 503 (1966)

17 TIMMERMANS, J., *Physico-chemical Constants of Pure Organic Molecules*, 2 vols., Elsevier, Amsterdam (1950, 1965)

18 QUINCKE, G., *Annln Phys.*, **19**, 401 (1883)

19 MACDONALD, J. R., *Rev. mod. Phys.*, **41**, 316 (1969)

20 TAIT, P. G., *Voyage of H.M.S. Challenger*, Vol. 2, Part 4, 1–76 (esp. p. 33) (1889); *Scientific Papers*, Vol. 2, Papers 61 and 107, Cambridge University Press, Cambridge (1900)

21 TAMMANN, G., *Z. phys. Chem.*, **17**, 620 (1895)

22 KELL, G. S., Private communication (1964)

23 HAYWARD, A. T. J., *Br. J. appl. Phys.*, **18**, 965 (1967)

24 HUDLESTON, L. J., *Trans. Faraday Soc.*, **33**, 97 (1937); BETT, K. E., WEALE, K. E., and NEWITT, D. M., *Br. J. appl. Phys.*, **5**, 243 (1954)

25 AMAGAT, E-H., *Annls Chim. Phys.*, **29**, 505 (1893)

26 BRIDGMAN, P. W., *Proc. Am. Acad. Arts Sci.*, **49**, 1 (1913)

27 BRIDGMAN, P. W., *Proc. Am. Acad. Arts Sci.*, **66**, 185 (1931); **67**, 1 (1932); **68**, 1 (1933)

28 BRIDGMAN, P. W., *Proc. Am. Acad. Arts Sci.*, **74**, 399 (1942); *see also idem, Physics of High Pressure*, Bell, London (1949), with *Suppl.* and *Collected Papers*, Pergamon, Oxford (1966)

29 GIBSON, R. E., and KINCAID, J. F., *J. Am. chem. Soc.*, **60**, 511 (1938); GIBSON, R. E., and LOEFFLER, O. H., *J. Am. chem. Soc.*, **61**, 2515, 2877 (1939); **63**, 898 (1941); *J. phys. Chem., Ithaca*, **43**, 207 (1939)

30 DOOLITTLE, A. K., SIMON, I., and CORNISH, R. M., *A.I.Ch.E.Jl.*, **6**, 150 (1960); DOOLITTLE, A. K., and D. B., *ibid.*, 153, 157; DOOLITTLE, A. K., *Chem. Engng Prog. Symp. Ser.*, **59**, No. 44, 1 (1963)

31 JESSUP, R. S., *J. Res. natn. Bur. Stand.*, **5**, 985 (1930); DOW, R. B., and FENSKE, M. R., *Ind. Engng Chem.*, **27**, 165 (1935); DOW, R. B., and FINK, C. E., *J. appl. Phys.*, **11**, 353 (1940); SMITH, L. B., BEATTIE, J. A., and KAY, W. C., *J. Am. chem. Soc.*, **59**, 1587 (1937); FELSING, W. A., and WATSON, G. M., *J. Am. chem. Soc.*, **64**, 1822 (1942); **65**, 780 (1943); EDULJEE, H. E., NEWITT, D. M., and WEALE, K. E., *J. chem. Soc.*, 3086, 3092 (1951); CUTLER, W. G., McMICKLE, R. H., WEBB, W., and SCHIESSLER, R. W., *J. chem. Phys.*, **29**, 727 (1958); LOWITZ, D. A., SPENCER, J. W., WEBB, W., and SCHIESSLER, R. W., *J. chem. Phys.*, **30**, 73 (1959); DIAZ PEÑA, M., and McGLASHAN, M. L., *Trans. Faraday Soc.*, **55**, 2018 (1959); **57**, 1511 (1961); DIAZ PEÑA, M., and TARDAJOS, G., *J. chem. Thermodyn.*, **10**, 19 (1978); **11**, 441 (1979); GEHRIG, M., and LENTZ, H., *J. chem. Thermodyn.*, **9**, 445 (1977); **11**, 291 (1979)

32 KELL, G. S., and WHALLEY, E., *Phil. Trans. R. Soc.*, **A258**, 565 (1965); KELL, G. S., *J. chem. Engng Data*, **12**, 66 (1967)

33 WHALLEY, E., in *Experimental Thermodynamics*, Eds. Le Neindre, B., and Vodar, B., Vol. 2. Chap. 8, Butterworths, London (1975)

34 WESTWATER, W., FRANTZ, H. W., and HILDEBRAND, J. H., *Phys. Rev.*, **31**. 135 (1928); HILDEBRAND, J. H., *Phys. Rev.*, **34**, 649 (1929); HILDEBRAND, J. H., and

CARTER, J. M., *J. Am. chem. Soc.*, **54**, 3592 (1932); ALDER, B. J., HAYCOCK, E. W., HILDEBRAND, J. H., and WATTS, H., *J. chem. Phys.*, **22**, 1060 (1954); BENNINGA, H., and SCOTT, R. L., *J. chem. Phys.*, **23**, 1911 (1955); SMITH, E. B., and HILDEBRAND, J. H., *J. chem. Phys.*, **31**, 145 (1959)

35 ALLEN, G., GEE, G., and MANGARAJ, D., *Polymer*, **1**, 467 (1960); MALCOLM, G. N., and RITCHIE, G. L. D., *J. phys. Chem., Ithaca*, **66**, 852 (1962); BIANCHI, U., AGABIO, G., and TURTURRO, A., *J. phys. Chem., Ithaca*, **69**, 4392 (1965); ORWOLL, R. A., and FLORY, P. J., *J. Am. chem. Soc.*, **89**, 6814 (1967); MACDONALD, D. D., HYNE, J. B., and SWINTON, F. L., *J. Am chem. Soc.*, **92**, 6355 (1970); MACDONALD, D. D., and HYNE, J. B., *Can. J. Chem.*, **49**, 611 (1971); LEE, I., and HYNE, J. B., *Can. J. Chem.*, **51**, 1855 (1973); BAGLEY, E. B., NELSON, T. P. CHEN, S.-A., and BARLOW, J. W., *Ind. Eng. Chem., Fundam.*, **9**, 93 (1970); **10**, 27 (1971)

36 BRYANT, M. O., and JONES, G. O., *Proc. phys. Soc.*, **B66**, 421 (1953); JONES, G. O., and WALKER, P. A., in *Conférence de Physique des Basses Températures*, Paris, (1955); *Proc. phys. Soc.*, **B69**, 1348 (1956); WALKER, P. A., *PhD Thesis*, Univ. London (1956)

37 OSBORNE, N. S., and VAN DUSEN, M. S., *Bull. natn. Bur. Stand.*, **14**, 397 (1918)

38 BABCOCK, H. A., *Proc. Am. Acad. Arts Sci.*, **55**, 323 (1920) [see p. 392]

39 HOGE, H. J., *J. Res. natn. Bur. Stand.*, **36**, 111 (1946)

40 McCULLOUGH, J. P., and SCOTT, D. W. (Eds.), *Experimental Thermodynamics*, Vol. 1: *Calorimetry of Non-reacting Systems*, Butterworths, London (1968); MARTIN, J. F., in *Chemical Thermodynamics, Specialist Periodical Reports*, Ed. McGlashan, M. L., Vol. 1, Chap. 4, Chemical Society, London (1973)

41 TYRER, D., *J. chem. Soc.*, **103**, 1675 (1913); **105**, 2534 (1914)

42 STAVELEY, L. A. K., and PARHAM, D. N., in *Changements de Phases*, Société de Chimie Physique, Paris (1952); STAVELEY, L. A. K., HART, K. R., and TUPMAN, W. I., *Discuss. Faraday Soc.*, **15**, 130 (1953); STAVELEY, L. A. K., TUPMAN, W. I., and HART, K. R., *Trans. Faraday Soc.*, **51**, 323 (1955)

43 HARRISON, D., and MOELWYN-HUGHES, E. A., *Proc. R. Soc.*, **A239**, 230 (1957)

44 FREYER, E. B., HUBBARD, J. C., and ANDREWS, D. H., *J. Am. chem. Soc.*, **51**, 759 (1929)

45 BARONE, A., *Handbuch der Physik*, Ed. Flügge, S., Vol. 11, Part 2, Springer, Berlin (1962); SCHAAFFS, W., *Molekularakustik*, Springer, Berlin (1963); MASON, W. P. (Ed.), *Physical Acoustics, Principles and Methods*, Vol. 1, Part A, Academic Press, New York (1964); SETTE, D., in *The Physics of Simple Liquids*, Eds. Temperley, H. N. V., Rowlinson, J. S., and Rushbrooke, G. S., North-Holland, Amsterdam (1968); MATHESON, A. J., *Molecular Acoustics*, Wiley, New York (1971); VAN DAEL, W., in *Experimental Thermodynamics*, Eds. Le Neindre, B., and Vodar, B., Vol. 2, Chap. 11, Butterworths, London (1975)

46 HERZFELD, K. F., and LITOVITZ, T. A., *Absorption and Dispersion of Ultrasonic Waves*, Academic Press, New York (1959); SETTE, D., in *Handbuch der Physik*, Ed. Flügge, S., Vol. 11, Part 1, Springer, Berlin (1961); BAUER, H. J. (p. 48) and LAMB, J. (p. 203), in *Physical Acoustics, Principles and Methods*, Ed. Mason, W. P., Vol. 2, Part A, Academic Press, New York (1964); BHATIA, A. B., *Ultrasonic Absorption*, Oxford University Press, Oxford (1967)

47 NAUGLE, D. G., and SQUIRE, C. F., *J. chem. Phys.*, **42**, 3725 (1965); **44**, 741 (1966); NAUGLE, D. G., LUNSFORD, J. H., and SINGER, J. R., *J. chem. Phys.*, **45**, 4669 (1966)

48 BRILLOUIN, L., *Annls Phys.*, **17**, 88 (1922)

49 CHIAO, R. Y., and STOICHEFF, B. P., *J. opt. Soc. Am.*, **54**, 1286 (1964); RANK, D. H., KIESS, E. M., FINK, U., and WIGGINS, T. A., *J. opt. Soc. Am.*, **55**, 925 (1965); CHIAO, R. Y., and FLEURY, P., in *Physics of Quantum Electronics*, Eds. Kelley, P. L., Lax, B., and Tannenwald, P. E., McGraw-Hill, New York (1966); MOUNTAIN, R. D., *J. Res. natn. Bur. Stand.*, **70A**, 207 (1966); *Rev. mod. Phys.*, **38**, 205 (1966); FIGGINS, R., *Contemp. Phys.*, **12**, 283 (1971)

50 HAKIM, S. E. A., and COMLEY, W. J., *Nature, Lond.*, **208**, 1082 (1965); COMLEY, W. J., *Br. J. appl. Phys.*, **17**, 1375 (1966)

51 HEASALL, E. L., and LAMB, J., *Proc. phys. Soc.*, **B69**, 869 (1956); *idem, Proc. R. Soc.*, **A237**, 233 (1956); KREBS, K., and LAMB, J., *Proc. R. Soc.*, **A244**, 558 (1958)

52 YOUNG, J. M., and PETRAUSKAS, A. A., *J. chem. Phys.*, **25**, 943 (1956)

53 KARPOVICH. J., *J. chem. Phys.*, **22**, 1767 (1954)

54 MOEN, C. J., *J. acoust. Soc. Am.*, **23**, 62 (1951)

55 ANDREAE, J. H., HEASELL, E. L., and LAMB, J., *Proc. phys. Soc.*, **B69**, 625 (1956)

56 LAMB, J., and PINKERTON, J. M. M., *Proc. R. Soc.*, **A199**, 114 (1949); PIERCY, J. E., and LAMB, J., *Trans. Faraday Soc.*, **52**, 930 (1956); MAIER, W., and RUDOLPH, H. D., *Z. phys. Chem. Frankf. Ausg.*, **10**, 83 (1957) [benzoic acid]
57 ANDREAE, J. H., *Proc. phys. Soc.*, **B70**, 71 (1957)
58 BLANDAMER, M. J., CLARKE, D. E., HIDDEN, N. J., and SYMONS, M. C. R., *Trans. Faraday Soc.*, **63**, 66 (1967)
59 ANGUS, S., and ARMSTRONG, B., *International Thermodynamic Tables of the Fluid State, Argon, 1971*, Butterworths, London (1972)
60 RABINOVICH, V. A., *Thermophysical Properties of Neon, Argon, Krypton and Xenon*, GS SSD Monograph, Moscow (1976) [In Russian]
61 PAVESE, F., *Metrologia*, **14**, 93 (1978); STAVELEY, L. A. K., LOBO, L. Q., and CALADO, J. C. G., *Cryogenics*, **21**, 131 (1981)
62 LIEPMANN, H. W., *Helv. phys. Acta*, **12**, 421 (1939); GAIT, J. K., *J. chem. Phys.*, **16**, 505 (1948); DOBBS, E. R., and FINEGOLD, L., *J. acoust. Soc. Am.*, **32**, 1215 (1960); VAN DAEL, W., VAN ITTERBEEK, A., COPS, A., and THOEN, J., *Physica*, **32**, 611 (1966)
63 MICHELS, A., LEVELT, J. M., and DE GRAAFF, W., *Physica*, **24**, 659 (1958); MICHELS, A., LEVELT, J. M., and WOLKERS, G. J., *ibid.*, 769
64 VAN ITTERBEEK, A., and VERBEKE, O., *Physica*, **26**, 931 (1960); VAN ITTERBEEK, A., VERBEKE, O., and STAES, K., *Physica*, **29**, 742 (1963)
65 ANGUS, A., DE REUCK, K. M. and ARMSTRONG, B., *International Thermodynamic Tables of the Fluid State. Nitrogen*, Pergamon, Oxford (1979)
66 CLUSIUS, K., *Z. phys. Chem.*, **B3**, 41 (1929); WIEBE, R., and BREVOORT, M. J., *J. Am. chem. Soc.*, **52**, 622 (1930); GIAUQUE, W. F., and CLAYTON, J. O., *J. Am. chem. Soc.*, **55**, 4875 (1933); ARMSTRONG, G. T., *J. Res. natn. Bur. Stand.*, **53**, 263 (1954)
67 BLAGOI, Y. P., BUTKO, A. E., MIKHAILENKO, S. A., and YAKUBA, V. V., *Sov. Phys. Acoust.*, **12**, 355, 359 (1967); PINE, A. S., *J. chem. Phys.*, **51**, 5171 (1969)
68 *Gmelins Handbuch der anorganischen Chemie*, Syst. No. 6, Chlor, Teil B, Lief. 1, Verlag Chemie, Weinheim (1968)
69 KLEIN, M. L., and McDONALD, I. R., *Molec. Phys.*, **42**, 243 (1981); SOPER, A. K., and EGELSTAFF, P. A., *ibid.*, 399
70 NUNES DA PONTE, M., and STAVELEY, L. A. K., *J. chem. Thermodyn.*, **13**, 179 (1981)
71 THOMAS, W., in *Prog. Int. Res. Thermodynamic Transport Props., 2nd Symp. Thermophys. Props.*, 166, American Society of Mechanical Engineers, Academic Press, New York (1962)
72 *Selected Values of Properties of Chemical Compounds*, TRC Data Project Publications, College Station, Texas (1973)
73 CHIHARA, H., and INABA, A., *J. chem. Thermodyn.*, **8**, 915 (1976)
74 GIAUQUE, W. F., and WIEBE, R., *J. Am. chem. Soc.*, **50**, 101 (1928)
75 EUCKEN, A., and KARWAT, M., *Z. phys. Chem.*, **112**, 472 (1924); CLUSIUS, K., *Z. phys. Chem.*, **B3**, 41 (1929)
76 FRANCK, E. U., BROSE, M. and MANGOLD, K., *Prog. Int. Res. Thermodynamic Transport Props., 2nd Symp. Thermophys. Props.*, 159, American Society of Mechanical Engineers, Academic Press, New York (1962)
77 ANGUS, S., ARMSTRONG, B., and DE REUCK, K. M., *International Thermodynamic Tables of the Fluid State, Methane*, Butterworths, London (1978)
78 VAN DAEL, W., VAN ITTERBEEK, A., THOEN, J., and COPS, A., *Physica*, **31**, 1643 (1965); **35**, 162 (1967)
79 BLAGOI, Y. P., BUTKO, A. E., MIKHAILENKO, S. A., and YAKUBA, V. V., *Russ. J. phys. Chem.*, **41**, 908 (1967)
80 BOUBLIK, T., FRIED, V., and HALA, E., *The Vapour Pressure of Pure Substances*, Elsevier, Amsterdam (1973)
81 CAMPBELL, A. N., and CHATTERJEE, R. M., *Can. J. Chem.*, **46**, 575 (1968); **47**, 3893 (1969)
82 HICKS, J. F. G., HOOLEY, J. G., and STEPHENSON, C. C., *J. Am. chem. Soc.*, **66**, 1064 (1944); BOELHOUWER, J. W. M., *Physica*, **26**, 1021 (1960); WILHELM, E., SACKMANN, H., and ZETTLER, M., *Ber. Bunsenges. phys. Chem.*, **78**, 795 (1974); GROLIER, J.-P. E., WILHELM, E., and HAMEDI, M. H., *Ber. Bunsenges. phys. Chem.*, **82**, 1282 (1978) [benzene and carbon tetrachloride]
83 PELLAM, J. R., and GALT, J. K., *J. chem. Phys.*, **14**, 608 (1946); LAGEMANN, R. T., McMILLAN, D. R., and WOOLF, W. E., *J. chem. Phys.*, **17**, 369 (1949); SACKMANN, H., AND BOCZEK, A., *Z. phys. Chem., Frankf. Ausg.*, **29**, 329 (1961); BOBIK, M., NIEPMANN, R., and MARIUS, W., *J. chem. Thermodyn.*, **11**, 351 (1979)

84 AMBROSE, D., SPRAKE, C. H. S., and TOWNSEND, R., *J. chem. Thermodyn.*, **1**, 499 (1969)
85 HOLDER, G. A., and WHALLEY, E., *Trans. Faraday. Soc.*, **58**, 2095 (1962); FORT, R. J., and MOORE, W. R., *Trans. Faraday Soc.*, **61**, 2102 (1965)
86 OLIVER, D. G., EATON, M., and HUFFMAN, H. M., *J. Am. chem. Soc.*, **70**, 1502 (1948) [benzene]; HUFFMAN, H. M., GROSS, M. E., SCOTT, D. W., and McCULLOUGH, J. P., *J. phys. Chem., Ithaca*, **65**, 495 (1961) [n-heptane]
87 BOBIK, M., *J. chem. Thermodyn.*, **10**, 1137 (1978)
88 VAN MITTENBURG, J. C., *J. chem. Thermodyn.*, **4**, 773 (1972)
89 DOUGLAS, T. B., FURUKAWA, G. T., McCOSKEY, R. E., and BALL, A. F., *J. Res. natn. Bur. Stand.*, **53**, 139 (1954)
90 AMIRKHANOV, K. I., ALIBEKOV, B. G., VIKHOV, D. I., MIRSKAYA, V. A., and LEVINA, L. N., *High Temp.*, **9**, 1211 (1971); **11**, 59 (1973)
91 ZOTOV, V. V., NERUCHEV, Y. A., and OTPUSHCHENNIKOV, N. F., *Russ, J. Engng Phys.*, **15**, 1098 (1968)
92 KELL, G. S., *J. chem. Engng Data*, **20**, 97 (1975)
93 AMBROSE, D., and LAWRENSON, I. J., *J. chem. Thermodyn.*, **4**, 755 (1972)
94 DE HAAS, J., *P.-v. Séanc. Com. int. Poids Més.*, Ser. 2, **12**, (1950)
95 DEL GROSSO, V. A., and MADER, C. W., *J. acoust. Soc. Am.*, **52**, 1442 (1972); CHEN, C.-T., and MILLERO, F. J., *J. acoust. Soc. Am.*, **60**, 1270 (1976)
96 BALFOUR, F. W., SENGERS, J. V., and LEVELT SENGERS, J. M. H., Paper at *9th Int. Conf. on Properties of Steam*, Munich (1979)
97 MICHELS, A., GELDERMANS, M., and DE GROOT, S. R., *Physica*, **12**, 105 (1946)
98 ROWLINSON, J. S., in *Handbuch der Physik*, Ed. Flügge, S., Vol. 12, Chap. 1, Springer, Berlin (1958)
99 FRANKISS, S. G., and GREEN, J. H. S., in *Chemical Thermodynamics, Specialist Periodical Reports*, Ed. McGlashan, M. L., Vol. 1, Chap. 8, Chemical Society, London (1973)
100 HILSENRATH, J., and ZIEGLER, G. G., *Tables of Einstein Functions, Natn. Bur. Stand. Monogr.*, No. 49 (1962); ABRAMOWITZ, M., and STEGUN, I. A., *Handbook of Mathematical Functions*, 999, National Bureau of Standards, Washington DC (1964)
101 NATIONAL ENGINEERING LABORATORY, *Steam Tables, 1964*, Ed. Bain, R. W., HMSO, Edinburgh (1964)
102 ROWLINSON, J. S., *The Perfect Gas*, 59–61, Pergamon, Oxford (1963); WOOLLEY, H. W., Paper at *9th Int. Conf. on Properties of Steam*, Munich (1979)
103 ANGUS, S., ARMSTRONG, B., and DE REUCK, K. M., *International Thermodynamic Tables of the Fluid State, Carbon Dioxide*, Pergamon, Oxford (1976)
104 ANGUS, S., ARMSTRONG, B., and DE REUCK, K. M., *International Thermodynamic Tables of the Fluid State, Ethylene*, Butterworths, London (1974)
105 DIN, F. (Ed.), *The Thermodynamic Functions of Gases*, 3 vols., Butterworths, London (1956–61)
106 VAN DER WAALS, J. D., *Die Kontinuität des gasformigen und flüssigen Zustandes*, 2nd Edn. Vol. 1: *Single Component Systems*, Barth, Leipzig (1899–1900); Engl. transl. 1st Edn Vol. 1 by Threlfall, R., and Adair, J. F., in *Physical Memoirs*, Vol. 1, Part 3, Physical Society, London (1890)
107 RIGBY, M., *Q Rev. chem. Soc.*, **24**, 416 (1970); ROWLINSON, J. S., *Nature, Lond.*, **244**, 414 (1973); DE BOER, J., *Physica*, **73**, 1 (1974); KLEIN, M. J., *ibid.*, 28; LEVELT SENGERS, J. M. H., *ibid.*, 73; **82A**, 319 (1976)
108 WILSON, E. B., *Commentary on the Scientific Writings of J. Willard Gibbs*, Eds. Donnan, F. G., and Haas, A., Vol. 1, 41, Yale University Press, New Haven (1936)
109 HILDEBRAND, J. H., and SCOTT, R. L., *The Solubility of Non-electrolytes*, 96–101, Reinhold, New York (1951); *Regular Solutions*, 77–79, Prentice-Hall, Englewood Cliffs, NJ (1962)
110 REDLICH, O., and KWONG, J. N. S., *Chem. Rev.*, **44**, 233 (1949); REDLICH, O., and DUNLOP, A. K., *Chem. Engng Prog. Symp. Ser.*, **59**, No. 44, 95 (1963)

Chapter 3
The critical state

3.1 Thermodynamics of the critical point

The critical state of a fluid is represented by the point on the (p, V, T) surface where the volumes of the gas and liquid phases become identical. The mechanical stability of this state is of a lower order than that required by the inequalities of Section 2.1. Such a point lies on the border separating the stable and unstable parts of a continuous (p, V, T) surface. A preliminary discussion of the behaviour of thermodynamic functions at and near this point may therefore be based on the (A, V, T) surface. It is impossible to use the (G, p, T) surface, since the condition of mechanical stability cannot be derived from it.

Liquid and gas can exist together at equilibrium if the temperatures, pressures and chemical potentials of the two phases are equal. These equalities may be written in terms of the molar Helmholtz free energy of a one-component system:

$$T^l = T^g \tag{3.1}$$

$$(\partial a / \partial v)^l_T = (\partial a / \partial v)^g_T \tag{3.2}$$

$$a^l - v^l(\partial a / \partial v)^l_T = a^g - v^g(\partial a / \partial v)^g_T \tag{3.3}$$

Thus a graph of a as a function of v at constant temperature is represented by the full line in *Figure 3.1*. The volumes of the coexistent phases, A and D, are connected by a straight tie-line of slope $(-p)$ which, by equations (3.2) and (3.3), must be the common tangent to both branches of the curve. Thus the experimental free energy is not a continuous differentiable function of volume at the dew and bubble points. It is, however, convenient to treat it as such a function of the form shown by the dashed curve $ABCD$. This lies above the full curve and is therefore a less stable state.

The instability of the curve $ABCD$ is of two kinds; between B and C the derivative $(\partial^2 a / \partial v^2)_T$ is negative, or $(\partial p / \partial v)_T$ is positive. Such a system is mechanically unstable; the parts of the curve between A and B and between C and D are mechanically stable but are metastable with respect to the two-phase system. Although the equilibrium curve is the discontinuous one and although an exact statistical treatment of the molar free energy of an

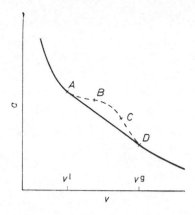

Figure 3.1 The Helmholtz free energy as a function of volume

infinitely large assembly would lead to this curve, it is useful to suppose that *a* may be represented by the continuous curve for the following reasons:

(1) The results obtained from this hypothesis and from the rules for constructing the tie-line, equations (3.2) and (3.3), agree qualitatively with the experimental results.
(2) Approximate statistical theories and all empirical equations of state lead to the continuous curve.
(3) The metastable regions of compressed vapour and expanded liquid are observable experimentally, although their theoretical status is not yet entirely clear[1].

The curve on the (p, V, T) surface which corresponds to the dashed curve in *Figure 3.1* is shown in *Figure 3.2*. The equality of the chemical potential of the two phases requires that (cf. equation (2.4))

$$\int_{ABCD} v\,dp = 0 \text{ (Maxwell's construction)} \tag{3.4}$$

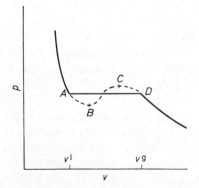

Figure 3.2 The pressure as a function of volume

Thus the straight line AD cuts the continuous curve to form two equal areas above and below the transition pressure.

A critical point occurs on a free-energy surface if the tangent AD becomes vanishingly small, so that the four points A, B, C and D coincide. At such a point

$$\left(\frac{\partial a}{\partial v}\right)_T < 0 \qquad \left(\frac{\partial^2 a}{\partial v^2}\right)_T = 0 \qquad \left(\frac{\partial^3 a}{\partial v^3}\right)_T = 0 \qquad \left(\frac{\partial^4 a}{\partial v^4}\right)_T > 0 \qquad (3.5)$$

that is,

$$p > 0 \qquad \left(\frac{\partial p}{\partial v}\right)_T = 0 \qquad \left(\frac{\partial^2 p}{\partial v^2}\right)_T = 0 \qquad \left(\frac{\partial^3 p}{\partial v^3}\right)_T < 0 \qquad (3.6)$$

The second and third of these conditions are necessary if B and C are to coincide, and the fourth is the condition that the fluid should be stable at volumes immediately above and below the critical point. It is so at this point, since the pressure falls with increasing volume, but the stability is of a lower order than that of other states of the fluid.

A fluid whose critical point is characterized by these conditions is often called a van der Waals fluid, since the conditions are satisfied by his equation of state. Their consequences were first explored in his classic work[2], *Die Kontinuität des gasförmigen und flüssigen Zustandes*. Experiment shows that the first three derivatives of a satisfy (3.5), the fourth is almost certainly zero and there is doubt about the behaviour of the higher derivatives. These doubts can be illustrated most clearly by discussing first, and in some detail, the consequences of the classical conditions.

The Helmholtz free energy can be written as a double Taylor expansion in $(v - v^c)$ and $(T - T^c)$ about its value at the critical point; this was the original approach of van der Waals and Verschaffelt, although the latter was aware of the limitations of the classical equations[3]. However, here we expand not the free energy itself but its density, that is, the function $\psi = A/V$, following the treatment of Sengers and Levelt Sengers[4]. The choice between A and ψ is arbitrary at this stage, but we shall see later that there are several advantages in choosing ψ. From equation (2.3)

$$d(A/V) = -(A + pV)\,dV/V^2 - S\,dT/V \qquad (3.7)$$

or

$$d\psi = \mu\,d\rho - \eta\,dT \qquad (3.8)$$

where $\rho = n/V$ is the molar density and $\eta = S/V$ is the entropy density. Hence the appropriate variables in which to expand ψ about its value at the critical point are $\delta\rho = \rho - \rho^c$ and $\delta T = T - T^c$.

$$\psi = \sum_{m=0}^{\infty} \sum_{n=0}^{\infty} \frac{1}{m!\,n!} \psi_{mn}^c (\delta\rho)^m (\delta T)^n \qquad (3.9)$$

where

$$\psi_{mn}^{c} = \left(\frac{\partial^{m+n}\psi}{\partial(\delta\rho)^{m}\partial(\delta T)^{n}} \right)^{c} \tag{3.10}$$

From equation (3.8) and the identity

$$\rho\mu = \psi + p \qquad \text{or} \qquad -p = \psi - \rho(\partial\psi/\partial\rho)_{T} \tag{3.11}$$

we have

$$\psi_{10} = \mu \qquad \psi_{20} = \frac{1}{\rho}\left(\frac{\partial p}{\partial\rho} \right)_{T} \qquad \psi_{30} = \left(\frac{\partial}{\partial\rho} \right)_{T}\left[\frac{1}{\rho}\left(\frac{\partial p}{\partial\rho} \right)_{T} \right] \quad \text{etc.} \tag{3.12}$$

and so from (3.6)

$$\psi_{20}^{c} = 0 \qquad \psi_{30}^{c} = 0 \qquad \psi_{40}^{c} > 0 \tag{3.13}$$

Differentiation of equation (3.9) leads to a similar expansion for $\mu = (\partial\psi/\partial\rho)_{T}$, of which the leading terms are

$$\mu - \mu^{c} = \mu_{01}^{c}(\delta T) + \mu_{11}^{c}(\delta\rho)(\delta T) + \tfrac{1}{6}\mu_{30}^{c}(\delta\rho)^{3} \tag{3.14}$$

since the terms in $\mu_{10}^{c} = \psi_{20}^{c}$ and $\mu_{20}^{c} = \psi_{30}^{c}$ are zero. Two equations of this kind can be written: one for μ^{l} in powers of δT and $\delta\rho^{l}$ and for μ^{g} in powers of δT and $\delta\rho^{g}$. Since $\mu^{l} = \mu^{g}$, the two can be subtracted to give

$$0 = \mu_{11}^{c}\delta T + \tfrac{1}{6}\mu_{30}^{c}[(\delta\rho^{l})^{2} + (\delta\rho^{l})(\delta\rho^{g}) + (\delta\rho^{g})^{2}] \tag{3.15}$$

and can be added to give

$$\mu^{l, g} - \mu^{c} = \mu_{01}^{c}\delta T + \tfrac{1}{2}(\delta\rho^{l} + \delta\rho^{g})$$
$$\times [\mu_{11}^{c}\delta T + \tfrac{1}{6}\mu_{30}^{c}((\delta\rho^{l})^{2} - (\delta\rho^{l})(\delta\rho^{g}) + (\delta\rho^{g})^{2})] \tag{3.16}$$

From equation (3.15) it follows that $\delta\rho^{l}$ and $\delta\rho^{g}$ are of leading order $|\delta T|^{1/2}$. Thus, as shown by van Laar[5], the classical conditions (3.5) lead to an expansion of $\delta\rho$ in powers of $|\delta T|^{1/2}$,

$$\delta\rho^{l, g} = B_{1}^{\pm}|\delta T|^{1/2} + B_{2}^{\pm}|\delta T| + B_{3}^{\pm}|\delta T|^{3/2} + \cdots \tag{3.17}$$

where the plus sign denotes the liquid and the minus sign the gas.

An expansion of the pressure about p^{c} leads to equations similar to equations (3.15) and (3.16) but with coefficients p_{11}^{c} and p_{30}^{c}. However, from equations (3.11) and (3.13) we have $p_{11}^{c} = \mu_{11}^{c}$ and $p_{30}^{c} = \mu_{30}^{c}$, and so equality of chemical potential and equality of pressure are, to leading order, equivalent statements for orthobaric fluids near the critical point. By taking the Taylor expansion (3.14) to order $(\delta T)^{2}$, and equating coefficients in $\mu^{l} = \mu^{g}$, it can be shown that[6]

$$B_{1}^{+} = -B_{1}^{-} \equiv B_{1} = (6\mu_{11}^{c}/\mu_{30}^{c})^{1/2} = (6p_{11}^{c}/p_{30}^{c})^{1/2} \tag{3.18}$$

and that $B_{2}^{+} = B_{2}^{-} \equiv B_{2}$ can be similarly expressed in terms of $\mu_{30}^{c}, \mu_{40}^{c}, \mu_{11}^{c}$

and μ_{21}^c. Thus

$$\delta\rho^{l,g} = \pm B_1|\delta T|^{1/2} + B_2|\delta T| + \cdots \tag{3.19}$$

and, to lowest order,

$$\tfrac{1}{2}(\rho^l - \rho^g) = B_1|\delta T|^{1/2}$$
$$\tfrac{1}{2}(\rho^l + \rho^g) = \rho^c + B_2|\delta T| \tag{3.20}$$

To the same order, equation (3.16) becomes

$$\mu^{l,g} - \mu^c = \mu_{01}^c(\delta T) \tag{3.21}$$

Hence the slope $(\partial\mu/\partial T)_\sigma^c$ below the critical temperature is equal to the slope $(\partial\mu/\partial T)_\rho^c$ on the isochore above the critical temperature.

Parallel results follow from the expansion of $a(v, T)$ instead of $\mu(\rho, T)$. The equations have the same form but with p for μ and δv^l and δv^g for $\delta\rho^l$ and $\delta\rho^g$ in equations (3.17)–(3.20). The equation equivalent to (3.21) implies that the limiting slope of the vapour pressure curve $(\partial p/\partial T)_\sigma^c$ is equal to that of the critical isochore $(\partial p/\partial T)_V^c$, or $\gamma_\sigma^c = \gamma_V^c$. To the lowest order the two sets of equations are equivalent, but this equivalence breaks down as soon as δT ceases to be small. Thus, equation (3.20) implies that ρ^c is the mean of ρ^l and ρ^g when $(\delta T)^{1/2}$ is small, while the corresponding equation for the volume implies that v^c is the mean of v^l and v^u. Both statements cannot be true except in the limit $\delta T = 0$, and it is a matter of experimental fact that the first is more correct than the second for small and moderate values of δT. This result can be generalized to say that if ψ is decomposed into two parts, first a smoothly varying function of ρ and T which plays no part in determining the behaviour near the critical point, and, second, a singular part ψ_s in which all the critical properties are contained, then ψ_s is a symmetric function of $\delta\rho$:

$$\psi_s(-\delta\rho, \delta T) = \psi_s(\delta\rho, \delta T) \tag{3.22}$$

It follows that μ contains a singular part that is antisymmetric in $\delta\rho$. Levelt Sengers and her colleagues[4,6,7] have reviewed fully the experimental evidence on these points. Theoretical reasons for using $\psi(\rho, T)$ rather than $a(v, T)$ will appear later in this and the next chapter.

Thus the physical consequence of the assumption that ψ is a continuous differentiable function of ρ and T is that the coexistence curve has a rounded top which is quadratic in the density (3.20). The order of the curve depends on that of the first non-vanishing derivative of ψ with respect to density. If the fourth derivative is assumed to vanish, then the fifth must do likewise and presumably the sixth would be non-zero and positive. Mechanical stability requires that the first non-vanishing derivative shall be even but does not indicate its order. If this derivative is of the order $2n$, then an analysis[8] similar to that above shows that $(\delta\rho)$ is proportional to

$(\delta T)^{1/(2n-2)}$. It is shown below that such proportionality is not in agreement with experiment, which suggests that $(\delta\rho)$ is proportional to $(\delta T)^{0.3}$. Such an exponent is inconsistent with an integral value of n. However, the following qualitative consequences of this treatment have been confirmed by experiment:

(1) The orthobaric curves meet at T^c as a continuous curve and not at a sharp intersection, and so
(2) The critical density is the mean of the orthobaric densities near the critical point.
(3) The vapour-pressure curve is collinear with the critical isochore.

It follows from equation (3.20) that α_σ, c_σ and $(\partial h/\partial T)_\sigma$ are all proportional to $(\delta T)^{-1/2}$ near T^c for a classical fluid, and so are an order of magnitude smaller than α_p, β_T and c_p, which are proportional to $(\delta T)^{-1}$. Again this conclusion is confirmed qualitatively by experiment.

All liquids pass through a Joule–Thomson inversion point some way below the critical point, at the temperature at which α_p is equal to $(1/T)$. The first of the inequalities (2.44) is reversed at this point, and c_p is greater than $(\partial h/\partial T)_\sigma$ from here to the critical point. The isothermal Joule–Thomson coefficient becomes infinite at the critical point but the adiabatic coefficient remains finite and is equal to $(1/\gamma_V^c)$.

In order to show that a thermal instability, that is, a zero in (T/C_V) as in equation (2.61), plays no essential part in determining the gas–liquid critical point, it is necessary to use the less familiar (U, S, V) surface[a], since A is not a differentiable function of V and T, nor ψ of ρ and T, if (T/C_V) is anywhere zero. The three equations (3.1)–(3.3) become on this surface

$$(\partial U/\partial S)_V^l = (\partial U/\partial S)_V^g \tag{3.23}$$

$$(\partial U/\partial V)_S^l = (\partial U/\partial V)_S^g \tag{3.24}$$

$$U^l - S^l(\partial U/\partial S)_V^l - V^l(\partial U/\partial V)_S^l = U^g - S^g(\partial U/\partial S)_V^g - V^g(\partial U/\partial V)_S^g \tag{3.25}$$

They are more symmetrical on this surface, where all the variables are extensive, than on the (A, V, T) surface. They are the conditions that the tangent planes to the (U, S, V) surface at the gas and liquid points should coincide. As the common tangent plane rolls over the surface, its two points of contact trace out the orthobaric curves which eventually meet at the critical point. These curves form a line along which U, S and V all increase monotonically on passing from liquid, through the critical point, to the gas[b]. The isotherms and isobars on this surface are the loci of the points of equal slope parallel to the principal axes, equations (3.23) and (3.24). The

[a] Photographs of a model of the $(-U, S, V)$ surface made by Maxwell are shown by Wilson[9]. A second model was made by Onnes and Happel[10]; the (U, V, T) surface is discussed by Wood[11].
[b] If C_σ for the gas becomes positive at low temperatures, there is a maximum in S on this branch. It is everywhere negative for most simple substances.

region of stability is given by equation (2.59) as

$$D > 0, \text{ where } D \text{ is the determinant } \begin{vmatrix} U_{2S} & U_{SV} \\ U_{SV} & U_{2V} \end{vmatrix} \tag{3.26}$$

The determinant is zero on the boundary curve which passes through the critical point and which separates the regions of stability and instability. Hence

$$D^c = 0 \quad \text{or} \quad \left(\frac{T}{C_V}\right)^c \left(-\frac{\partial p}{\partial V}\right)_T^c = 0 \tag{3.27}$$

$$(\partial p/\partial V)_S > (\partial p/\partial V)_T \geqslant 0 \tag{3.28}$$

$$(T/C_V) > (T/C_p) \geqslant 0 \tag{3.29}$$

3.2 Inequalities at the critical point

The assumption that A is an analytic function of V and T or ψ of ρ and T, at and near the critical point, leads to a full description of the ways in which $(\rho^1 - \rho^g)$, $(\partial p/\partial V)_T$, (T/C_p), etc., become zero as the critical point is approached; but it is, of course, irreconcilable with the requirement that the compressibility is everywhere positive. This is clear from *Figure 3.1*. The difficulty is overcome by the device of the common tangent (*Figure 3.1*) or Maxwell's rule (*Figure 3.2*). However, the description remains conceptually unsatisfactory and, as is shown below, is also quantitatively inaccurate.

The requirements of thermodynamic stability, and in particular that (T/C_V) be positive, lead to some powerful inequalities between the rates at which different functions become zero as the critical point is approached along various paths. These inequalities do not depend upon A being everywhere an analytic function, but do require that A, U, S and p are continuous functions of V and T. They are valuable both for the insight they give into the possible algebraic forms of A and for checking experimental results. In fact, the margins by which real fluids satisfy them are so small that it is now assumed that they are true equalities, and on this assumption rests the description of the critical point given in Section 3.5.

These results can be obtained in two ways. The requirement that C_V is positive means that A is a function of T which is everywhere convex upwards. Griffiths[12] used such conditions of convexity to obtain many inequalities, some of which are of great practical value. However, many, and possibly all, of these inequalities can be derived from exact equations from which one or more terms have been discarded. The omitted terms are functions of heat capacities and so are of known sign. This method of derivation is due to Rushbrooke[13] who first obtained one of these inequalities for the Curie point of a ferromagnet. His derivation was extended to the critical point of a fluid by Fisher[14]. A variant of this method is followed here which leads, in one operation, to the three most useful inequalities. Since it requires no Taylor expansion and involves no considerations of symmetry, the argument is most conveniently expressed in the familiar variables, p, V, and T.

The inequalities are between the rates of approach to zero of different functions as $(T - T^c)$ and $(V - V^c)$ tend to zero. Consider some function $X(V, T)$ and define an exponent χ by

$$\chi^{\pm} = \lim_{T \to T_{\pm}^c} \{\ln X(V, T)/\ln[\pm(T - T^c)]\} \tag{3.30}$$

This exponent defines the rate of approach of X to zero, or to infinity if χ is negative. The path to be followed in taking the limit must be specified and is often $V = V^c$. The exponents χ^+ and χ^- can differ, and χ^- can sometimes be divided further into χ_1^- and χ_2^- according as the approach to the critical point is made along a path in the one-phase or two-phase fluid. The superscripts and subscripts are omitted when no confusion can arise.

The exponents are defined in *Table 3.1*, in which the symbols are those used in this field but which must not be confused with α, the coefficient of thermal expansion, β, the compressibility, etc., introduced in Chapter 2. It is

Table 3.1

Property	Exponent	Comments
C_V	$-\alpha^+$	along critical isochore, $T > T^c$
	$-\alpha_1^-$	through orthobaric states of homogeneous fluid
	$-\alpha_2^-$	along critical isochore, $T < T^c$
$(V^g - V^c)$, $(V^c - V^l)$	β	through orthobaric states
$-(\partial p/\partial V)_T$	γ^+	along critical isochore, $T > T^c$
	γ_1^-	through orthobaric states of homogeneous fluid
$|p - p^c|$ as function of $(V - V^c)$	δ	along critical isotherm
$(\partial^2 p/\partial T^2)_\sigma$	$-\theta$	curvature of vapour-pressure line

possible to subdivide further the exponents α_1^-, β, γ_1^- and δ if the coexistence curve and the critical isotherm are of different degree for $V > V^c$ and $V < V^c$. Rushbrooke[13] and Griffiths[12] consider such possibilities, but there is no experimental evidence for such differences and so they are ignored here.

An exponent of zero is ambiguous. It can mean that the function has no zero or infinity (but may have a discontinuity), or else that there is a logarithmic infinity.

We now express the results of the last section for the generalized classical fluid in terms of these exponents. If the first non-vanishing derivative of A with respect of V at the critical point is of order $2n$, where n is necessarily integral, then

$$\alpha^+ = \alpha_1^- = 0 \qquad \alpha_2^- = (n-2)/(n-1) \tag{3.31}$$

$$\beta = 1/2(n-1) \qquad \gamma^+ = \gamma_1^- = 1 \tag{3.32}$$

$$\delta = 2n - 1 \qquad \theta = (n-2)/(n-1) \tag{3.33}$$

The classical van der Waals fluid ($n = 2$) has a simple discontinuity in C_V, that is $\alpha_2^- = 0$.

The required inequalities between these exponents can now be obtained by considering the function $Y(V, T)$, defined by

$$Y = U - T^c S \tag{3.34}$$

whose derivatives are

$$\left(\frac{\partial Y}{\partial T}\right)_V = -\frac{T^c - T}{T} C_V \qquad \left(\frac{\partial Y}{\partial V}\right)_T = -(T^c - T)\left(\frac{\partial p}{\partial T}\right)_V - p \tag{3.35}$$

The difference

$$\Delta Y = Y(V^c, T) - Y(V^c, T^c) \qquad (T < T^c) \tag{3.36}$$

can now be obtained in three different ways. Equating these results, two at a time, gives three equations, from each of which, by omission of suitable terms, we obtain a different inequality. The difference is found by adding the changes in Y along each of the three paths shown in *Figure 3.3*. These are: (a) from C to D, (b) from C to A^l, A^g to B^l, B^g, and (c) from C to B^l, B^g through the orthobaric states. Each mole of fluid at D corresponds to x^l

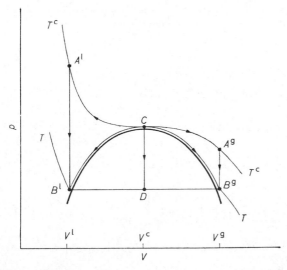

Figure 3.3 The change ΔY of equation (3.36) is that from point C to point D and may be found in three ways by considering changes of Y along the three paths shown

moles in state B^l and x^g moles in B^g, where

$$x^l = \frac{V^g - V^c}{V^g - V^l} \qquad x^g = \frac{V^c - V^l}{V^g - V^l} \tag{3.37}$$

Here, V^g and V^l are the volumes at B^g and B^l, and in what follows these are taken to be fixed at those appropriate to the final temperature T; they do not change as the fluid is taken along paths such as A^g to B^g, C to B^g, etc. Quantities that change along the path C to B^g are denoted by the subscript σ; $V = V_\sigma^g$.

The changes in Y are

(a) From C to D

$$\Delta Y = \int_T^{T^c} \frac{T^c - T}{T} (C_V)_{V=V^c} \, dT \tag{3.38}$$

(b) From C to A^l, A^g to B^l, B^g.

Here, each mole of the fluid at C is split into two samples of x^l and x^g moles each, where these are the mole fractions appropriate to the final temperature T. The first sample is taken along the path CA^lB^l and the second along CA^gB^g:

$$\Delta Y = \sum_{l,g} x^{l,g} \left\{ \int_{V^{l,g}}^{V^c} p_{T=T^c} \, dV + \int_T^{T^c} \frac{T^c - T}{T} (C_V)_{V=V^{l,g}} \, dT \right\} \tag{3.39}$$

$$= \sum_{l,g} x^{l,g} \left\{ \int_{V^{l,g}}^{V^c} (p - p^c)_{T=T^c} \, dV + \int_T^{T^c} \frac{T^c - T}{T} (C_V)_{V=V^{l,g}} \, dT \right\} \tag{3.40}$$

where the summation is over the two samples with superscript l and g.

(c) From C to B^lB^g

$$\Delta Y = -\sum_{l,g} x^{l,g} \int_T^{T^c} \left(\frac{\partial Y}{\partial T} \right)_{V=V_\sigma^{l,g}} dT \tag{3.41}$$

$$= -\sum_{l,g} x^{l,g} \int_T^{T^c} \left[\left(\frac{\partial Y}{\partial T} \right)_V + \left(\frac{\partial Y}{\partial V} \right)_T \left(\frac{\partial V}{\partial T} \right)_\sigma \right]_{V=V_\sigma^{l,g}} dT \tag{3.42}$$

$$= \sum_{l,g} x^{l,g} \left\{ \int_T^{T^c} \frac{T^c - T}{T} (C_V)_{V=V_\sigma^{l,g}} \, dT \right.$$

$$+ \int_T^{T^c} \left[\left(\frac{\partial V}{\partial T} \right)_\sigma \left(p_\sigma + (T^c - T) \left(\frac{\partial p}{\partial T} \right)_\sigma \right) \right.$$

$$\left. \left. - (T^c - T) \left(\left(\frac{\partial p}{\partial V} \right)_{T,\sigma} \left(\frac{\partial V}{\partial T} \right)_\sigma^2 \right) \right]_{V=V_\sigma^{l,g}} dT \right\} \tag{3.43}$$

Now, if β is the same on both sides of the critical point[13]

$$\sum_{l,g} x^{l,g} \int_T^{T^c} \left[\left(\frac{\partial V}{\partial T} \right)_\sigma \right]_{V=V_\sigma^{l,g}} \left[p_\sigma + (T^c - T) \left(\frac{\partial p}{\partial T} \right)_\sigma \right] dT = 0 \tag{3.44}$$

since $(\partial V / \partial T)_\sigma$ is positive on the liquid branch and negative on the gas branch, and since $[p_\sigma + (T^c - T)(\partial p / \partial T)_\sigma]$ is the same on both branches. Hence equation (3.43) becomes

$$\Delta Y = \sum_{1, g} x^{1, g} \left\{ \int_T^{T^c} \frac{T^c - T}{T} (C_V)_{V = V_\sigma^{1, g}} \, dT \right.$$

$$\left. - \int_T^{T^c} (T^c - T) \left[\left(\frac{\partial p}{\partial V} \right)_T \left(\frac{\partial V}{\partial T} \right)_\sigma^2 \right]_{V = V_\sigma^{1, g}} dT \right\} \tag{3.45}$$

We have three equations for ΔY, (3.38), (3.40) and (3.45). Each term in these is positive and, in particular, the two pairs of integrals (l and g)

$$\int_T^{T^c} \frac{T^c - T}{T} (C_V)_{V = V^{1, g}} \, dT > 0$$

and $\hspace{10cm}$ (3.46)

$$\int_T^{T^c} \frac{T^c - T}{T} (C_V)_{V = V_\sigma^{1, g}} \, dT > 0$$

First, equate (3.38) and (3.40) and omit the first pair of integrals in (3.46). The resulting inequality can be turned into one in the exponents of *Table 3.1* by substituting for $(C_V)_{V = V^c}$, etc., the expression $(T^c - T)^{-\alpha_2}$, etc. The larger quantity has an exponent equal to or lower than that of the smaller, and so we obtain

$$\alpha_2^- + \beta(\delta + 1) \geqslant 2 \tag{3.47}$$

Similarly, by equating (3.38) and (3.45) and omitting the second pair of integrals in (3.46), we have

$$\alpha_2^- + 2\beta + \gamma_1^- \geqslant 2 \tag{3.48}$$

A third inequality is obtained by equating (3.40) and (3.45). The equation so formed contains terms that are differences of the integrals in (3.46), that is, of the form

$$\int_T^{T^c} \frac{T^c - T}{T} [(C_V)_{V = V_\sigma^{1, g}} - (C_V)_{V = V^{1, g}}] \, dT \tag{3.49}$$

There is good experimental evidence that C_V increases on approaching the orthobaric boundary from either side along isotherms just below the critical point. This behaviour is also required by the presence of an infinity in C_V for the one-phase fluid (see below). Hence expression (3.49) is necessarily positive, and its neglect leads to a third inequality, if again β is the same on each side of the critical point,

$$\gamma_1^- - \beta(\delta - 1) > 0 \tag{3.50}$$

This result was put forward by Griffiths for the Curie point of a ferromagnet; Liberman[15] derived it for a fluid.

The only other inequality discussed below is a much weaker one, obtained by both Griffiths and Rushbrooke, namely,

$$\alpha_2^- + \beta - \theta > 0 \qquad (3.51)$$

There are analogous inequalities, as has been mentioned, for other 'critical points', such as the Curie point of a ferromagnet, the Néel point of an antiferromagnet, and for the corresponding points on many lattice models of these phenomena—the lattice gas, the Ising model and the Heisenberg model. These inequalities are sometimes easier to derive than those for fluids, since there is often an essential symmetry in the ordering parameter with respect to the sign of the field which is not present in a fluid between $(V^g - V^c)$ and $(V^c - V^l)$ or even between $(\rho^l - \rho^c)$ and $(\rho^c - \rho^g)$. This book is not the place to explore the analogies between critical points of different kinds, or between these and the transitions to superfluid or superconducting states, but it may be said that, despite the diverse physical nature of these phenomena, the underlying singularities in the appropriate thermodynamic functions are remarkably similar. (There is now an immense body of work on critical points of all kinds, for a full account of which the reader must turn to other books[16].) The dimensionality of the system and the range of the interactions are of greater importance in determining the nature of the singularities than the details of interactions between the elements of the systems. Critical points in fluid mixtures are the closest to the gas–liquid points of this chapter, and these are discussed in Chapters 5 and 6.

The inequalities of this section become equations if A is an analytic function. This follows at once from equations (3.31)–(3.33). A more realistic representation of A as a function of V and T is to require it to be analytic only in the homogeneous region but not on the phase boundary. This is the minimal assumption needed to justify the 'common tangent' (or Maxwell's) rule, as may be seen by equating chemical potentials in two orthobaric states along a path lying wholly within the homogeneous fluid; a formal proof has been given by Griffiths[17].

3.3 The measurement of critical constants

The classical view of the critical point is that it is the state at which the densities of the coexisting phases are equal, and is also the highest temperature and pressure at which $(\partial p/\partial V)_T$ is zero. It has not been easy to establish experimentally that these two conditions define the same state of the fluid. The difficulties arise from the highly unusual mechanical, thermal and optical properties of the fluid in this region: the compressibility is infinite, and so the earth's gravitational field is large enough to produce large gradients of density in a vessel a few centimetres high; the infinities of C_p and C_V make it hard to reach thermal equilibrium. Even slow changes of pressure are more nearly adiabatic than isothermal, and cyclic changes lead to hysteresis.

As the densities of the two phases approach each other, the dividing meniscus becomes at first faint and then very hazy, and the measurement of the exact temperature of its disappearance needs great care. Observation of the system is made harder by a strong scattering of light—the *critical opalescence*—which can lead to complete opacity at the critical point.

All these effects make precise equilibrium measurements unusually difficult, and good agreement between critical constants measured in different ways is hard to achieve. Levelt Sengers[18] has reviewed experimental techniques in this field.

The critical temperature is the easiest of the three constants to measure. It is usually found by observing the temperature at which the meniscus vanishes in a system maintained at an overall density approximately equal to the critical. If a sealed tube containing a liquid and its vapour is heated uniformly, then one of three things may happen. If the overall density is less than the critical, the meniscus falls as the liquid evaporates. The equation for the rate of change with temperature of the volume of either phase in a closed system follows directly from equation (2.89)

$$\frac{dV^l}{dT} = -\frac{dV^g}{dT} = \frac{n^l\alpha_\sigma^l + n^g\alpha_\sigma^g}{(1/v^l) - (1/v^g)} \tag{3.52}$$

The coefficients of thermal expansion, α_σ^l and α_σ^g, are positive and negative, respectively, and so the right-hand side of this equation is negative if the overall density is low (n^l small). The volume of the liquid decreases until it has all evaporated, after which further heating produces a less rapid rise of pressure than that along the saturation curve.

The second possibility is that the density exceeds the critical (n^l large) when the meniscus rises on heating until no vapour remains. Further heating produces a rapid rise in pressure along a liquid isochore.

If, however, the density is close to the critical, then the meniscus rises slowly until it is near the centre of the tube where it becomes flat and faint and eventually vanishes at the critical temperature.

No great care is needed to load the tube exactly to the critical density for measurements of moderate precision (say, 0.05–0.1 K), for a slight error only causes the meniscus to vanish a little above or below the centre of the tube, at a height at which the local density is equal to the critical. However, for the best work (measurements to 0.01 K or better) the density must be within 1 per cent of the critical, the tube must be short and well-stirred, and the final heating must be carried out very slowly in an accurately controlled thermostat. Conversely, the lack of sensitivity to the loading density means that this method cannot be used for any but the roughest measurements of the critical volume.

If the tube is not sealed but open at its lower end to a reservoir of mercury, then the density may be altered at will and the pressure measured at the same time. The critical pressure may be measured with little more trouble than the temperature, as it is equally insensitive to small changes of density at this point. This was the apparatus used in the original work of Andrews[19] and later by Young[20]. It has not been changed significantly since.

The critical volume is the most difficult of the three constants to measure accurately. The method which is probably the best, and certainly the most commonly used, is to extrapolate the mean of the orthobaric liquid and gas densities up to the critical temperature. It was shown earlier, equations (3.20), that the classical description of the critical region requires that the critical density should be equal to the mean of the orthobaric densities at temperatures just below the critical. This was shown to be so experimentally by Cailletet and Mathias[21] in 1886, and the statement is now incorporated in what is called the *law of rectilinear diameters*. They found that if they drew diameters (*AB, Figure 3.4*) across the graph of orthobaric

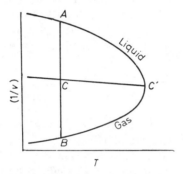

Figure 3.4 The law of rectilinear diameters. The line CC' is the locus of the mid-points of diameters such as AB

densities as functions of the temperature, then the mid-points of these diameters, *C*, etc., lay on a straight line[a] which passed through the critical density *C'*. They thought at first that this line was parallel to the temperature axis, but it is now known that it has a slight slope and a small but usually negligible curvature. The fact that the law appears to hold even close to the critical point is evidence that the exponent β of the last section is the same on both branches of the orthobaric curve.

This empirical rule holds over the whole liquid range but is most commonly used to fit measurements of ρ^l and ρ^g from about 50 to 3 K below T^c, so as to obtain ρ^c by extrapolation. It reduces to equations (3.20) as T approaches T^c. *Figure 3.5* shows some moderately accurate measurements[22] of the density of nitrous oxide which conform well to this law and from which a value of ρ^c may be determined that is probably correct to 1 per cent.

---

[a] The term *rectilinear diameter* is now usually given to this locus of the mid-points and not to the lines, *AB*, etc.

The results plotted in *Figure 3.5* stop about $2\,K$ below the critical temperature, as do most measurements made before 1970. It was suggested that year that there were strong theoretical arguments, based on the behaviour of model systems[23], for supposing that the diameter was not truly rectilinear near T^c but has a weak singularity in its slope; that is, that $(\rho^l + \rho^g - 2\rho^c)$ behaved as $|\delta T|^{1-\alpha}$, where $-\alpha$ is the same exponent that governs the behaviour of C_V (*Table 3.1*). We see below that C_V has a logarithmic (or stronger) singularity in a real fluid, and the implication from the results for the models is that real fluids should therefore have also a singularity in the slope of the rectilinear diameter. Some recent measurements[24] claim to have detected this, but their analysis is far from simple[25]. It was pointed out by Buckingham[26] and emphasized by Scott[26] that if one particular function, e.g., ρ, has a $|\delta T|^{1-\alpha}$ singularity, then any less symmetric function, e.g., $v = \rho^{-1}$, or ρ^2, behaves as $|\delta T|^{2\beta}$, and so the sought-for effect will be missed unless the correct function is chosen. There are strong, but not conclusive arguments for supposing that ρ is the appropriate function to choose.

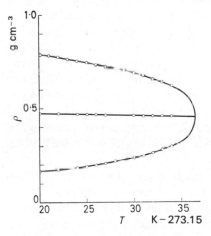

Figure 3.5 *The orthobaric densities of nitrous oxide[22]. The straight line through the mean densities has the equation*

$$(\rho^l + \rho^g)/2 = [0.452 + 0.00132(T^c - T)/K]\ \text{g cm}^{-3}$$

The measurement of the critical constants by finding the point of zero slope and of inflection on the critical isotherm is much more difficult, and such attempts have in the past led to some notable disagreements. Schneider and his colleagues[27] have shown by measurements which are more accurate than most that true equilibrium can be achieved only by keeping the temperature constant to within $0.001\,K$ and by continuously stirring the fluid for many hours. Under these conditions, the coexistence curves of xenon and sulphur hexafluoride have rounded tops. The effect of gravity was eliminated by studying the curve with the tube in both vertical and horizontal positions. The coexistence curves found in the former position are flatter than those in the latter and lie outside them. The latter are undoubtedly the more 'correct' curves, that is, they represent the behaviour of the fluids in a negligible gravitational field, since no density gradient can be set up in a tube of zero height.

Schneider was further able to show, by using [127]Xe as a tracer, that the density gradient in a vertical tube of xenon was no greater than that to be expected for the earth's gravitational field when a correct allowance was made for the very great compressibility in this region. This conclusion is confirmed by optical measurements[28-30].

The heat capacity at constant volume of a fluid near its critical point has been measured in two ways. First, the residual heat capacity may be found from an accurate knowledge of the pressure as a function of volume and temperature. From equations (2.16) and (2.97)

$$c_v^* = -T \int_v^\infty (\partial^2 p/\partial T^2)_V \, dv \qquad (3.53)$$

The integrand is the curvature of the isochore, or the rate of change of γ_V with temperature. It is negative both in the gas phase and in the saturated liquid at low temperatures. It is always small and never easy to measure.

Secondly, it is possible to measure the heat capacity directly for a fluid in a sealed metal container near the critical point, where γ_V is never more than about $2 \, \text{bar} \, \text{K}^{-1}$. However, the heat capacity of the container is much greater than that of the fluid if it is strong enough to withstand the critical pressure. A further difficulty with this second method is the maintenance of equilibrium. Slow changes of temperature are apparently necessary if accurate results are to be obtained. This is always feasible in a mechanical measurement but rarely in a thermal one.

Young[31] first showed that $(\partial^2 p/\partial T^2)_V$ changes sign at or near the critical point by some careful measurements of the isotherms of isopentane. More recently, Michels, Bijl and Michels[32] have shown that carbon dioxide behaves similarly. At high densities there is a second change of sign for this gas where C_V passes through a minimum. The maximum in C_V decreases with increasing temperature but is still detectable in carbon dioxide 100 K above the critical temperature. This method of observing the maximum in C_V is, however, not sufficiently precise for determining the exponent α^+ of *Table 3.1*. The differentiation and subsequent integration needed in equation (3.53) smooth out any true infinity into a high but smooth maximum.

The calorimetric measurement of C_V is made by heating the two-phase system through the critical boundary curve and into the homogeneous state beyond it. Early measurements of C_V along the critical isochore confirmed that there was a sharp rise as the critical point was approached from above or below, and also showed that the heat capacity below the critical temperature exceeded that above for a given difference $|T - T^c|$. Michels and Strijland[32] showed that there was almost certainly an infinity below T^c, but their results, together with their study of the curvature of the iso-chores[32], do not prove the existence of an infinity on the high-temperature side. It is the work of Voronel and his colleagues[33,34] on argon, oxygen, nitrogen and ethane, of Buckingham[35] on xenon and carbon dioxide, and of Moldover[36] on helium which suggest most strongly that there are infinities in C_V both above and below T^c. They appear to be logarithmic or sharper.

If the heat capacity of the one-phase system is finite at the critical point, then it may be shown that the equilibrium speed of sound is non-zero. Equation (2.50) becomes indeterminate at the critical point, but by sub-stituting for β_S in equation (2.46) the following expression is obtained:

$$W^2 = \frac{v}{\mathscr{M}}\left[\frac{1}{\beta_T} + \frac{Tv\gamma_V^2}{c_v}\right] \tag{3.54}$$

where $\mathscr{M}$ is the molar mass. At the critical point, this reduces to

$$W^2 = Tv^2(\gamma_\sigma^c)^2/\mathscr{M}c_v \tag{3.55}$$

where γ_σ^c is the limiting slope of the vapour-pressure curve. Similarly, the adiabatic compressibility is given by

$$\beta_S = c_v/Tv(\gamma_\sigma^c)^2 \tag{3.56}$$

Thus a maximum (or infinity) in the heat capacity will lead to a maximum (or infinity) in β_S and a minimum (or zero) in W at the critical point.

The experimental speeds of sound have all been measured at high frequencies (generally about 0.5 MHz) so that the wavelength can be kept small compared with the dimensions of the sample of fluid. Unfortunately, the speed at such frequencies is not the equilibrium speed. Voronel[34] has discussed the discrepancy between the observed and the 'thermodynamic' speeds.

3.4 Experimental values of the critical exponents

The difficulties of measuring thermodynamic functions in the critical region were set out in the last section. They make it hard to arrive at precise values of the critical exponents of Section 3.2, and several reviews [4,37–40] have been devoted to assessing the experimental evidence. These are readily available, and some are of great length. We do not repeat their assessments here, but instead summarize their conclusions and indicate the principal experimental sources on which they have relied.

The substances on which most of the best experimental work has been done are helium (^{3}He and ^{4}He), argon, xenon, oxygen, carbon dioxide, and sulphur hexafluoride, but there is no need to distinguish between them since the evidence is that all substances have the same critical exponents. These fluids have very different mass densities and the analysis of the measure-ments requires that full attention be given to gravitational corrections[4,40,41].

We consider the exponents in turn.

α^+ The measurements cited in the last section show that there is an infinity in C_V which is a little sharper than logarithmic, $\alpha^+ = 0.10 \pm 0.05$.

α_1^- No information.

α_2^- The infinity on the critical isochore below T^c appears to be the same as that above ($\alpha_2^- = \alpha^+$) although $(C_V)_2 > (C_V)_1$ for a given value of $|\delta T|$.

β This is the exponent that seems easiest to measure and was, for many years, taken to be 0.35 ± 0.01. The optical method introduced by Wilcox and his colleagues[30] has, however, allowed measurements to be made very close to T^c, that is, for $(\delta T/T^c) < 10^{-4}$, which is a range impossible to probe by mechanical measurements. The fringe numbers and angles of the Fraunhofer pattern from the density profile of a thin slab of fluid are determined by ρ, μ and the compressibility β_T. Such measurements[39] suggest that the exponent β falls to 0.32 ± 0.01 in the 'true' critical region, that is the temperature range in which the apparent values of the exponents have reached their limiting values.

γ^+ The best values, including those from optical measurements[39], are in the range 1.24 ± 0.05.

γ^- This is hard to measure but there is no reason to suppose that it is not equal to γ^+.

δ This exponent is certainly larger than its classical value of 3 and appears to be 4.8 ± 0.2. However, there is some doubt about this value, and the true accuracy may be less than the apparent precision. This is a consequence of the fact that the exponent is so large that the measurements on which it is based are necessarily well removed from the critical point.

θ This is hard to measure. If experimental vapour pressures are fitted to an empirical analytic equation, then it is commonly found that the deviation graph shows a sudden sweep upwards near the critical point. Unfortunately, the infinity (*Table 3.1*) is only in the second derivative of p_σ and can be fitted by values of θ as large[42] as 0.5. A recent analysis of measurements on ethylene[43] leads, however, to $\theta \sim 0.1$.

If we assume that all exponents of the same kind are equal, $\alpha^+ = \alpha_1^- = \alpha_2^-$, and $\gamma^+ = \gamma_1^-$—an assumption for which there are theoretical arguments and no contradictory experimental evidence—then we have

$$\alpha = 0.10 \pm 0.05 \qquad \beta = 0.32 \pm 0.01$$

$$\gamma = 1.24 \pm 0.05 \qquad \delta = 4.8 \pm 0.2$$

These results can be tested by means of the inequalities (3.47), (3.48) and (3.50); the fourth (3.51) is amply satisfied if $\alpha \sim \theta$. The figures above lead to

$$\alpha + \beta(\delta + 1) = 1.95 \pm 0.15$$

$$\alpha + 2\beta + \gamma = 2.00 \pm 0.10$$

$$\gamma - \beta(\delta - 1) = 0.0 \pm 0.1$$

The natural conclusion is that the experimental results satisfy the in-equalities as equations, as do the exponents obtained by assuming that $A(\delta V, \delta T)$ or $\psi(\delta\rho, \delta T)$ is an analytic function, (3.31)–(3.33). If this is so then only two of the four exponents α, β, γ and δ are independent, and this relation imposes a restriction on the possible (non-analytic) forms of $A(\delta V, \delta T)$ or $\psi(\delta\rho, \delta T)$ to which we now turn.

3.5 Scaling the free energy

The discussion above shows that the classical picture of the critical point is correct only in its broad outlines and not in its specific predictions. Two experimental simplifications have emerged from the comparison of the classical picture with reality; first, that ψ_s the singular part of the free energy density, is symmetric in $\delta\rho$, and secondly, that only two of the four exponents α, β, γ and δ, are independent. The other two are found by taking three of the inequalities of Section 3.2 to be equations. These two simplifications tell us much about the function $\psi(\rho, T)$.

The first point, symmetry, recalls Landau's discussion of higher-order phase transitions[44], which can be regarded as a generalization or abstraction of the classical treatment of Section 3.1. He supposed that the singular part of ψ could be written as a quadratic function of s^2, where s is an order-parameter which is a measure of the difference of properties of the coexisting phases. We return to Landau's treatment later, but here note that a natural candidate for an order-parameter is the density difference $\delta\rho^l = -\delta\rho^g$. Let us therefore take the Taylor expansion (3.9), stopping at the fourth power in $\delta\rho$, and see how it can be decomposed into a symmetrical singular part and a smooth background term.

$$\psi(\delta\rho, T) = \alpha_0 + \alpha_1(\delta\rho) + \alpha_2(\delta\rho)^2 + \alpha_3(\delta\rho)^3 + \alpha_4(\delta\rho)^4 \tag{3.57}$$

where α_i are functions of temperature. Stability requires that ψ_{40}^c and ψ_{21}^c are both positive, where ψ_{mn} are the coefficients defined in (3.10). These conditions are

$$\alpha_4 > 0 \quad \text{and} \quad \text{sgn } \alpha_2 = \text{sgn } \delta T \quad \text{at } \delta T \sim 0 \tag{3.58}$$

By differentiation of equation (3.57)

$$\mu = (\delta\psi/\delta\rho)_T = \alpha_1 + 2\alpha_2(\delta\rho) + 3\alpha_3(\delta\rho)^2 + 4\alpha_4(\delta\rho)^3 \tag{3.59}$$

Let us define a difference $\Delta\mu$ by

$$\Delta\mu = \mu(\rho, T) - \mu(\rho^c, T) \tag{3.60}$$

which we require to be the antisymmetric part of μ, in accord with the experimental evidence[4,6,7]. The second term, $\mu(\rho^c, T)$ is α_1, and so, from equation (3.59)

$$\Delta\mu = 2\alpha_2(\delta\rho) + 3\alpha_3(\delta\rho)^2 + 4\alpha_4(\delta\rho)^3 \tag{3.61}$$

This function must be antisymmetric, and be the derivative of ψ_s, the singular part of ψ. The antisymmetry requires that $\alpha_3 = 0$ and it is conventional to scale μ and the coefficients so that the essentially positive coefficient α_4 becomes unity. Hence

$$\Delta\mu = 2\alpha_2(\delta\rho) + 4(\delta\rho)^3 \tag{3.62}$$

The two parts of ψ are therefore the background term $[\alpha_0 + \alpha_1(\delta\rho)]$, and the singular part whose derivative is $\Delta\mu$:

$$\psi_s(\delta\rho, T) = \alpha_2(\delta\rho)^2 + (\delta\rho)^4 \tag{3.63}$$

$$(\partial\psi_s/\partial\delta\rho)_T = \Delta\mu \tag{3.64}$$

The background term $[\alpha_0 + \alpha_1(\delta\rho)]$ is $[\psi(\rho^c, T) + \mu(\rho^c, T)\delta\rho]$; that is, it comprises a term which is a function only of T, $[\psi(\rho^c, T) - \mu(\rho^c, T)\rho^c]$, and a term which is linear in ρ, namely $[\mu(\rho^c, T)\rho]$. The distinction between smooth and singular parts of ψ is somewhat artificial in this purely classical description, but its value becomes evident later in this section.

The equations above reduce the classical description to its simplest terms. If $\alpha_2 > 0$, then $\Delta\mu$ is a monotonic function of $\delta\rho$, and so there is one phase. If $\alpha_2 < 0$, then there are two phases whose properties are given by the equation $\Delta\mu = 0$, or, from equation (3.62),

$$\delta\rho^l = -\delta\rho^g = (-\tfrac{1}{2}\alpha_2)^{1/2} \tag{3.65}$$

Clearly, α_2 is equal (or proportional) to δT, and this equation is the same as (3.20). The critical point is at $\alpha_2 = 0$, $\delta\rho = 0$.

The representation of ψ in equation (3.57) is in terms of its canonical variables ρ and T, (3.8), but it is often convenient to use instead intensive variables that take the same values in both phases, that is, μ, p, and T, or linear combinations of these. Let such variables, or some as yet unspecified combination of them, be denoted by a set of coefficients a_i. It is the essence of Landau's treatment that $\psi(a_i)$ be found by writing a more general free-energy $\Psi(a_i, s)$ formally as an expansion of powers of an order-parameter s, and then minimizing Ψ with respect to s. The function Ψ is called a generalized free energy since it is defined for values of s other than the equilibrium value of this parameter. We have, therefore,

$$\psi(a_i) = \min_s \Psi(a_i, s)$$
$$\Psi(a_i, s) = a_1 s + a_2 s^2 + s^4 \tag{3.66}$$

(For complete generality we should add a background term to $\psi(a_i)$ but since it plays no part in determining the phase behaviour near the critical point, we omit it.) The form of Ψ in equation (3.66) is quite general, for if there were a term in s^3 it could be removed by changing the origin from which the so-far undefined parameter s is measured; that is, by replacing s by $(s + s_0)$. Similarly, the coefficient a_4, which is essentially positive, can be put equal to unity by choosing a suitable scale. The minimization of Ψ

requires that $s = s_m$, where

$$\partial \Psi / \partial s = a_1 + 2a_2 s_m + 4s_m^3 = 0 \qquad (3.67)$$

With this value of a_1, (3.66) becomes

$$\psi = -a_2 s_m^2 - 3s_m^4 \qquad (3.68)$$

which is

$$\psi_s - \Delta\mu.\delta\rho = -\alpha_2(\delta\rho)^2 - 3(\delta\rho)^4 \qquad (3.69)$$

if we identify $\delta\rho$ with s_m, the equilibrium value of the order parameter, α_2 or δT with a_2, and so $\Delta\mu$ with $-a_1$.

The minimization of Ψ to give ψ as a function of the intensive variables (or fields), a_i, can be done by inspection[45] if a_1 and a_2 are replaced by two parameters c and r defined by

$$a_1 = -2cr^2 \qquad a_2 = r^2 - 2c^2 \qquad (3.70)$$

when equation (3.66) becomes

$$\Psi = (s - c)^2[(s + c)^2 + r^2] - c^2(c^2 + r^2) \qquad (3.71)$$

Clearly, Ψ is at a minimum at $s = c$ and, if $r^2 = 0$, also at $s = -c$. Hence the free energy density ψ of equation (3.66) is

$$\psi(a_1, a_2) = -c^2(c^2 + r^2) \qquad (3.72)$$

where a_1 and a_2 are given parametrically in terms of c and r by equation (3.70).

The phase diagram is again readily constructed. If $r^2 = 0$, there are two phases, $s = \pm c$, and the two variables a_1 and a_2 are given by $a_1 = 0$, $a_2 \leqslant 0$. From equation (3.72), the order parameter takes the two values $s_m = \pm(-\frac{1}{2}a_2)^{1/2}$. If $r^2 > 0$, then $a_2 > 0$ and there is only one phase $s_m = c$ for each pair of values (a_1, a_2). Thus in (a_1, a_2)-space the phase boundary is a straight line on the negative half of the a_1 axis (i.e., $a_1 = 0$, $a_2 \leqslant 0$) which ends at a critical point at $a_1 = a_2 = 0$.

Landau's description is a classical one since the exponents are $\alpha = 0$, $\beta = \frac{1}{2}$, $\gamma = 1$ and $\delta = 3$, but it is useful for two reasons. First, it concentrates attention on the part of ψ that has the useful property of *scaling*, which is discussed below. In practice, equations (3.64) and (3.66) define the relevant part of ψ, namely,

$$\psi_s(\rho, T) = \psi(\rho, T) - \psi(\rho^c, T) - (\rho - \rho^c)\mu(\rho^c, T) \qquad (3.73)$$

Secondly, we shall see in the next chapter that equation (3.66) has a natural generalization which is useful for the discussion of critical points of higher order that occur in mixtures.

To describe accurately the critical point of a fluid, we therefore require a function $\psi_s(\delta\rho, \delta T)$, defined by equation (3.73), which satisfies equations (3.22) and (3.64), which leads to exponents that satisfy the equations,

$$\alpha + \beta(\delta + 1) = 2 \qquad \alpha + 2\beta + \gamma = 2 \qquad \gamma - \beta(\delta + 1) = 0, \qquad (3.74)$$

but which is not restricted to the classical Taylor expansions so far discussed. The way to construct such a function was shown first[46] by Widom for fluids and by Domb and Hunter for the parallel problem of the Curie point of a ferromagnet. The function ψ_s has to be a *generalized homogeneous function* of its arguments[4,16]. A homogeneous function of two variables, x and y, satisfies the equation

$$F(\lambda^\xi x, \lambda^\xi y) = \lambda F(x, y) \tag{3.75}$$

for all values of x, y and λ. A generalized homogeneous function has two independent (but fixed) exponents

$$F(\lambda^\xi x, \lambda^\eta y) = \lambda F(x, y) \tag{3.76}$$

Let

$$\psi_s(\lambda^\xi \delta\rho, \eta^\xi \delta T) = \lambda \psi_s(\delta\rho, \delta T) \tag{3.77}$$

and differentiate both sides twice with respect to $\delta\rho$. The first differentiation gives

$$\lambda^\xi \cdot \Delta\mu(\lambda^\xi \delta\rho, \lambda^\eta \delta T) = \lambda \cdot \Delta\mu(\delta\rho, \delta T) \tag{3.78}$$

and the second

$$\lambda^{2\xi} \cdot \beta_T(\lambda^\xi \delta\rho, \lambda^\xi \delta T) = \lambda \cdot \beta_T(\delta\rho, \delta T) \tag{3.79}$$

where β_T is the isothermal compressibility. The equation holds for all values of $\delta\rho$, δT and λ, so let

$$\delta\rho = 0 \qquad \lambda = (\delta T/T^c)^{-1/\eta} \tag{3.80}$$

so that

$$(\delta T/T^c)^{(1-2\xi)/\eta} \cdot \beta_T(\delta\rho = 0, \delta T = T^c) = \beta_T(\delta\rho = 0, \delta T) \tag{3.81}$$

where $\beta_T(\delta\rho = 0, \delta T = T^c)$ is a particular value of the function $\beta_T(\delta\rho, \delta T)$; there is no implication that the form of ψ_s under discussion is a useful representation of the free energy at $\delta T = T^c$, or $T = 2T^c$, but merely that the function β_T has an extrapolated value at that point.

The rate at which β_T^{-1} goes to zero with δT on the critical isochore $\delta\rho = 0$, defines the exponent γ, so from (3.81)

$$\gamma = (1 - 2\xi)/\eta \tag{3.82}$$

It follows from identity (3.11) that on the critical isotherm

$$p - p^c = \rho(\mu - \mu^c) \qquad (\delta T = 0) \tag{3.83}$$

so that the exponent δ can be found from equation (3.78) by putting

$$\delta T = 0 \qquad \lambda = (\delta\rho/\rho^c)^{-1/\xi} \tag{3.84}$$

to give

$$(\delta\rho/\rho^c)^{(1-\xi)/\xi} \cdot \Delta\mu(\delta\rho = \rho^c, \delta T = 0) = \Delta\mu(\delta\rho, \delta T = 0) \tag{3.85}$$

or

$$\delta = (1 - \xi)/\xi \tag{3.86}$$

The same argument for the singular part of the heat capacity density, $-(\partial^2\psi_s/\partial(\delta T)^2)_\rho$, yields

$$\alpha = (2\xi - 1)/\eta \tag{3.87}$$

To obtain the fourth exponent β, we substitute

$$\lambda = (-\delta T/T^c)^{-1/\eta} \qquad (\delta T < 0) \tag{3.88}$$

into equation (3.78) to give

$$(-\delta T/T^c)^{(1-\zeta)/\eta}\cdot\Delta\mu((-\delta T/T^c)^{-\zeta/\eta}\delta\rho, \delta T - T^c) - \Delta\mu(\delta\rho, \delta T) \tag{3.89}$$

On the orthobaric curve $\Delta\mu = 0$ and $\delta T < 0$. Consider, therefore, a set of deviations from the critical point in which δT is coupled to $\delta\rho$ in such a way that the displacements all fall on the orthobaric curve, and so on $\Delta\mu = 0$. Since $\Delta\mu$ is a monotonic function of $\delta\rho$, it follows that the left-hand side of equation (3.89) can be zero only for one value of the argument; that is

$$(-\delta T/T^c)^{-\zeta/\eta}\delta\rho = \text{constant} \qquad (\Delta\mu = 0, \delta T < 0) \tag{3.90}$$

The change of $\delta\rho$ with δT along the orthobaric curve defines the exponent β, hence

$$\beta = \zeta/\eta \tag{3.91}$$

The four equations (3.82), (3.86), (3.87) and (3.91) yield

$$\xi^{-1} = 1 + \delta \qquad \eta^{-1} = 2 - \alpha \tag{3.92}$$

and values of β and γ related to α and δ by equation (3.74). Thus a function ψ_s, equation (3.77), symmetrical in $\delta\rho$, and with ξ and η determined by equation (3.92) is the sought-after form of the free-energy density. The classical form of ψ_s conforms to this equation with $\xi = \frac{1}{4}$ and $\eta = \frac{1}{2}$, but the experimental results require values of 0.17 and 0.53.

The assertion that ψ_s is a generalized homogeneous function with these values of ξ and η is called the *scaling law*, since it is an assertion that ψ_s depends not on $\delta\rho$ and δT separately but only on a scaled ratio of these variables. This can be seen by letting λ have the value given in (3.84), when equation (3.77) becomes

$$(\delta\rho/\rho^c)^{-1/(\delta+1)}\cdot\psi_s(\delta\rho, \delta T) = \psi_s(\delta\rho = \rho^c, t) \tag{3.93}$$

where t is the scaled ratio

$$t = \delta T/(\delta\rho/\rho^c)^{1/\beta} \tag{3.94}$$

Thus $\psi_s(\delta\rho, \delta T)$, when multiplied by $(\delta\rho)^{-1/(\delta+1)}$, is a function of this one variable only, as is shown in equation (3.93).

Scaling laws are, however, only asymptotically correct; away from the critical point *corrections to scaling* become important, as was shown in the last section in the discussion of the apparent values of the exponent β.

There are good theoretical reasons[47] for supposing that if a property F varies, according to the scaling law, as $(\delta T)^{-\lambda}$ near T^c, then away from T^c it will vary as

$$F = A(\delta T)^{-\lambda}[1 + a(\delta T/T^c)^{\Delta} + O(\delta T/T^c)^{2\Delta}] \qquad (3.95)$$

where both the coefficient a and the exponent Δ are of the order of 0.5.

Several forms of ψ_s or its derivative $\Delta\mu$ have been put forward, which conform to the scaling law and have the required symmetry, and which are good fits to the experimental results. The choice between them is largely a matter of convenience and here the parametric representations of Schofield and Josephson have been found to be the easiest to handle[48]. They suggested that $\delta\rho$ and δT be replaced by 'polar coordinates', r and θ, the first of which represents the 'distance' from the critical point, and the second of which represents the 'direction' of the displacement. The variation of $\Delta\mu$ with θ is analytic, its dependence on r contains the singularities. Let

$$\delta T/T^c = r \cdot T(\theta) \qquad \delta\rho/\rho^c = r^{\beta} \cdot M(\theta) \qquad (3.96)$$

so that the scaling variable t of equation (3.94) is a function of θ only. Furthermore let

$$\Delta\mu(\rho^c/p^c) = r^{\beta\delta} \cdot H(\theta) \qquad (3.97)$$

so that $\Delta\mu/|\delta\rho|^{\delta}$ is also a function of θ alone, and hence of t. These equations satisfy (3.53) and (3.94); it remains only to choose $T(\theta)$, $M(\theta)$, and $H(\theta)$ to fit the experimental results. The common choice is the *cubic model*

$$T(\theta) = 1 - b^2\theta^2 \qquad M(\theta) = k\theta(1 + c\theta^2) \qquad H(\theta) = a\theta(1 - \theta^2) \qquad (3.98)$$

or the simpler *linear model* obtained by putting $c = 0$. The line $\theta = 0$ is the critical isochore for $T > T^c$, the two lines $\theta = \pm b^{-1}$ are the two branches of the critical isotherm, and the two lines $\theta = \pm 1$ are the two orthobaric curves, $+$ for the liquid, and $-$ for the gas.

All other thermodynamic functions can be obtained[4] from $\Delta\mu(r, \theta)$ in terms of the adjustable parameters a, b, c, and k, but it is found that b and c (if used) change little from one substance to another. That is, not only are the four critical exponents, α–δ, apparently the same for all fluids but so also are the scaling functions themselves, except for the two remaining scale factors a and k of equation (3.98). This reduction in the number of adjustable parameters gives what are called the *restricted linear* and *restricted cubic models*. Both have two adjustable parameters but the latter has the better functional form. In it b^2 and c are set equal to $3/(3 - 2\beta)$ and $(2\beta\delta - 3)/(3 - 2\beta)$, respectively, for all fluids. With this restriction, a contour of fixed r around the critical point is also a contour of fixed β_T, with $\beta_T^{-1} \sim r^{\gamma}$.

The hypothesis that all fluids have the same exponents and that there are only two adjustable parameters in the scaling functions is the hypothesis of *universality*. The evidence[4,38] for it is too extensive to reproduce here but it has been tested for a wide range of fluids from helium to water. What is still uncertain is whether fluids, as a class, have the same universal behaviour as the three-dimensional Ising model of ferromagnetism, although the best results suggest that this is probably so.

It is natural to ask if the great simplification introduced by the use of universal exponents and scaling functions has behind it any clear theoretical interpretation. Even to sketch an answer requires that we go beyond the purely thermodynamic properties of fluids and consider their structure, internal correlations and fluctuations.

Away from the critical point a fluid can be treated as a continuum on a scale of length large compared with the size of the molecules, say 10 nm or more. It is only when we come to a scale of less than 1 nm (10 Å) that we have to recognize the local inhomogeneities caused by the presence or absence of molecules at a given point and time. We can describe this by introducing a *correlation length* ξ which is a measure of the range of correlations of density fluctuations. In a non-critical dense fluid $\xi \sim 0.5$ nm, but it grows rapidly as the critical point is approached; indeed, its divergence at that point is described by introducing another critical exponent v; on the critical isochore $\xi \sim (\delta T)^{-v}$. The classical value of v is $\frac{1}{2}$ but the experimental value is about 0.627. When δT is sufficiently small for ξ to be of the order of 300 nm, or the wavelength of visible light, then the fluid shows the characteristic critical opalescence.

The universal behaviour of the critical singularities is a consequence of ξ greatly exceeding the range of the intermolecular forces. At a critical point, it is the range and magnitude of the density fluctuations that control the thermodynamic behaviour, the range and magnitude of the intermolecular forces have become irrelevant except in the trivial sense that they determine the absolute size of T^c and ρ^c. As long as the intermolecular forces are of short range, and as long as we consider only three-dimensional systems, then universal behaviour is to be expected. It probably does not matter whether the system is continuous or a crystal lattice; as long as the range of the forces and the lattice spacing (with nearest-neighbour forces) are both small compared with ξ in the critical region, then the same universal class of behaviour is to be expected. A change of dimensionality takes a system out of this universality class, and if we go to systems of dimensionality of four or above, or to a model with forces of infinite range, then we recover the classical behaviour of the van der Waals–Landau treatment. Indeed, there are further equations, not discussed here, which relate structural exponents such as v to the dimensionality and the thermodynamic exponents.

The simple ideas of the last paragraph have served as the basis for much more sophisticated and powerful theories, which are now beginning to yield purely theoretical values of the exponents that agree well with experiments. These theoretical ideas were formulated first in a qualitative way by Kadanoff[49] and then put on a much deeper basis by Wilson and others by the introduction of the technique of the renormalization group[50]. These subjects need, and have been given, books of their own[16], and no short summary is attempted here.

References

1 TEMPERLEY, H. N. V., *Changes of State*, 41 et seq., Cleaver-Hume, London (1956); TEMPERLEY, H. N. V., and TREVENA, D. H., *Liquids and their Properties*, Chap. 8, Horwood, Chichester (1978); FISHER, M. E., *Physics*, **3**, 255 (1967); LANGER, J. S., *Annln Phys.*, **41**, 1 (1967)

2 VAN DER WAALS, J. D., *Die Kontinuität des gasförmigen und flüssigen Zustandes*, 2nd Edn, Vol. 1: *Single Component Systems* (1899); Vol. 2: *Binary Mixtures* (1900), Barth, Leipzig; Eng. transl. 1st Edn Vol. 1 by Threlfall, R., and Adair, J. F., in *Physical Memoirs*, Vol. 1, Part 3, Physical Society, London (1890)

3 VERSCHAFFELT, J. E., *Communs phys. Lab. Univ. Leiden*, No. 55 (1900), No. 81 (1902); LEVELT SENGERS, J. M. H., *Physica*, **82A**, 319 (1976)

4 SENGERS, J. V., and LEVELT SENGERS, J. M. H., in *Progress in Liquid Physics*, Ed. Croxton, C. A., Chap. 4, Wiley, Chichester (1978)

5 VAN LAAR, J. J., *Proc. Sect. Sci. K. ned. Akad. Wet.*, **14**, 1091 (1912)

6 LEVELT SENGERS, J. M. H., *Ind. & Eng. Chem., Fundam.*, **9**, 470 (1970)

7 VICENTINI-MISSONI, M., LEVELT SENGERS, J. M. H., and GREEN, M. S., *J. Res. natn. Bur. Stand.*, **73A**, 563 (1969)

8 BAEHR, H. D., *Forsch. Ingenieurw.*, **29**, 143 (1963); *Brennst.-Wärme-Kraft*, **15**, 514 (1963)

9 WILSON, E. B., in *Commentary on the Scientific Writings of J. Willard Gibbs*, Eds. Donnan, F. G. and Haas, A., Vol. 1, 51, Yale University Press, New Haven (1936)

10 ONNES, H. K., and HAPPEL, H., *Communs phys. Lab. Univ. Leiden*, No. 86 (1903)

11 WOOD, S. E., *J. phys. Chem., Ithaca*, **66**, 600 (1962)

12 GRIFFITHS, R. B., *J. chem. Phys.*, **43**, 1958 (1965)

13 RUSHBROOKE, G. S., *J. chem. Phys.*, **39**, 842 (1963); **43**, 3439 (1965)

14 FISHER, M. E., *J. math. Phys.*, **5**, 944 (1964)

15 LIBERMAN, D. A., *J. chem. Phys.*, **44**, 419 (1966)

16 STANLEY, H. E., *Introduction to Phase Transitions and Critical Phenomena*, Clarendon, Oxford (1971); DOMB, C., and GREEN, M. S., (Eds.) *Phase Transitions and Critical Phenomena*, Vol. 1–6, Academic Press, London (1972–76); MA, S.-K., *Modern Theory of Critical Phenomena*, Benjamin, Reading, Mass. (1976)

17 GRIFFITHS, R. B., *Phys. Rev.*, **158**, 176 (1967); GREEN, M. S., VICENTINI-MISSONI, M., and LEVELT SENGERS, J. M. H., *Phys. Rev. Lett.*, **18**, 1113 (1967)

18 LEVELT SENGERS, J. M. H., in *Experimental Thermodynamics*, Vol. 2: *Experimental Thermodynamics of Non-reacting Fluids*, Eds. Le Neindre, B., and Vodar, B., Chap. 14, Butterworths, London (1975)

19 ANDREWS, T., *Phil. Trans.*, **159**, 575 (1869)

20 YOUNG, S., *Proc. R. Ir. Acad*, **12**, 374 (1909–10)

21 CAILLETET, L., and MATHIAS, E., *C. r. hebd. Séanc. Acad. Sci., Paris*, **102**, 1202 (1886); **104**, 1563 (1887)

22 COOK, D., *Trans. Faraday Soc.*, **49**, 716 (1953)

23 WIDOM, B., and ROWLINSON, J. S., *J. chem. Phys.*, **52**, 1670 (1970); MERMIN, N. D., and REHR, J. J., *Phys. Rev. Lett.*, **26**, 1155 (1971)

24 WEINER, J., LANGLEY, K. H., and FORD, N. C., *Phys. Rev. Lett.*, **32**, 879 (1974)

25 LEY-KOO, M., and GREEN, M. S., *Phys. Rev.*, **A16**, 2483 (1977)

26 BUCKINGHAM, M. J., in *Phase Transitions and Critical Phenomena*, Eds. Domb, C., and Green, M. S., Vol. 2, Academic Press, London (1972); SCOTT, R. L., in *Chemical Thermodynamics, Specialist Periodical Reports*, Ed. McGlashan, M. L., Vol. 2, Chap. 8., Chemical Society, London (1978)

27 SCHNEIDER, W. G., in *Changements de Phases*, 69, Société de Chimie Physique (1952); ATACK, D., and SCHNEIDER, W. G., *J. phys. Chem., Ithaca*, **54**, 1323 (1950); **55**, 532 (1951); MACCORMACK, K. E., and SCHNEIDER, W. G., *Can. J. Chem.*, **29**, 699 (1951); WEINBERGER, M. A., and SCHNEIDER, W. G., *Can. J. Chem.*, **30**, 422, 847 (1952); HABGOOD, H. W., and SCHNEIDER, W. G., *Can. J. Chem.*, **32**, 98, 164 (1954); SCHNEIDER, W. G., and HABGOOD, H. W., *J. chem. Phys.*, **21**, 2080 (1953)

28 PALMER, H. B., *J. chem. Phys.*, **22**, 625 (1954); LORENTZEN, H. L., *Acta chem. scand.*, **7**, 1335 (1953); **9**, 1724 (1955)

29 STRAUB, J., *Chemie-Ingr-Tech.*, 291 (1967); ARTYUKHOVSKAYA, L. M., SHIMANSKAYA, E. T., and SHIMANSKII, Y. I., *Soviet Phys. JETP*, **32**, 375 (1971)

30 WILCOX, L. R., and BALZARINI, D., *J. chem. Phys.*, **48**, 753 (1968); ESTLER, W. T., HOCKEN, R., CHARLTON, T., and WILCOX, L. R., *Phys. Rev.*, **A12**, 2118 (1975)

31 YOUNG, S., *Proc. phys. Soc. Lond.*, **13**, 602 (1895)
32 MICHELS, A., BIJL, A., and MICHELS, C., *Proc. R. Soc.*, **A160**, 376 (1937); MICHELS, A., BLAISSE, B., and MICHELS, C., *ibid.*, 358; MICHELS, A., and DE GROOT, S. R., *Appl. scient. Res.*, **A1**, 94 (1949); MICHELS, A. and STRIJLAND, J., *Physica*, **18**, 613 (1952).
33 VORONEL, A. V., and CHASHKIN, Y. R., *Soviet Phys. JETP*, **24**, 263 (1967); VORONEL, A. V., SMIRNOV, V. A., and CHASHKIN, Y. R., *Soviet Phys. JETP Lett.*, **9**, 229 (1969); VORONEL, A. V. GORBUNOVA, V. G., SMIRNOV, V. A., SHMAKOV, N. G., and SHCHEKOCHIKHINA, V. V., *Soviet Phys. JETP*, **36**, 505 (1973)
34 VORONEL, A. V., in *Phase Transitions and Critical Phenomena*, Eds. Domb, C., and Green, M. S., Vol. 5B, Academic Press, London (1976)
35 EDWARDS, C., LIPA, J. A., and BUCKINGHAM, M. J., *Phys. Rev. Lett.*, **20**, 496 (1968); LIPA, J. A., EDWARDS, C., and BUCKINGHAM, M. J., *Phys. Rev. Lett.*, **25**, 1086 (1970); *Phys. Rev.*, **A15**, 778 (1977)
36 MOLDOVER, M. R., *Phys. Rev.*, **182**, 342 (1969)
37 FISHER, M. E., *Rep. Prog. Phys.*, **30**, 615 (1967); HELLER, P., *ibid.*, 731
38 LEVELT SENGERS, J. M. H., and SENGERS, J. V., *Phys. Rev.*, **A12**, 2622 (1975); LEVELT SENGERS, J. M. H., GREEN, W. L., and SENGERS, J. V., *J. phys. chem. ref. Data*, **5**, 1 (1976); SENGERS, J. V., in *Phase Transitions*, Eds. Levy, M., Le Guillou, J. C., and Zinn-Justin, J., Plenum, New York (1981)
39 HOCKEN, R., and MOLDOVER, M. R., *Phys. Rev. Lett.*, **37**, 29 (1976)
40 MOLDOVER, M. R., SENGERS, J. V., GAMMON, R. W., and HOCKEN, R. J., *Rev. mod. Phys.*, **51**, 79 (1979)
41 HOHENBERG, P. C., and BARMATZ, M., *Phys. Rev.*, **A6**, 289 (1972)
42 HALL, K. R., and EUBANK, P. T., *Ind. & Eng. Chem., Fundam.*, **15**, 323 (1976); BAEHR, H. D., GARNJOST, H., and POLLAK, R., *J. chem. Thermodyn.*, **8**, 113 (1976); WAGNER, W., EWERS, J., and PENTERMANN, W., *ibid.*, 1049
43 HASTINGS, J. R., and LEVELT SENGERS, J. M. H., in *7th Symposium on Thermophysical Properties*, Ed. Cezairliyan, A., 794, American Society of Mechanical Engineers, New York (1978)
44 LANDAU, L. D., (1937), in LANDAU, L. D., and LIFSHITZ, E. M., *Statistical Physics*, 2nd Edn, Section 138, Pergamon, Oxford (1969)
45 FOX, J. R., *J. stat. Phys.*, **21**, 243 (1979)
46 WIDOM, B., *J. chem. Phys.*, **43**, 3898 (1965); DOMB, C., and HUNTER, D. L., *Proc. phys. Soc.*, **86**, 1147 (1965)
47 AHARONY, A., and AHLERS, G., *Phys. Rev. Lett.*, **44**, 782 (1980)
48 SCHOFIELD, P., *Phys. Rev. Lett.*, **22**, 606 (1969); JOSEPHSON, B. D., *J. Phys.*, **C2**, 1113 (1969); SCHOFIELD, P., LITSTER, J. D., and HO, J. T., *Phys. Rev. Lett.*, **23**, 1098 (1969); HO, J. T., and LITSTER, J. D., *Phys. Rev.*, **B2**, 4523 (1970)
49 KADANOFF, L. P., *Physics*, **2**, 263 (1966); in *Phase Transitions and Critical Phenomena*, Eds. Domb, C. and Green, M. S., Vol. 5A, Chap. 1, Academic Press, London (1976)
50 WILSON, K. G., *Phys. Rev.*, **B4**, 3174, 3184 (1971); WILSON, K. G., and FISHER, M. E., *Phys. Rev. Lett.*, **28**, 240 (1972); WILSON, K. G., and KOGUT, J. B., *Phys. Rep.*, **12C**, 75 (1974)

Chapter 4

The thermodynamics of liquid mixtures

4.1 Introduction

There are two principal reasons for the great amount of experimental and theoretical work that has been done on the properties of liquid mixtures.

The first is that they provide one way of studying the physical forces acting between two molecules of different species. Liquid mixtures are not, however, the most direct source of such information, owing to the difficulties of the interpretation of the properties of liquids in terms of the intermolecular forces. The equilibrium and, in some cases, the transport properties of dilute gases together with the results of molecular beam scattering experiments and the spectroscopic study of gaseous dimers are our principal sources of this information[1], but the use of liquid mixtures for this purpose is now becoming more important with the advance of their theoretical study. The use of liquid mixtures for the study of strong specific forces between different molecules is of much longer standing.

There is now less emphasis than hitherto on attempts to explain the properties of mixtures solely from a knowledge of those of the pure components. Such attempts rest upon the fallacy that the forces $(\alpha-\beta)$ between two molecules of species α and β are always determinable from the strengths of the forces $(\alpha-\alpha)$ and $(\beta-\beta)$. If it were true that the $(\alpha-\beta)$ forces were always some 'average' of the $(\alpha-\alpha)$ and $(\beta-\beta)$ forces, then the properties of a binary mixture would be predictable, in principle, solely from a knowledge of those of the two pure components.

However, such averaging is not universally valid. It is true that for very simple substances and for the prediction of relatively crude properties there are suitable averages of the intermolecular forces. These are considered further in Chapter 8. However, such averaging is unsatisfactory for many classes of substances and inadequate for the detailed interpretation even of the simplest mixtures. One should rather take the observed properties of a binary mixture as an experimental source of information about the $(\alpha-\beta)$ forces.

The possibilities of *a priori* prediction are greater in a multicomponent system. The energy of an array of, say, three molecules α, β and γ is approximately given by the sum of the energies of the three pairs $(\alpha-\beta)$, $(\beta-\gamma)$

and (γ–α) considered separately. Any small departure of the total energy from this sum can probably be represented adequately by an average of the corresponding departures in the pure components. It should be possible, therefore, to predict the properties of a multicomponent mixture from a knowledge of those of the pure substances and of all the binary mixtures. However, even these predictions can be made only by appealing to our knowledge of the intermolecular forces and so are outside the scope of classical thermodynamics.

The second reason for the study of mixtures is the appearance of new phenomena which are not present in pure substances. The most interesting of these are new types of phase equilibrium which arise from the extra degrees of freedom introduced by the possibility of varying the proportions of the components. The number of degrees of freedom may be calculated from the phase rule of Gibbs. A system of one component and one phase has two degrees of freedom. That is, the two intensive properties, pressure and temperature, may both be changed (within limits) without causing any new phase to appear. A two-component system of one phase has three degrees of freedom, for the composition may also be freely varied. A one-component system of two phases has one degree of freedom. If the temperature is fixed arbitrarily, then there is only one value of the pressure for which the two phases can exist together in equilibrium.

A graph of the coexisting pressures and temperatures defines a vapour-pressure or other similar phase-boundary curve. There is, in general, no such unique curve for a mixture, since the pressure is a function of composition as well as of temperature although azeotropes are exceptions to this statement.

Three coexisting fluid phases are not found in one-component systems, except for the unusual behaviour of liquid helium at its λ-point, and of those substances that form liquid crystals. They are common in binary fluid mixtures, and it is the critical points of such systems that have been responsible for much of the interest in mixtures.

The basic thermodynamic relationships that are applicable to binary liquid mixtures at temperatures and pressures well below those at the gas–liquid critical point of either component are derived in the first sections of this chapter. The thermodynamic equations relevant to wholly miscible systems are derived in Sections 4.2–4.6 and are followed by Section 4.8 on partially-miscible mixtures. The final sections are a discussion of critical and tricritical points in binary and multicomponent mixtures. The additional effects that arise at high pressures and that can lead to phase behaviour of great diversity are discussed later in Chapter 6.

4.2 Partial molar quantities

The composition of a mixture may be expressed in several ways, some of which are symmetrical in the components and some of which are not. The only system used here is a symmetrical one in which the amount of each component (expressed in moles) is denoted n_1, n_2, etc., and in which the relative amount of each component is described by its mole fraction, x_1, x_2, etc., where

$$x_1 = n_1 \bigg/ \sum_\alpha n_\alpha \quad (\alpha = 1, 2, 3, ...) \tag{4.1}$$

If V is the volume of a binary mixture of n_1 moles of component 1 and n_2 of component 2, it is interesting to enquire how much of this volume is to be ascribed to the first and how much to the second component. For, in general, V is not equal to the sum of the volumes occupied by the separate components at the same pressure and temperature before mixing. It is, however, possible in principle to measure the increase in volume (δV) on adding an infinitesimal amount (δn_1) at constant pressure and temperature to a large amount of mixture of known composition. The ratio of (δV) to (δn_1) is the *partial molar volume* of species 1 in this mixture, here denoted v_1. More formally

$$v_1 = (\partial V/\partial n_1)_{n_2, p, T} \tag{4.2}$$

Partial molar quantities are intensive properties which depend only on the pressure, temperature and composition of the mixture. Thus if $(n_1 + n_2)$ moles of liquid are formed by repeatedly adding small quantities (δn_1) and (δn_2) of the two substances, with the ratio ($\delta n_1/\delta n_2$) a constant, then the total volume is given by

$$V = n_1 v_1 + n_2 v_2 \tag{4.3}$$

This follows at once from the fact that v_1 and v_2 are functions only of the relative composition of the mixture and not of its total amount. It is seen, therefore, that the partial molar properties of a system provide an answer to the question of how much of an extensive property, in this case the volume, is to be ascribed to each component.

If V is expressed as a function of temperature, pressure and amount of substance, $V = f(T, p, n)$, then, for a binary mixture,

$$dV = \left(\frac{\partial V}{\partial T}\right)_{p,n} dT + \left(\frac{\partial V}{\partial p}\right)_{T,n} dp + \left(\frac{\partial V}{\partial n_1}\right)_{T,p,n_2} dn_1 + \left(\frac{\partial V}{\partial n_2}\right)_{T,p,n_1} dn_2 \tag{4.4}$$

or, from equation (4.2)

$$dV = V\alpha_p\, dT - V\beta_T\, dp + v_1\, dn_1 + v_2\, dn_2 \tag{4.5}$$

Differentiation of equation (4.3) gives

$$dV = n_1\, dv_1 + n_2\, dv_2 + v_1\, dn_1 + v_2\, dn_2 \tag{4.6}$$

so, comparing equations (4.5) and (4.6),

$$n_1\, dv_1 + n_2\, dv_2 = V\alpha_p\, dT - V\beta_T\, dp = 0 \quad \text{(const. } T, p) \tag{4.7}$$

This is an important restriction on the simultaneous change of v_1 and v_2 in a mixture and is an example of a Gibbs–Duhem equation.

These equations may be generalized for any extensive property R, and for any number of components

$$r_1 = (\partial R/\partial n_1)_{n', p, T} \tag{4.8}$$

$$R = \sum_\alpha n_\alpha r_\alpha \tag{4.9}$$

$$0 = \sum_\alpha n_\alpha \, dr_\alpha \quad \text{(Gibbs–Duhem equation, const. } p, T\text{)} \tag{4.10}$$

where the prime indicates all components but the one with respect to which the differentiation is performed. $(\partial G/\partial n_1)_{n', p, T}$ is the chemical potential of component 1 in the mixture, and the potentials satisfy the most important form of the Gibbs–Duhem equation

$$\sum_\alpha n_\alpha d\mu_\alpha = -S \, dT + V \, dp \tag{4.11}$$

The sum on the left is zero for changes of composition at constant p and T.

It is usual to measure partial molar quantities not by means of the hypothetical experiment described at the beginning of this section but from a knowledge of the molar quantity r as a function of mole fraction x, where

$$r = R/n = \sum_\alpha x_\alpha r_\alpha \tag{4.12}$$

In order to do this for a binary mixture, it is necessary to consider the relation between the quantities, $(\partial R/\partial n_1)_{n_2, p, T}$ and $(\partial r/\partial x_1)_{p, T}$. The first is the rate of change of the extensive quantity R with n_1 at constant n_2, that is, it is the partial molar quantity. The second is the change of the mean molar quantity r with the mole fraction x_1. This differentiation is not performed at constant n_2 or at constant x_2, since it is clear that the sum of dx_1 and dx_2 must be zero in a binary mixture. By differentiation of equation (4.12)

$$\left(\frac{\partial r}{\partial x_1}\right) = \frac{\partial}{\partial x_1}[x_1 r_1 + (1 - x_1)r_2] \tag{4.13}$$

$$= r_1 - r_2 + x_1 \left(\frac{\partial r_1}{\partial x_1}\right) + x_2 \left(\frac{\partial r_2}{\partial x_1}\right) \tag{4.14}$$

The sum of the last two terms of equation (4.14) is zero, as is seen by multiplying by n, the total number of moles, to obtain the usual form of the Gibbs–Duhem equation, (4.10). Therefore,

$$(\partial r/\partial x_1) = -(\partial r/\partial x_2) = r_1 - r_2 \tag{4.15}$$

and, from equation (4.12),

$$r_1 = r - x_2(\partial r/\partial x_2) \tag{4.16}$$

$$r_2 = r - x_1(\partial r/\partial x_1) \tag{4.17}$$

These are the required relations between the partial molar quantities and the derivatives $(\partial r/\partial x)$.

Byers Brown[2] has extended these equations to multicomponent mixtures. From equation (4.12)

$$\left(\frac{\partial r}{\partial n_\alpha}\right)_{n'} = \frac{r_\alpha}{n} - \frac{R}{n^2} = \frac{r_\alpha - r}{n} \tag{4.18}$$

or

$$r_\alpha = r + n(\partial r/\partial n_\alpha)_{n'} \tag{4.19}$$

where the subscript n' indicates that all numbers of moles except n_α are to be kept constant. In a mixture of q components, r is a function of the mole fractions $x_1, x_2, \ldots, x_q$, but only $(q-1)$ of these are independent. Let these be $x_2, x_3, \ldots, x_q$, and let r be a function of these independent variables:

$$\left(\frac{\partial r}{\partial n_1}\right)_{n'} = \sum_{\beta=2}^{q}\left(\frac{\partial r}{\partial x_\beta}\right)\left(\frac{\partial x_\beta}{\partial n_1}\right)_{n'} = -\frac{1}{n}\sum_{\beta=2}^{q} x_\beta\left(\frac{\partial r}{\partial x_\beta}\right) \tag{4.20}$$

and, for α equal to $2, 3, \ldots, q$,

$$\left(\frac{\partial r}{\partial n_\alpha}\right)_{n'} = \left(\frac{\partial r}{\partial x_\alpha}\right)\left(\frac{\partial x_\alpha}{\partial n_\alpha}\right)_{n'} + \sum_{\beta=2}^{q}{}'\left(\frac{\partial r}{\partial x_\beta}\right)\left(\frac{\partial x_\beta}{\partial n_\alpha}\right) \tag{4.21}$$

$$= \left(\frac{1-x_\alpha}{n}\right)\left(\frac{\partial r}{\partial x_\alpha}\right) - \frac{1}{n}\sum_{\beta=2}^{q}{}' x_\beta\left(\frac{\partial r}{\partial x_\beta}\right) \tag{4.22}$$

$$= \frac{1}{n}\left[\left(\frac{\partial r}{\partial x_\alpha}\right) - \sum_{\beta=2}^{q} x_\beta\left(\frac{\partial r}{\partial x_\beta}\right)\right] \tag{4.23}$$

where the prime on a summation sign shows that the term with β equal to α is to be omitted. Hence, from equation (4.19)

$$r_1 = r - \sum_{\beta=2}^{q} x_\beta\left(\frac{\partial r}{\partial x_\beta}\right) \tag{4.24}$$

$$r_\alpha = r + \left(\frac{\partial r}{\partial x_\alpha}\right) - \sum_{\beta=2}^{q} x_\beta\left(\frac{\partial r}{\partial x_\beta}\right) \qquad (\alpha = 2, 3, \ldots, q) \tag{4.25}$$

In differentiation with respect to any mole fraction, say x_α, all the others except x_1 and x_α are to be kept constant, as they are all independent variables. These equations are, therefore, the extension of equations (4.16) and (4.17) to multicomponent mixtures. However, they are not the neatest form of the equations, as it is necessary to treat separately the dependent mole fraction, which was chosen here to be x_1. It is more convenient to use a differential operator (D/Dx_α) to denote differentiation with respect to x_α in which all other mole fractions are treated as independent variables and so are held constant. With this convention

$$
\left(\frac{\partial r}{\partial x_\alpha}\right) = \left(\frac{Dr}{Dx_\alpha}\right) + \left(\frac{Dr}{Dx_1}\right)\left(\frac{Dx_1}{Dx_\alpha}\right)
$$

$$
= \left(\frac{Dr}{Dx_\alpha}\right) - \left(\frac{Dr}{Dx_1}\right) \qquad (\alpha = 2, 3, ..., q) \tag{4.26}
$$

since

$$
\left(\frac{Dx_1}{Dx_\alpha}\right) = \frac{D}{Dx_\alpha}\left(1 - \sum_{\beta=2}^{q} x_\beta\right) = -1 \tag{4.27}
$$

Substitution in equations (4.24) and (4.25) gives

$$
r_1 = r - \sum_{\beta=2}^{q} x_\beta\left(\frac{Dr}{Dx_\beta}\right) + (1 - x_1)\left(\frac{Dr}{Dx_1}\right) \tag{4.28}
$$

$$
r_\alpha = r + \left(\frac{Dr}{Dx_\alpha}\right) - \sum_{\beta=2}^{q} x_\beta\left(\frac{Dr}{Dx_\beta}\right) - x_1\left(\frac{Dr}{Dx_1}\right) \qquad (\alpha = 2, 3, ..., q) \tag{4.29}
$$

These equations may be written in a common and more simple form:

$$
r_\alpha = r + \left(\frac{Dr}{Dx_\alpha}\right) - \sum_{\beta=1}^{q} x_\beta\left(\frac{Dr}{Dx_\beta}\right) \qquad (\alpha = 1, 2, ..., q) \tag{4.30}
$$

This equation satisfies equation (4.9) and is the most convenient way of finding r_α in a multicomponent system, where r is known as a function of the mole fractions. It is so used in Chapter 8. The differentiation is most simply performed if r is written as a symmetrical function of the mole fractions, but the equation is equally valid for any of the many alternative ways of expressing r as a function of either q or $(q - 1)$ variables.

Equations (4.16) and (4.17) are the basis of the commonly used *method of intercepts* for calculating partial molar quantities in binary mixtures. *Figure 4.1* shows a typical graph of a mean molar quantity (say, v, h or c_p) as a function of composition. It is clear from these equations that the intercepts of the tangent to the curve at any point, x_1, give the values of r_1 and r_2 for that composition. However, it is difficult to draw accurately the tangent to an experimental curve, and so this is not a recommended way of obtaining r_1 and r_2. It is usually better to start by defining an apparent molar quantity by

$$
r_1^{app} = (r - x_2 r_2^0)/x_1 \tag{4.31}
$$

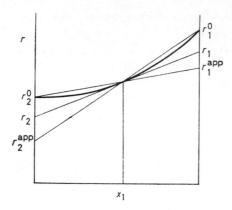

Figure 4.1 A mean molar quantity, r, as a function of the mole fraction, x_1. The curve from r_1^0 to r_2^0 is the value of r. A tangent to this curve at $r(x_1)$ has intercepts of $r_1(x_1)$ and $r_2(x_1)$ (equations (4.16) and (4.17)), and the chords from $r(x_1)$ to r_1^0 and r_2^0 have intercepts of r_2^{app} and r_1^{app}, respectively (equations (4.31)–(4.33))

where r_2^0 is the value in pure component 2. By differentiation of this equation

$$r_1 = r_1^0 - x_2^2(\partial r_2^{\mathrm{app}}/\partial x_2) \tag{4.32}$$

$$r_2 = r_2^0 - x_1^2(\partial r_1^{\mathrm{app}}/\partial x_1) \tag{4.33}$$

If r is close to a linear function of x, then the second terms in these equations are much smaller than the first, and even if r is not linear, it often happens that $(\partial r_1^{\mathrm{app}}/\partial x_1)$ is almost independent of x_1. Errors in determining slopes are therefore less serious in this method than in the method of intercepts.

Young and Vogel[3] give a useful summary of other ways of manipulating equations (4.16) and (4.17) in order to obtain accurate values of r_1 and r_2 from experimental results. Partial molar quantities are now often calculated, not from the measured values of the mean molar quantity r, but from the molar excess function, r^E. These functions are discussed in Section 4.4.

The relations between different partial molar quantities are the same as those between molar quantities of a pure substance. Thus

$$\left(\frac{\partial \mu_\alpha}{\partial p}\right)_{T,x} = v_\alpha \qquad \left(\frac{\partial \mu_\alpha}{\partial T}\right)_{p,x} = -s_\alpha \qquad \left(\frac{\partial h_\alpha}{\partial T}\right)_{p,x} = (c_p)_\alpha \quad \text{etc.} \tag{4.34}$$

4.3 The ideal mixture[a]

The ideal mixture is a hypothetical one whose properties are introduced into the thermodynamic description of real mixtures as convenient standards of

---

[a] The term *perfect mixture* is sometimes used for what is here called an *ideal mixture*. The word *perfect* is used in this book only to describe a gas which has the equation of state, $pv = RT$

normal behaviour. The term is not restricted to mixtures of two (or more) liquids but may be used, for example, for solutions of both solids and gases in liquids. In almost every case there is a certain lack of precision about the definition of an ideal mixture, since it necessarily makes use of the properties of phases, solid, liquid or gas, extrapolated across phase boundaries into regions of pressure and temperature where they do not exist. Fortunately, this imprecision is least serious for purely liquid mixtures with which this book is concerned[a], and becomes negligible at low vapour pressures.

There are several definitions of an ideal liquid mixture; one of the most convenient is that it is a mixture in which the chemical potentials of all components are given by the equations

$$\mu_\alpha(p, T, x) = \mu_\alpha^0(p, T) + RT \ln x_\alpha \qquad (\alpha = 1, 2, ..., q) \qquad (4.35)$$

where $\mu_\alpha^0(p, T)$ is the potential of pure component α at the same pressure and temperature as the mixture being studied. These equations are to be understood to hold over non-zero ranges of pressure and temperature about (p, T).

This definition is unexceptionable for a gas mixture and, indeed, a mixture of perfect gases is the simplest example of an ideal mixture[5]. However, it is not entirely satisfactory for a liquid mixture on its saturation curve. The vapour pressure of a mixture is usually lower than that of one or more of its components at the same temperature. Some of the $\mu_\alpha^0(p, T)$ therefore refer to liquid states extrapolated to pressures below their vapour pressures, and so have no precise physical meaning. Fortunately, the change of μ^0 with pressure is small for liquids near their triple points, since their molar volumes are much smaller than those of a perfect gas at the same pressure and temperature. The change is given by

$$\frac{1}{RT}\left(\frac{\partial \mu_\alpha^0}{\partial \ln p}\right) = \frac{p v_\alpha^0}{RT} \qquad (4.36)$$

Thus, at low vapour pressures, the extrapolations needed to obtain μ_α^0 are either negligible or at least small enough to be calculated with confidence. The use of the ideal mixture as a standard is therefore restricted to low vapour pressures, that is, below about 3 bar. Most precise work has been carried out at and below the normal boiling points and so may be compared with the ideal mixture, but it is unfortunate that there is no simple standard of normal behaviour at higher vapour pressures, and particularly for mixtures near the gas–liquid critical point.

The consequences of equations (4.35) are readily set out. In an ideal mixture

[a] For the definition and discussion of ideal solutions of solids and gases in liquids, see ref. 4.

$$G = \sum_\alpha n_\alpha \mu_\alpha^0 + RT \sum_\alpha n_\alpha \ln x_\alpha \qquad (4.37)$$

$$H = \sum_\alpha n_\alpha h_\alpha^0 \qquad (4.38)$$

$$S = \sum_\alpha n_\alpha s_\alpha^0 - R \sum_\alpha n_\alpha \ln x_\alpha \qquad (4.39)$$

$$V = \sum_\alpha n_\alpha v_\alpha^0 \qquad (4.40)$$

$$C_p = \sum_\alpha n_\alpha (c_p)_\alpha^0 \qquad (4.41)$$

where μ_α^0, h_α^0, s_α^0, v_α^0 and $(c_p)_\alpha^0$ are the molar properties of the pure components at the same pressure and temperature as those of the mixture.

It is seen that G and S are not linear functions of the composition. The sums of the logarithmic terms in equations (4.37) and (4.39) are, respectively, the ideal free energy and entropy of mixing. A thermodynamic function of mixing is the difference between the value of any function in the mixture and the sum of those for the same amount of unmixed components at the same pressure and temperature. Such functions are denoted here G^m, H^m, etc. The ideal free energy of mixing is always negative and the ideal entropy of mixing is always positive. Thus the formation of an ideal mixture from its components is a spontaneous irreversible process, whether carried out isothermally or adiabatically.

It is sometimes useful to discuss a mixture formed from its components at constant temperature and total volume. Functions of mixing in such circumstances are distinguished from those above, at constant pressure, by using V and p as subscripts[6]. V_p^m is zero in an ideal mixture and so it follows that

$$G_p^m = G_V^m = A_p^m = A_V^m \qquad S_p^m = S_V^m \qquad (4.42)$$

$$H_p^m = H_V^m = U_p^m = U_V^m = 0 \quad \text{etc.} \qquad (4.43)$$

Figure 4.2 shows graphs of G^m, H^m and TS^m for a binary ideal mixture. Although not apparent on this scale, it is important to note that, because of the functional forms of equations (4.37) and (4.39), both G^m and TS^m have infinite slopes (either positive or negative) at $x_1 = 0$ and $x_1 = 1$.

The most interesting property of an ideal mixture is the isothermal change with composition of its vapour pressure. An expression for this change may be derived by equating the chemical potentials of each component in gas and liquid phases. Consider a binary mixture:

Liquid

$$\mu_1(\text{liq}, p, x_1) = \mu_1^0(\text{liq}, p^*) + RT \ln x_1 + (p - p^*)v_1^0 \qquad (4.44)$$

$$\mu_2(\text{liq}, p, x_2) = \mu_2^0(\text{liq}, p^*) + RT \ln x_2 + (p - p^*)v_2^0 \qquad (4.45)$$

where p^* is a small but arbitrary pressure.

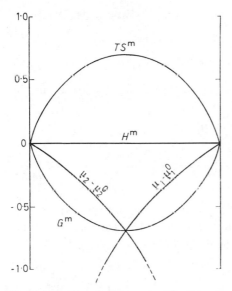

Figure 4.2 The ideal mixture. The free energy, heat and entropy of mixing and the chemical potentials, as functions of x_1. All are in units of RT

Gas[5]

$$\mu_1(\text{gas}, p, y_1) = \mu_1^0(\text{gas}, p^*) + RT \ln(p y_1/p^*) \\ + (p - p^*)B_{11} + 2p(\delta B)_{12} y_2^2 \tag{4.46}$$

$$\mu_2(\text{gas}, p, y_2) = \mu_2^0(\text{gas}, p^*) + RT \ln(p y_2/p^*) \\ + (p - p^*)B_{22} + 2p(\delta B)_{12} y_1^2 \tag{4.47}$$

where y denotes a mole fraction in the gas phase, and where B_{11}, B_{12} and B_{22} are the second virial coefficients for the interactions of pairs of molecules in collisions 1–1, 1–2, and 2–2, respectively. The difference $(\delta B)_{12}$ is given by

$$(\delta B)_{12} = B_{12} - \tfrac{1}{2}B_{11} - \tfrac{1}{2}B_{22} \tag{4.48}$$

Higher terms in the virial expansions for the potentials in the gas phase may be neglected at all pressures at which the concept of an ideal mixture is useful. The arbitrary pressure p^* is now set equal to p_1^0, the vapour pressure of pure 1, in equations (4.44) and (4.46) and equal to p_2^0 in equations (4.45) and (4.47). Equating μ_1 and μ_2 in each phase gives

$$\mu_1^0(\text{liq}, p_1^0) + RT \ln x_1 + (p - p_1^0)v_1^0 \\ = \mu_1^0(\text{gas}, p_1^0) + RT \ln(p y_1/p_1^0) + (p - p_1^0)B_{11} + 2p(\delta B)_{12} y_2^2 \tag{4.49}$$

$$\mu_2^0(\text{liq}, p_2^0) + RT \ln x_2 + (p - p_2^0)v_2^0 \\ = \mu_2^0(\text{gas}, p_2^0) + RT \ln(p y_2/p_2^0) + (p - p_2^0)B_{22} + 2p(\delta B)_{12} y_1^2 \tag{4.50}$$

Now

$$\mu_1^0(\text{liq}, p_1^0) = \mu_1^0(\text{gas}, p_1^0) \qquad \mu_2^0(\text{liq}, p_2^0) = \mu_2^0(\text{gas}, p_2^0) \tag{4.51}$$

since p_1^0 and p_2^0 are the vapour pressures of the pure components. Hence

$$RT \ln p| = RT \ln (x_1 p_1^0/y_1) + (p - p_1^0)(v_1^0 - B_{11}) - 2p(\delta B)_{12} y_2^2 \qquad (4.52)$$

$$= RT \ln (x_2 p_2^0/y_2) + (p - p_2^0)(v_2^0 - B_{22}) - 2p(\delta B)_{12} y_1^2 \qquad (4.53)$$

By combining these equations and substituting $(1 - y_1)$ for y_2 in the logarithmic term of equation (4.53)

$$p = x_1 p_1^0 \exp\left[\frac{(p - p_1^0)(v_1^0 - B_{11}) - 2p(\delta B)_{12} y_2^2}{RT}\right]$$
$$+ x_2 p_2^0 \exp\left[\frac{(p - p_2^0)(v_2^0 - B_{22}) - 2p(\delta B)_{12} y_1^2}{RT}\right] \qquad (4.54)$$

This equation is greatly simplified if it is legitimate to put the exponential terms equal to unity; that is, if the vapour is a perfect gas and the molar volumes of the liquids are negligibly small. This gives

$$p = x_1 p_1^0 + x_2 p_2^0 \qquad (4.55)$$

That is, the vapour pressure of an ideal mixture of negligible volume in equilibrium with a perfect gas mixture is a linear function of the mole fractions. This is a statement of *Raoult's law* and may be generalized to ideal multicomponent mixtures.

This law is sometimes made the definition of an ideal mixture, in place of equations (4.35), but such a starting point is not so satisfactory as that chosen here, since Raoult's law follows from equation (4.35) only after making the assumptions which led from equation (4.54) to (4.55). These assumptions become more correct the lower the temperature, and are the same as those that were necessary for the derivation of the Clapeyron–Clausius equation (2.76) from Clapeyron's equation (2.74).

A more realistic approximation is the assumption that the vapour is not a perfect gas mixture but an ideal mixture of imperfect gases[5]. This assumption requires that $(\delta B)_{12}$ is zero, and is generally more nearly correct than that B_{11}, B_{12} and B_{22} are all zero. It gives the following expression for the vapour pressure after expanding the exponential terms in equation (4.54) and neglecting all powers of p beyond the first

$$p = x_1 p_1^0 + x_2 p_2^0 - x_1 x_2 (p_1^0 - p_2^0)\left[\frac{p_1^0(v_1^0 - B_{11}) - p_2^0(v_2^0 - B_{22})}{RT}\right] \qquad (4.56)$$

The second virial coefficient is always negative for a saturated vapour, and so $(v^0 - B)$ is positive. An ideal mixture in equilibrium with an ideal but imperfect gas mixture therefore shows negative deviations from Raoult's law unless B_{11} and B_{22} are very different in size. Conversely, a system which is observed to obey Raoult's law is generally not an ideal mixture as defined by equations (4.35), but has a small positive excess free energy. The system benzene + ethylene dichloride is an example of this behaviour (*see* Section 5.6).

The partial pressure of a component in a vapour mixture is defined as the product of the mole fraction and the total pressure. It is not a partial molar quantity in the sense of Section 4.2. The partial pressures of a system which obeys Raoult's law are given by

$$p_\alpha = y_\alpha p = x_\alpha p_\alpha^0 \tag{4.57}$$

The bubble-point line on a pressure–composition graph (*Figure 4.3*) shows p as a function of x, and is here a straight line. The dew-point line is p as a function of y and is concave upwards. The equation of this curve in a binary mixture is

$$p = p_1^0 p_2^0 / [p_1^0 - y_1(p_1^0 - p_2^0)] \tag{4.58}$$

and so the curve is part of a rectangular hyperbola.

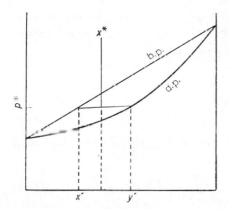

Figure 4.3 *The bubble-point and dew-point lines of an ideal mixture at constant temperature*

If a liquid mixture of composition x^* (*Figure 4.3*) is slowly decompressed, then gas first appears where the vertical line through x^* cuts the bubble-point curve. At a pressure p^* the system consists of gas and liquid in equilibrium. The compositions of the phases are, respectively, y', x', and the ratio of the number of moles of liquid to the number of gas is $(y' - x^*)/(x^* - x')$. The (T, x) graph, or boiling-point diagram at constant pressure, may be calculated from the (p, x) diagram if the variation of the vapour pressure with temperature is known. Neither dew- nor bubble-point line is linear for a mixture that obeys Raoult's law, as pressure is not a linear function of temperature[7].

Two liquids may be readily separated by distillation if their *volatility ratio* is large. This ratio, α_{21}, is defined as that of (y_2/y_1) to (x_2/x_1). It is a measure of the relative difference of composition of the gas and liquid phases in equilibrium. For an ideal mixture

$$\alpha_{12}(\neq \alpha_{21}) \equiv K_1/K_2 = p_1^0/p_2^0 \quad \text{where } K_1 = y_1/x_1 \tag{4.59}$$

thus showing, as is well known, that two liquids are the more readily separable by distillation the greater the difference in vapour pressure or, at constant total pressure, the greater the difference in boiling point.

4.4 Thermodynamics of non-ideal mixtures

Few real mixtures are ideal. In this section, the formal thermodynamic treatment of non-ideal mixtures is developed as a necessary preliminary to the description of the behaviour of actual systems in the following chapter.

The equations defining an ideal mixture may be modified so that they are formally valid for any real mixture at low pressure by writing

$$\mu_\alpha(p, T, x) - \mu_\alpha^0(p, T) = RT \ln (x_\alpha \gamma_\alpha) \tag{4.60}$$

This equation defines the *activity coefficients*, γ_α. They are all unity in ideal mixtures, and in non-ideal mixtures, functions of pressure, temperature and composition. The partial molar quantities are given by

$$h_\alpha = h_\alpha^0 - RT^2(\partial \ln \gamma_\alpha/\partial T)_p \tag{4.61}$$

$$s_\alpha = s_\alpha^0 - R \ln (x_\alpha \gamma_\alpha) - RT(\partial \ln \gamma_\alpha/\partial T)_p \tag{4.62}$$

$$v_\alpha = v_\alpha^0 + RT(\partial \ln \gamma_\alpha/\partial p)_T \tag{4.63}$$

The excess thermodynamic functions are obtained by subtracting from these the ideal quantities to give

$$\mu_\alpha^E = RT \ln \gamma_\alpha \qquad G^E = RT \sum_\alpha n_\alpha \ln \gamma_\alpha \tag{4.64}$$

$$h_\alpha^E = - RT^2(\partial \ln \gamma_\alpha/\partial T)_p \qquad H^E = - RT^2 \sum_\alpha n_\alpha(\partial \ln \gamma_\alpha/\partial T)_p \quad \text{etc.} \tag{4.65}$$

Excess functions are related by equations of the usual type

$$G^E = H^E - TS^E = H^E + T(\partial G^E/\partial T)_p \tag{4.66}$$

$$(C_p)^E = T(\partial S^E/\partial T)_p = (\partial H^E/\partial T)_p \quad \text{etc.} \tag{4.67}$$

The relation of V^E to the other excess functions is purely formal, for in practice there is no convenient method of measuring the pressure derivative of γ_α or of G^E other than that of measuring the excess volume.

The principal excess partial molar quantities can be written as the differences between the partial molar quantities for the non-ideal mixture and those of an ideal mixture,

$$\mu_\alpha^E = \mu_\alpha - (\mu_\alpha^0 + RT \ln x_\alpha) \tag{4.68}$$

$$h_\alpha^E = h_\alpha - h_\alpha^0 \tag{4.69}$$

$$s_\alpha^E = s_\alpha - (s_\alpha^0 - R \ln x_\alpha) \tag{4.70}$$

$$v_\alpha^E = v_\alpha - v_\alpha^0 \tag{4.71}$$

The excess partial molar quantities can be obtained from the measured values of the molar excess functions in exactly the same way that the partial molar quantities are calculated from the total thermodynamic functions. Some of these methods were described in Section 4.2. The use of excess functions leads to a considerable increase in the precision of the calculated total partial molar quantities and, nowadays, this is the usual method. Other ways of manipulating excess functions to obtain partial molar excess functions have been discussed by Franks and Smith[8] and by Van Ness and Mrazek[9].

The excess Helmholtz free energy and the excess energy at constant pressure are given by

$$A_p^E - G_p^E = U_p^E - H_p^E = -pV_p^E \simeq 0 \tag{4.72}$$

They are little used. The excess functions and functions of mixing at constant total volume are more interesting. They may be related to the constant pressure functions as follows[6].

Consider a binary mixture of mole fraction x formed from its components at constant, and essentially zero, pressure and at constant temperature

$$G_p^m = G(x, 0) - (1 - x)G(0, 0) - xG(1, 0) \tag{4.73}$$

where the two variables are x and p. If the mixing is carried out at constant total volume, then the pressure will change to p_V^m. This will be large and positive if V_p^m is positive, and negative if V_p^m is negative. Hence

$$G_V^m = G(x, p_V^m) - (1 - x)G(0, 0) - xG(1, 0) \tag{4.74}$$

or

$$G_V^m - G_p^m = G(x, p_V^m) - G(x, 0) \tag{4.75}$$

$$= \int_0^{p_V^m} \left(\frac{\partial G(x)}{\partial p} \right)_{T,x} dp = \int_0^{p_V^m} V(x) \, dp \tag{4.76}$$

Similarly,

$$V_p^m = V(x, 0) - V(x, p_V^m) = -\int_0^{p_V^m} \left(\frac{\partial V(x)}{\partial p} \right)_{T,x} dp \tag{4.77}$$

$V(x, p)$ may be expressed as a Taylor series in p

$$V(x, p) = V(x, 0) + \left(\frac{\partial V(x)}{\partial p} \right)_{p=0} p + \cdots \tag{4.78}$$

and, to a first approximation, all terms beyond the second may be neglected—an approximation which is similar to assuming that β_T is independent of pressure. Substitution from this equation into equations (4.76) and (4.77) gives

$$V_p^m = -(\partial V(x)/\partial p)_{p=0} p_V^m + \cdots \tag{4.79}$$

$$G_V^m - G_p^m = -\left[\frac{V(x)}{\partial V(x)/\partial p} \right]_{p=0} V_p^m + \frac{1}{2} \left[\frac{1}{\partial V(x)/\partial p} \right]_{p=0} (V_p^m)^2 + \cdots \tag{4.80}$$

and

$$G_V^m - A_V^m = V(x, p_V^m)p_V^m \tag{4.81}$$

$$= -\left[\frac{V(x)}{\partial V(x)/\partial p}\right]_{p=0} V_p^m + \left[\frac{1}{\partial V(x)/\partial p}\right]_{p=0} (V_p^m)^2 + \cdots \tag{4.82}$$

Whence

$$G_p^m - A_V^m = G_p^E - A_V^E = \tfrac{1}{2}\left[\frac{1}{\partial V(x)/\partial p}\right]_{p=0} (V_p^m)^2 \tag{4.83}$$

$$= \tfrac{1}{2}\frac{1}{V\beta_T}(V_p^m)^2 \tag{4.84}$$

The inclusion of higher terms in the Taylor expansion for the volume shows that the next non-vanishing term in equation (4.83) is of the order of $(V_p^m)^3$. Similarly,

$$S_p^m - S_V^m = S(x, 0) - S(x, p_V^m) \tag{4.85}$$

$$= \int_0^{p_V^m}\left(\frac{\partial V(x)}{\partial T}\right)_p dp \tag{4.86}$$

which gives

$$TS_p^m - TS_V^m = T\left(\frac{V(x)}{T}\right)_p p_V^m = \left[T\left(\frac{\partial p}{\partial T}\right)_V\right]_{p=0} V_p^m + O(V_p^m)^2 \tag{4.87}$$

By addition to equation (4.77)

$$H_p^m - U_V^m = \left[T\left(\frac{\partial p}{\partial T}\right)_V\right]_{p=0} V_p^m + O(V_p^m)^2 \tag{4.88}$$

Thus the differences in energy and entropy are of the order of V_p^m and that in the free energy of $(V_p^m)^2$. The latter difference is negligible compared to G_p^m and small compared to G_p^E, but the former is comparable with H_p^m and TS_p^m. Typical values for an equimolar mixture at room temperature are

$$g_p^m = -1500\,\mathrm{J\,mol^{-1}} \qquad g_p^E = 200\,\mathrm{J\,mol^{-1}} \qquad h_p^E = 300\,\mathrm{J\,mol^{-1}},$$

$$v_p^E = 1\,\mathrm{cm^3\,mol^{-1}} \qquad v = 100\,\mathrm{cm^3\,mol^{-1}} \qquad \beta_T = 10^{-4}\,\mathrm{bar^{-1}}$$

$$T\gamma_V = 3000\,\mathrm{bar}$$

These figures give

$$g_p^m - a_V^m = 5\,\mathrm{J\,mol^{-1}}$$

$$h_p^m - u_V^m \approx TS_p^m - TS_V^m = 300\,\mathrm{J\,mol^{-1}}$$

Thus the excess heat and entropy of a solution have very different values at constant total volume and at constant pressure. Only the latter are measured directly, and the former are calculated from them by using the equations above. TS_V^E is sometimes much smaller than TS_p^E, but it is doubtful if excess functions at constant volume are more closely related to intermolecular forces than those at constant pressure.

The variation of the vapour pressure with composition at constant temperature is closely related to G_p^E. The condition of equilibrium between liquid and vapour in a binary mixture is, as before, the equality of the chemical potentials of both species in each phase. The potentials of the gas are given by equations (4.46) and (4.47) and of the liquid by equation (4.60). Hence

$$\mu_1^E(p) = RT \ln (y_1 p / x_1 p_1^0) - (p - p_1^0)(v_1^0 - B_{11}) + 2p(\delta B)_{12} y_2^2 \qquad (4.89)$$

and similarly for component 2. The molar excess free energy is given by

$$g_p^E(p^*) = x_1 \mu_1^E(p) + x_2 \mu_2^E(p) + (p^* - p)v_p^E \qquad (4.90)$$

where p^* is a small arbitrary pressure. The last term is negligible if p^* is of the order of atmospheric pressure, and so g_p^E is almost independent of p^*.

It is seen that the activity coefficients are found most directly by measuring the vapour pressure and composition (p, y_1, y_2) in equilibrium with a liquid mixture of known composition (x_1, x_2) at a fixed temperature, T. Subsidiary measurements are needed to obtain the molar volumes of the liquids (v_1^0, v_2^0) and the second virial coefficients of the vapour (B_{11}, B_{12}, B_{22}).

4.5 Tests for consistency

The excess chemical potentials in a mixture are not wholly independent but, like all partial molar quantities, are related through the Gibbs–Duhem equation (4.10). This equation enables various tests to be made of the consistency of the experimental measurements and calculations which have led to the evaluation of the partial molar quantities. Such tests are commonly applied to chemical potentials and to partial molar volumes.

When the Gibbs–Duhem equation is written in the form

$$x_1 (\partial r_1 / \partial x_1)_T = x_2 (\partial r_2 / \partial x_2)_T \qquad (4.91)$$

it is obvious that, if $(\partial r_1 / \partial x_1)_T$ is zero at a maximum of r_1, then $(\partial r_2 / \partial x_2)_T$ is zero at a minimum of r_2. Such behaviour is shown, for example, by the partial molar volumes in aqueous solutions of ethanol and of 1,4-dioxan[10] (*Figure 4.4*). However, Griffiths[11] has reported partial molar volumes for the latter system in which a cusp in one curve is opposed to a minimum in the other and which therefore do not satisfy equation (4.91). The partial molar volumes of Schott[11] have improbable extrema.

The corresponding equation for the activity coefficients is obtained by substituting $(\mu_1 - \mu_1^0)$ for r_1 in equation (4.91)

$$\frac{(\partial x_1 \gamma_1 / \partial x_1)}{\gamma_1} = \frac{(\partial x_2 \gamma_2 / \partial x_2)}{\gamma_2} \qquad (4.92)$$

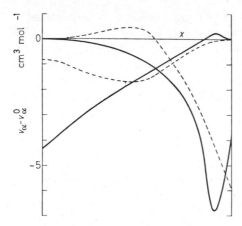

Figure 4.4 Maxima and minima in partial molar volumes[10]. Full lines, ethanol + water at 273 K; dashed lines, dioxan + water at 298 K. The mole fraction, x, is that of water

If the vapour is a perfect gas mixture, then γ_1 is equal to $(y_1 p / x_1 p_1^0)$. Hence

$$\frac{(\partial p_1 / \partial x_1)}{p_1 / x_1} = \frac{(\partial p_2 / \partial x_2)}{p_2 / x_2} \tag{4.93}$$

This form of the Gibbs–Duhem equation is usually called the Duhem–Margules equation, but its practical value in tests for consistency is not great, since it holds only for a system in which the vapour is a perfect gas mixture. However, gross violations of equation (4.93) are always evidence of inconsistency. A bad example is the partial pressure measurements of Smyth and Engel[12] on the system ethanol + n-heptane. Here, the partial pressure of ethanol at 50 °C is reported to show a maximum whilst that of

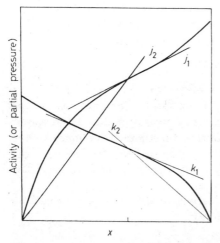

Figure 4.5 The test for consistency of Beatty and Calingaert[13]. The ratio of the slope of the tangent j_1 to the chord j_2 must be the same as the ratio of k_1 to k_2

heptane behaves normally. Such a maximum not only violates equation (4.93) but would lead to material instability and so to the separation of the mixture into two phases (*see* Section 4.6).

Beatty and Calingaert[13] have formulated neatly the consequences of equation (4.93). They observe that each side of the equation is the ratio of the slope of the partial-pressure curve to that of the line joining the partial pressure to the origin (*Figure 4.5*). The equation requires that these ratios be the same for both components at all compositions. In particular, if the line through the origin becomes a tangent to the partial-pressure curve for one component, then the same relationship must hold for the second.

Another kind of test, once much used but now recognized to be of limited utility, tests the consistency of simultaneous measurements of p, x and y through the use of the volatility ratio defined by equation (4.59). By subtraction of equation (4.89) from the corresponding equation for component 2 and substitution from equation (4.15),

$$\mu_2^E - \mu_1^E = RT \ln (\gamma_2/\gamma_1) = (\partial g^E/\partial x_2)_T \tag{4.94}$$

$$= RT \ln \left(\frac{\alpha p_1^0}{p_2^0}\right) + (p - p_1^0)(v_1^0 - B_{11})$$

$$- (p - p_2^0)(v_2^0 - B_{22}) + 2p(\delta B)_{12}(y_1^2 - y_2^2) \tag{4.95}$$

This equation may be integrated over x_2 from 0 to 1 to give

$$\int_0^1 \ln \left(\frac{\alpha p_1^0}{p_2^0}\right) dx_2 = \frac{1}{RT} \int_0^1 [(p - p_2^0)(v_2^0 - B_{22})$$

$$- (p - p_1^0)(v_1^0 - B_{11}) - 2p(\delta B)_{12}(y_1^2 - y_2^2)] dx_2 \tag{4.96}$$

since g^E is zero both at $x_2 = 0$ and at $x_2 = 1$. The following statements follow directly from these equations:

(1) The integrands on both sides of equation (4.96) are zero for an ideal liquid mixture in equilibrium with a perfect gas mixture.
(2) The volatility ratio is equal to (p_2^0/p_1^0) at the composition at which g^E has a maximum or a minimum in a non-ideal mixture in equilibrium with a perfect gas. In such a system, the integrand on the left-hand side of equation (4.96) takes on both positive and negative values and the integral vanishes.
(3) The function $\ln(\alpha p_1^0/p_2^0)$ is a linear function of x in a *quadratic*[a] liquid mixture in equilibrium with a perfect gas mixture.

The second of these conclusions is the test for consistency proposed[14] independently by Coulson and Herington and by Redlich and Kister. They plotted $\ln (\alpha p_1^0/p_2^0)$ against x_2 and, by graphical integration, found whether or not the areas above and below the x-axis were equal.

This simple form of the test is adequate only if the vapour pressures of

[a] *See* Section 5.2

the components are close, since it neglects imperfections of the vapour. They and others have considered the modifications necessary to account for these imperfections. One simple form of equation (4.96) which takes adequate account of gas imperfection, except in the most extreme cases, can be obtained as follows. First, it is found that the integrand on the right-hand side of equation (4.96) is a well-behaved function of x to which the last term contributes very little if $(\delta B)_{12}$ is much smaller than either B_{11} or B_{22}. The first approximation is, therefore, the neglect of this term. The second is the replacement of p by the value required by Raoult's law, namely $(x_1 p_1^0 + x_2 p_2^0)$. This allows the integration to be performed analytically to give

$$\int_0^1 \ln\left(\frac{\alpha p_1^0}{p_2^0}\right) dx_2 = \frac{(p_1^0 - p_2^0)(v_1^0 + v_2^0 - B_{11} - B_{22})}{2RT} \tag{4.97}$$

Further modifications of the equal area test have been discussed by Herington[15]. All such methods are limited by the fact that the left-hand sides of equations (4.96) and (4.97) are calculated from experimental values of x and y together with values of the vapour pressures of the two pure components. The experimental total pressures for the various mixtures are not required.

It is usually found that, in an experiment in which all three parameters x, y and p are determined experimentally, excess Gibbs functions of the highest precision can be obtained from p, x results only, using methods that are discussed in Chapter 5. A true test of thermodynamic consistency is then to compare the experimental values of y with the theoretical, calculated values.

4.6 Azeotropy

The (p, T, x) surface of a binary mixture can have varying shapes, some of which are related to distinctive physical properties such as azeotropy (the property of a liquid mixture of distilling without change of composition) or the presence of liquid–liquid and gas–liquid critical points. The discussion of these requires expressions for bounding surfaces in a (p, T, x) diagram.

Consider a mole of a binary mixture which, at a given p and T, separates into two phases of composition x_2' and x_2''. If the phases are gas and liquid then the (p, T, x) surface is formed of two sheets, as shown in *Figure 4.6*. Equilibrium between *phase'* and *phase"* requires equality of pressure and temperature, and of the chemical potentials:

$$\mu_1' = \mu_1'' \qquad \mu_2' = \mu_2'' \tag{4.98}$$

Let p, T, x_2' and x_2'' be a state which satisfies these equations. Consider a neighbouring state of the system $(p + \delta p)$, $(T + \delta T)$, $(x_2' + \delta x_2')$, $(x_2'' + \delta x_2'')$, in

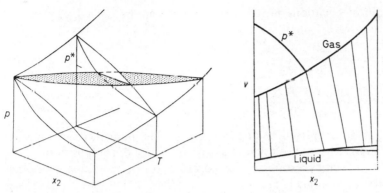

Figure 4.6 *The* (p, T, x) *surface and the* (v, x) *projection of the* (p, v, x) *surface (at constant temperature) of a simple binary mixture. The shaded cut on the* (p, T, x) *surface shows the boiling and condensation points at a given pressure. The isobar* p^* *on the* (v, x) *diagram is shown in the gas, liquid and two-phase regions, the other isobars only in the last*

which it is also at equilibrium. If the potentials are still to satisfy conditions (4.98), then

$$(\partial \mu_1/\partial p)'_{T,x}\delta p + (\partial \mu_1/\partial T)'_{p,x}\delta T + (\partial \mu_1/\partial x_2)'_{p,T}\delta x'_2$$
$$= (\partial \mu_1/\partial p)''_{T,x}\delta p + (\partial \mu_1/\partial T)''_{p,x}\delta T + (\partial \mu_1/\partial x_2)''_{p,T}\delta x''_2 \tag{4.99}$$

and similarly for component 2. Since μ_1 is the partial Gibbs free energy, we have from equation (4.16)

$$\mu_1 = g - x_2(\partial g/\partial x_2)_{p,T} \tag{4.100}$$

and differentiation gives

$$x_2^{-1}(\partial \mu_1/\partial x_2)_{p,T} = -(\partial^2 g/\partial x_2^2)_{p,T} \tag{4.101}$$
$$= -(\partial^2 g/\partial x_1^2)_{p,T} \tag{4.102}$$
$$= x_1^{-1}(\partial \mu_2/\partial x_1)_{p,T} \tag{4.103}$$

Subtraction of the two sides of equation (4.99) and the companion equation for component 2 gives

$$\Delta v_1 \delta p - \Delta s_1 \delta T - x'_2(\partial^2 g/\partial x^2)'_{p,T}\delta x'_2 + x''_2(\partial^2 g/\partial x^2)''_{p,T}\delta x''_2 = 0 \tag{4.104}$$
$$\Delta v_2 \delta p - \Delta s_2 \delta T + x'_1(\partial^2 g/\partial x^2)'_{p,T}\delta x'_2 - x''_1(\partial^2 g/\partial x^2)''_{p,T}\delta x''_2 = 0 \tag{4.105}$$

where

$$\Delta v_1 = v'_1 - v''_1 \quad \text{etc.} \tag{4.106}$$

Since

$$\Delta \mu_1 = \Delta h_1 - T\Delta s_1 = 0 \tag{4.107}$$

then

$$\Delta s_1 = \Delta h_1/T \tag{4.108}$$

The pair of equations (4.104) and (4.105) can be solved at constant pressure to give

$$\left(\frac{\partial x_2}{\partial T}\right)_p' = -\frac{\Delta h_1 x_1'' + \Delta h_2 x_2''}{T(g_{2x}')\Delta x_2} \tag{4.109}$$

$$\left(\frac{\partial x_2}{\partial T}\right)_p'' = -\frac{\Delta h_1 x_1' + \Delta h_2 x_2'}{T(g_{2x}'')\Delta x_2} \tag{4.110}$$

where g_{2x} is an abbreviation for $(\partial^2 g/\partial x^2)_{p,T}$. The numerator of equation (4.109) is a molar heat of solution of a drop of *phase"* of composition (x_1'', x_2'') in *phase'* at constant pressure and temperature. It can be written in another form by using equation (4.16)

$$\Delta h_1 x_1'' + \Delta h_2 x_2'' = \Delta h - \Delta x_2 (\partial h/\partial x_2)_{p,T}' \tag{4.111}$$

Hence

$$\left(\frac{\partial x_2}{\partial T}\right)_{p,\sigma}' = \frac{1}{T g_{2x}'}\left[\left(\frac{\partial h}{\partial x_2}\right)_{p,T}' - \frac{\Delta h}{\Delta x_2}\right] \tag{4.112}$$

and similarly for the second phase. By solving equations (4.104) and (4.105) at constant temperature, we obtain

$$\left(\frac{\partial x_2}{\partial p}\right)_{T,\sigma}' = \frac{\Delta v_1 x_1'' + \Delta v_2 x_2''}{g_{2x}'\Delta x_2} = -\frac{1}{g_{2x}'}\left[\left(\frac{\partial v}{\partial x_2}\right)_{p,T}' - \frac{\Delta v}{\Delta x_2}\right] \tag{4.113}$$

The equations (4.112) and (4.113) are the slopes of one side of a (p, T, x) surface in directions perpendicular to the p and T axes (*Figure 4.6*). The slopes of the other surface (*phase"*) are found by substituting x'' for x', etc. A third pair of equations, for slopes at constant x' and x'', may be obtained from these and the corresponding equations for *phase"*. The first of this pair is

$$\left(\frac{\partial p}{\partial T}\right)_{x_2,\sigma}' = \frac{\Delta h_1 x_1'' + \Delta h_2 x_2''}{T(\Delta v_1 x_1'' + \Delta v_2 x_2'')} \tag{4.114}$$

The compositions of the two phases can become equal in two different ways. If x' approaches x'', as h' approaches h'' and v' approaches v'', etc., then the point of identity is a critical point. If, however, x' becomes equal to x'' without h' being equal to h'', etc., then the system is at an *azeotropic point*. Azeotropic systems are discussed in this section and critical systems in Section 4.8.

It follows directly from equation (4.109) and the first part of equation (4.113) that if x' is equal to x'', whilst Δh and Δv, etc., remain non-zero, then the slopes $(\partial p/\partial x_2)_\sigma'$, $(\partial p/\partial x_2)_\sigma''$, $(\partial T/\partial x_2)_\sigma'$ and $(\partial T/\partial x_2)_\sigma''$ are all zero. That is, the vapour pressure (at constant temperature) and the boiling point (at constant pressure) are either maxima or minima with respect to changes in composition. A maximum in the vapour pressure is always accompanied by a minimum in the boiling point, since Δh and Δv have the same signs. It also follows from these equations that $(\partial p/\partial x_2)_\sigma'$ and $(\partial p/\partial x_2)_\sigma''$ become equal as they approach zero, and similarly with the temperature derivatives.

These deductions are known collectively as the *Gibbs–Konowalow laws*. It is conventional to describe an azeotrope with a maximum vapour pressure and a minimum boiling point as a positive azeotrope, and the converse as a negative azeotrope. The former are much the more common, and *Figure 4.7* shows the (p, T, x) surface and the (v, x) diagram for this azeotrope. The minimum volume of the gas phase is at the azeotropic composition if the vapour is a perfect gas mixture.

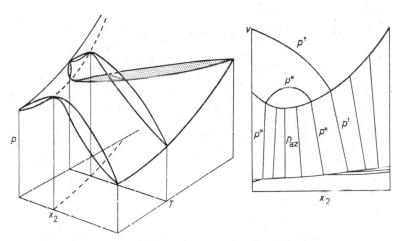

Figure 4.7 The (p, T, x) surface and the (v, x) projection of the (p, v, x) surface (at constant temperature) for a mixture that forms a positive azeotrope. The conventions are as in Figure 4.6. The dashed line is the locus of the azeotrope. The order of increasing pressure of the isobars on the (v, x) diagram is $p_2^0 < p^\dagger < p_1^0 < p^* < p_{az}$

A system of two phases and two components has two degrees of freedom. Hence, for a displacement on a (p, T, x) surface

$$dp = (\partial p/\partial T)_{x,\sigma}\, dT + (\partial p/\partial x)_{T,\sigma}\, dx \tag{4.115}$$

The last term is zero for an azeotrope and so, putting x' equal to x'' in equation (4.114)

$$\frac{dp_{az}}{dT} = \left(\frac{\partial p}{\partial T}\right)_{x,\sigma} = \frac{\Delta h}{T\Delta v} = \frac{\Delta s}{\Delta v} \tag{4.116}$$

This equation is the same as Clapeyron's equation for a system of one component, and indeed, an azeotrope has many of the characteristics of a pure substance. The logarithm of the azeotropic pressure is almost a linear function of $(1/T)$. An approximate derivation of this statement follows from equation (4.116) in the same way as equation (2.76) follows from equation (2.74). *Figure 4.8* shows the azeotropic pressure and composition for the system ethanol + water[16]. The agreement of the several observers is good for the vapour pressure but poor for the less readily measurable azeotropic composition.

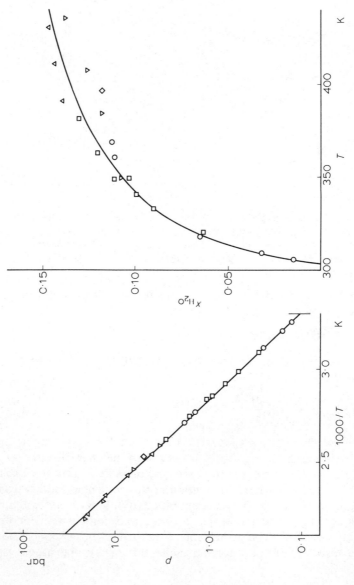

Figure 4.8 The vapour pressure and composition of the azeotrope of ethanol + water[16]. Circles, Wade and Merriman (0.13–1.93 bar); squares, Larkin and Pemberton (0.30–3.16 bar); triangles, Kleinert (4.32–19.72 bar); inverted triangles, Otsaki and Williams (1.01–20.68 bar); diamonds, Othmer and Levy (5.17 bar)

Successive or fractional distillation of a mixture which forms a positive azeotrope leads to a distillate of the composition of the latter and a residue of the pure component that was initially present in excess. Straight distillation of a mixture which forms a negative azeotrope leads eventually to a residue of the azeotropic composition and a distillate from which one of the pure components can be separated by fractionation. Thus a negative azeotrope can be separated by straight distillation from a mixture of arbitrary composition, but a positive azeotrope only by fractional distillation.

The former, although less common, were therefore studied and analysed about 50 years before there was any systematic study of the latter. The earliest workers thought that negative azeotropes were chemical compounds because of their low vapour pressures and reproducible compositions. This was disproved by Roscoe and Dittmar[17] in 1860 when they showed that the composition of the azeotrope formed from hydrogen chloride and water changed with the pressure under which the distillation was performed. Their results agree remarkably well with the best modern work of Bonner and his colleagues[18] (*Table 4.1*).

Table 4.1 Composition and boiling points of the hydrogen chloride + water azeotrope

p/mmHg	50	760	1000	2500
$t/°C$ (Bonner & Wallace)	48.72	108.58_4	116.18_5	—
Wt% HCl (Roscoe & Dittmar)	23.2	20.24	19.7	18.0
Wt% HCl (Bonner & Titus)	23.42	20.222	19.734	—

A positive azeotrope is at a maximum of the p–x graph and is, therefore, at low vapour pressures, a mixture of positive excess free energy. Conversely, a negative azeotrope has a negative excess free energy. There is, however, no direct correlation between the sizes of the excess free energies and the occurrence or absence of azeotropy, since this depends also on the ratio of the vapour pressures of the pure components. Elementary reasoning from the properties of the (p, T, x) surface leads to the following generalizations.

(1) The closer the vapour pressures of the components the more likely is azeotropy. It is inevitable at any temperature at which the vapour pressures are the same. (Such a temperature and pressure is often called a *Bancroft point*.)
(2) If the excess free energy is symmetrical in the composition, then a positive azeotrope is richer in the component of higher vapour pressure, and conversely for a negative azeotrope.
(3) An increase of temperature and of vapour pressure in a positive azeotrope increases the mole fraction of the component whose vapour

pressure increases the more rapidly with temperature. The converse holds for a negative azeotrope.
(4) The closer the vapour pressures of the pure components, the more rapidly does the azeotropic composition change with temperature.

These generalizations are not without exceptions but are usually an adequate guide. If a mixture is *quadratic* (*see* Section 5.2), and if its vapour is a perfect gas mixture, then the azeotropic pressure and composition at a given temperature can be calculated from the vapour pressures of the pure components and the excess free energy. Conversely, a knowledge of the composition and pressure at a given temperature (or even of the composition and temperature at a given pressure) is sufficient for an estimate of the excess free energy[19]. Such estimates are unlikely to be badly in error, since the excess free energy is the most symmetric of the excess functions. The composition of an azeotrope formed by a quadratic mixture is approximately a linear function of the temperature[19,20].

There have been few measurements of the pressure, temperature and composition of an azeotrope over the whole of the liquid range, but it is often possible to obtain an adequate knowledge of a system by combining several sets of measurements. There are four possibilities. A binary mixture can form an azeotrope (1) over the whole of the liquid range, (2) over a range bounded above by a return to normal behaviour, (3) over a range similarly bounded below, or (4) over a range bounded above and below. The first type of behaviour is called *absolute azeotropy* and the others *limited azeotropy*.

We enquire next into the shape of (p, T, x) and similar surfaces near critical points in binary mixtures. These surfaces have similarities to, and differences from those for critical points of one-component systems. A proper comparison of mixtures and pure substances requires an extension of the apparatus of classical thermodynamics to which we turn in the next section.

4.7 Fields and densities

For many purposes, the size of a thermodynamic system is irrelevant and is often eliminated by working, for example, in terms of molar functions rather than extensive functions. This elimination was formalized by Griffiths and Wheeler in a way that is particularly useful for the analysis of phase equilibria. After a short introduction, they open their paper[21] on 'Critical points in multicomponent systems', as follows (we make a slight change of notation).

The customary division of thermodynamic variables into *intensive* and *extensive* is a very important one, though the reasons for this division are seldom clearly stated, and the terms themselves can be misleading.

One knows, for example, that temperature belongs in the first category and entropy in the second. But what about entropy per mole, s, a quantity which remains unchanged upon the combination to two identical systems (and consequent doubling of extensive variables)? We shall use the term *density* for such a variable (a more precise definition of density is given below), which though in one sense intensive plays quite a different role in thermodynamic equations than temperature and pressure, which we shall refer to as *fields*. Other examples of fields are the chemical potentials of different components in fluid mixtures, and magnetic and electric fields in paramagnets and dielectrics, respectively,

The fields (in contrast with densities) have the property that they take on identical values in two phases which are in thermodynamic equilibrium with each other. A system with q independent thermodynamic variables ($q = 2$ for a pure fluid) may be characterized by $q + 1$ fields $f_0, f_1, ..., f_q$ with one of these, say f_0, regarded as a function of the rest. The dependent field ... will be called the *potential*. With this particular choice of independent and dependent variables, the q densities d_i are defined by

$$d_i = -(\partial f_0 / \partial f_i) \tag{4.117}$$

For the thermodynamic systems we wish to consider, it is always possible to choose the potential as a concave function of the other fields, and the statement that f_0 is a concave function of all the [other] f_i together comprehends all the requirements of thermodynamic stability.

There is a wide choice of variables but it is usual to define the densities on a molar or on a volumetric basis. Thus division of equation (2.4) by the amount of substance, n, gives for a one-component system,

$$d\mu = vdp - sdT \tag{4.118}$$

so that if μ is the potential, the two independent fields are $-p$ and T, and the densities v and s. The generalization to a multicomponent system is the Gibbs–Duhem equation, cf. equation (4.11),

$$\sum_i x_i \, d\mu_i = vdp - sdT \tag{4.119}$$

or

$$d\mu_1 = vdp - sdT - \sum_{i>1} x_i \, d\mu_{i1} \tag{4.120}$$

where μ_{i1} is $(\mu_i - \mu_1)$. The additional fields are these differences, μ_{i1}, and the conjugate densities, the mole fraction x_i. Chemical potential is often more conveniently replaced by the activity λ,

$$\lambda_i = \exp(\mu_i / RT) \tag{4.121}$$

In these variables equation (4.120) becomes

$$\mathrm{d}\ln\lambda_i = v\mathrm{d}(p/RT) + u\mathrm{d}(1/RT) - \sum_{i>1} x_i \,\mathrm{d}\ln(\lambda_i/\lambda_1) \tag{4.122}$$

so that if $\ln\lambda_1$ is the potential, the $q = c + 1$ independent fields are $-p/RT$, $-1/RT$, and $\ln(\lambda_i/\lambda_1)$ with $i > 1$.

In Section 3.1 we have met volumetric densities, namely, the free-energy density, $\psi = A/V$, as a function of the molar densities $\rho_i = n_i/V$ and the temperature. The Gibbs–Duhem equation in these variables is obtained by dividing equation (4.119) by $v = V/n$,

$$\mathrm{d}p = \eta\mathrm{d}T + \sum_i \rho_i \,\mathrm{d}\mu_i \tag{4.123}$$

where $\eta = S/V$ is the entropy density. This equation can be re-written in terms of the energy density $\phi = U/V$,

$$\mathrm{d}(p/RT) = -\phi\mathrm{d}(1/RT) + \sum_i \rho_i \,\mathrm{d}(\mu_i/RT) \tag{4.124}$$

$$= -\phi\mathrm{d}(1/RT) + \sum_i \rho_i \,\mathrm{d}\ln\lambda_i \tag{4.125}$$

In a binary system we have, therefore, several canonical choices of independent fields; e.g., $(-p/RT)$, $(-1/RT)$ and $\ln(\lambda_2/\lambda_1)$ from equation (4.122), or $(1/RT)$, $\ln(1/\lambda_1)$ and $\ln(1/\lambda_2)$, from equation (4.125). The first of these trios is a close match for the traditional set, p, T, and x_2, used in plotting phase diagrams. The important difference, as we shall see, is that the variables of equation (4.122) are all fields, whilst the traditional set is mixed—two fields and one density. A change from the variables of equation (4.122) to the set (p, T, λ_2) does not change the topology of a phase-diagram, and is often made in practice. If the standard state from which μ_2 is measured is chosen appropriately then λ_2 can run from zero at $x_2 = 0$ to unity at $x_2 = 1$. It is therefore a convenient field with which to record the composition.

In a system of c components, and so $q = c + 1$ independent fields, the equilibrium between two phases is determined by the $q + 1$ field equations

$$f_i' = f_i'' \qquad i = 0, 1, \ldots, q \tag{4.126}$$

These equations define a hypersurface of dimension $q - 1$ in the space of the q independent fields. The hypersurface can terminate in several ways.

(1) A field can reach the limit of its definition, e.g., $\lambda_i = 0$, when component i is no longer present. The hypersurface of dimension $q - 1$ usually then becomes a hypersurface of dimension $q - 2$ in a reduced field space of $q - 1$ dimensions. Thus, in a binary system $(q = 3)$, liquid–vapour equilibrium is a surface in a (p, T, λ_2)-space, which becomes a vapour pressure line in a (p, T)-space at $\lambda_2 = 0$.
(2) The coexistence hypersurface can cut another at a 'triple-point' hypersurface of $(q - 2)$ dimensions. Thus, in a binary system a liquid–gas surface meets a solid–liquid surface at a line of triple points.

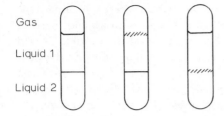

Gas

Liquid 1

Liquid 2

Figure 4.9 Critical end-points at which the meniscus vanishes either between a gas and a liquid in the presence of a second liquid phase, or between two liquids in the presence of a gas phase

(3) The coexistence hypersurface can end in a critical-point hypersurface of $(q - 2)$ dimensions. In a binary system this is a line of critical points, which can generally be called unambiguously a line of gas-liquid or of liquid–liquid critical points. (Ambiguous cases are deferred to Chapter 6.) Thus, if we have two partially immiscible liquids in equilibrium with mixed vapour, we can reduce the number of fluid phases to two by changing the fields until a critical point is reached. If this is marked by the vanishing of the meniscus between the vapour and one of the liquids, then it is a gas–liquid critical point; if by the vanishing of the meniscus between the two liquids, then it is a liquid–liquid critical point (*Figure 4.9*). Changing (say) the pressure will show that the critical point lies on a line in a (p, T, λ) field space. We return to the topology of these lines in the next section and in Chapter 6; here we note only that there is no thermodynamic distinction between the two kinds of critical lines, and that the discussion in this and the next section applies to both.

The essential point of the geometric analysis of Griffiths and Wheeler[21] is shown by a comparison of *Figures 4.10*(a) and (b), which are the gas–liquid surfaces for a binary mixture, first in a (p, T, x)-space, and secondly in a (p, T, λ)-space. The pair of sheets in the first become a single sheet in the

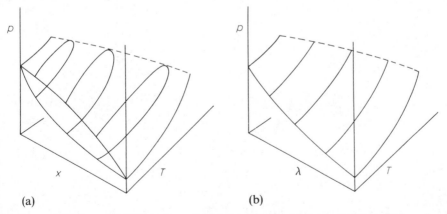

(a)   (b)

Figure 4.10 The gas–liquid surface of a binary mixture (a) in the space (p, T, x), and (b) in (p, T, λ). The critical lines are shown by dashes

second, since $\lambda_2^g = \lambda_2^l$. The two (p, T, x) sheets meet smoothly along the critical line, while the single (p, T, λ) sheet merely comes to an end there. A section of *Figure 4.10*(a) by a plane containing the two field variables, p and T, yields topologically quite a different curve for a binary mixture than the same section for a pure substance ($x_2 = 0$ or $x_2 = 1$). In *Figure 4.10*(b), on the other hand, such a section is topologically the same whether the section is taken at $\lambda_2 = 0$ or $\lambda_2 = 1$, or at some intermediate value. Indeed, the same kind of curve, ending in a critical point, is the result of sections parallel to any pairs of field variables, (p, T), (p, λ_2) or (T, λ_2), or, indeed, by (almost) any arbitrary section.

The general deduction that Griffiths and Wheeler draw from such arguments is that if we go from a c-component system to one of $(c + 1)$, or $(c + 2)$, etc. components, then the thermodynamic behaviour at and near critical points is essentially the same in all systems provided that the thermodynamic space is extended only by adding further field variables. If we add a density, as in going from (p, T) for a one-component system to (p, T, x) for a two-component system, then the character of the thermodynamic surface changes drastically. This powerful result enables us to generalize all that has been learnt about critical points of one-component systems (Chapter 3) to give us a full description of critical points in multicomponent systems.

It is not necessary to use all-field spaces to reap this advantage; all that is required is that on changing the number of components there is no change in the number of densities; that is, the added variables should all be fields. Thus the rounded ends of the (p, x) or (T, x) sections of *Figure 4.10*(a) are in essence the same as the rounded orthobaric curves, (T, v) or (T, ρ), or (p, v) or (p, ρ), near a one-component critical point. In each case the system is described in a space of one density (x, or v, or ρ) and one or more fields (p, or T, or p and T).

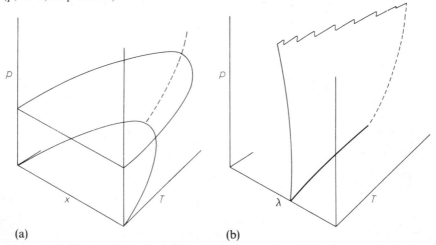

(a)   (b)

Figure 4.11 The liquid–liquid surface of a binary mixture (a) *in the space* (p, T, x), *and* (b) *in* (p, T, λ). *The critical lines are shown by dashes*

Two partially immiscible liquids generally become more miscible as the temperature rises and may reach a liquid–liquid critical point, as shown in *Figures 4.11*(a) and (b), which are topologically the same as *Figures 4.10*(a) and (b). In the next section we analyse these surfaces, using the results of Chapter 3 and the principle of Griffiths and Wheeler. We shall not abandon the use of mixed spaces since they are more familiar than the all-field spaces, and contain more information.

4.8 Thermodynamics of partially miscible liquids

The most convenient experimental variables for a binary mixture are p, T and x, with which we naturally associate the thermodynamic function $g(p, T, x)$. This function is in a space of one density and so is legitimately compared with the functions $a(T, v)$ or $\psi(T, \rho)$ of Chapter 3. Scott[22,23] gives dictionaries of translations of the kind $a(T, v) \equiv g(p, T, x)$, for systems of one, two and three components.

We saw in the last chapter that there were strong empirical reasons for using ψ rather than a for a one-component system, since it generates more symmetrical orthobaric (and similar) curves. This symmetry ensures that the scaling laws hold over wider thermodynamic domains for $\psi(T, \rho)$ than for $a(T, v)$. For mixtures there is no compelling evidence to show if x is the best density to use as the order parameter. For some systems[24] (e.g., $CH_4 + CF_4$) the volume fraction ϕ_2

$$\phi_1 = x_1 v_1 / (x_1 v_1 + x_2 v_2) \tag{4.127}$$

leads to more symmetric curves than x_2, but there are difficulties in using this function since the molar volumes v_i have no obvious definition. They may be taken to be those of the pure components, v_i^0, but then the volume fractions will not sum to unity unless $v^E = 0$. If the v_i are the partial volumes (Section 4.2) then they are extremely difficult to measure near critical points. In this section we therefore consider $g(p, T, x)$ and compare it with $a(T, v)$.

Chapter 3 opened with a sketch (*Figure 3.1*) of the isothermal section of the $a(T, v)$ surface in which was included an analytic continuation of $a(v)$ through the metastable and unstable states between v^l and v^g. *Figure 4.12* shows the equivalent isothermal–isobaric section of the $g(p, T, x)$-space for two partially miscible liquids of compositions x' and x''. It follows from equation (4.16), and equations (4.100)–(4.103) that the common tangent at A and D satisfies the condition of equilibrium, $\mu_1' = \mu_1''$ and $\mu_2' = \mu_2''$. The points of inflection, B and C, are the points at which $(\partial^2 g / \partial x^2)_{p, T}$ is zero. This derivative is negative between B and C and so this part of the isotherm is unstable. This kind of instability, which is called *material* or *diffusional instability*, is a third class to add to the thermal and mechanical instabilities of Chapters 2 and 3. A system on BC is in a state in which spontaneous

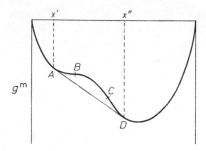

Figure 4.12 The molar free energy of mixing as a function of mole fraction, when g is a continuous function of x

fluctuations of composition would lead to a decrease of g. Such a change is called spinodal decomposition into separate phases, for reasons that will become clear in Chapter 6.

In a binary system

$$dg = vdp - sdT + \mu_1 dx_1 + \mu_2 dx_2$$
$$= vdp - sdT + \mu_{21} dx_2 \tag{4.128}$$

where μ_{21} is again $(\mu_2 - \mu_1)$. The parallel equation for a one-component system is

$$da = -sdT - pdv \tag{4.129}$$

The terms with densities as the independent variables are $\mu_{21}dx_2$ and $-pdv$. Thus the equations that define a one-component critical point

$$\left(\frac{\partial^2 a}{\partial v^2}\right)_T = -\left(\frac{\partial p}{\partial v}\right)_T = 0 \qquad \left(\frac{\partial^3 a}{\partial v^3}\right)_T = -\left(\frac{\partial^2 p}{\partial v^2}\right)_T = 0 \tag{4.130}$$

have as their parallel

$$\left(\frac{\partial^2 g}{\partial x_2^2}\right)_{p,T} = \left(\frac{\partial \mu_{21}}{\partial x_2}\right)_{p,T} = 0 \qquad \left(\frac{\partial^3 g}{\partial x_2^3}\right)_{p,T} = \left(\frac{\partial^2 \mu_{21}}{\partial x_2^2}\right)_{p,T} = 0 \tag{4.131}$$

Classical stability requires that $(\partial^4 g/\partial x_2^4)_{p,T} > 0$. If we were to follow precisely the notation in Chapter 3 then $(\partial^2 g/\partial x_2^2)_{p,T}$ would be written g_{200}, but we assume for the moment that p and T are fixed and retain the more transparent abbreviation g_{2x}. Thus the classical conditions for a critical point are

$$g_{2x}^c = 0 \qquad g_{3x}^c = 0 \qquad g_{4x}^c > 0 \tag{4.132}$$

These conditions determine the limiting form of equation (4.112) near a critical point. Let g be represented by a Taylor expansion in T and x_2, at constant pressure, about its value at the critical point. The leading terms are

$$g = g^c + g_x^c(\delta x)$$
$$= s^c(\delta T) - s_x^c(\delta x)(\delta T) - \tfrac{1}{2}s_{2x}^c(\delta x)^2(\delta T) + \tfrac{1}{24}g_{4x}^c(\delta x)^4 + \cdots \tag{4.133}$$

where (δx) now denotes $(x_2 - x_2^c)$ and g_{2x}^c and g_{3x}^c have been put equal to zero. Higher powers of (δT) may be neglected. The differential coefficients of g with respect to x and T, that is, g_x, g_{2x}, s and s_x, are easily obtained from equation (4.133). Equilibrium between two phases requires the equality of both g_x and $(g - (\delta x)g_x)$ in each phase, that is, that there is a common tangent to the curve $g(x)$ at $\delta x'$ and $\delta x''$. This equality may be expressed in terms of the derivatives of equation (4.133)

$$-s_{2x}^c(\delta T)(\delta x' - \delta x'') + \tfrac{1}{6}g_{4x}^c[(\delta x')^3 - (\delta x'')^3] = 0 \tag{4.134}$$

$$g_{4x}^c[(\delta x')^2 - (\delta x'')^2][(\delta x') - (\delta x'')]^2 = 0 \tag{4.135}$$

The second equation gives at once

$$\delta x' = -\delta x'' = \tfrac{1}{2}\Delta x \tag{4.136}$$

The first then gives

$$\tfrac{1}{6}g_{4x}^c(\delta x')^2 = \tfrac{1}{6}g_{4x}^c(\delta x'')^2 = s_{2x}^c(\delta T) \tag{4.137}$$

The right-hand side of equation (4.111) may be expressed in terms of the derivatives of g

$$\frac{1}{T}\left[\left(\frac{\partial h}{\partial x}\right)'_{p,\,T} - \frac{\Delta h}{\Delta x}\right] = \left(\frac{\partial s}{\partial x}\right)'_{p,\,T} - \frac{\Delta s}{\Delta x} = \tfrac{1}{2}s_{2x}^c\Delta x \tag{4.138}$$

whence

$$\left(\frac{\partial x_2}{\partial T}\right)'_p = -\left(\frac{\partial x_2}{\partial T}\right)''_p - \frac{6s_{2x}^c}{g_{4x}^c\Delta x_2} \tag{4.139}$$

This equation shows that x is a quadratic function of $(T - T^c)$ near the critical point for a mixture of constant (or negligible) pressure. The sign of the right-hand side of equation (4.139) is determined by that of s_{2x}^c, since g_{4x}^c is necessarily positive. The sign of s_{2x}^c is the same as that of h_{2x}^c, since g_{2x}^c is zero. If $x_2'' > x_2^c > x_2'$, then $(\partial x_2/\partial T)'_p$ is positive, and $(\partial x_2/\partial T)''_p$ is negative, if s_{2x}^c and h_{2x}^c are negative. These signs define an upper critical solution point[2] (*Figure 4.13*). The inequalities at upper and lower points may be expressed concisely by using the excess thermodynamic functions[a]

$$g_{2x}^c = RT^c/x_1x_2 + (g_{2x}^E)^c = 0 \tag{4.140}$$

$$h_{2x}^c = (h_{2x}^E)^c < 0 \quad \text{(UCST)} \qquad \text{or} > 0 \quad \text{(LCST)} \tag{4.141}$$

$$s_{2x}^c = -R/x_1x_2 + (s_{2x}^E)^c < 0 \quad \text{(UCST)} \qquad \text{or} > 0 \quad \text{(LCST)} \tag{4.142}$$

The signs of the second derivatives of h and s determine those of the functions themselves only if the h–x and s–x graphs have no points of

---

[a] See, for example, Copp and Everett[25] and the discussion of their paper (ref. 25, pp. 267–278).

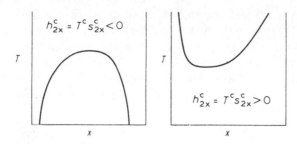

Figure 4.13 Typical sketches of the phase-boundary curves near upper and lower critical solution points (UCST and LCST)

inflection. This is true in a quadratic mixture (*see* Section 5.2) for which

$$x_1 x_2 g_{2x} = RT - 2g^E \tag{4.143}$$

$$x_1 x_2 h_{2x} = -2h^E \tag{4.144}$$

$$x_1 x_2 s_{2x} = -R - 2s^E \tag{4.145}$$

A quadratic mixture can have a UCST if h^E is positive, irrespective of the sign of s^E as long as it is greater than $(-R/2)$. It can have a LCST only if h^E and s^E are both negative and if the latter is smaller than $(-R/2)$. In both cases, the phase-boundary curve is symmetrical in x_1 and x_2 and the critical point is at the equimolar composition.

In practice, it is found that mixtures which are sufficiently non-ideal to show a LCST are rarely quadratic, but that these inequalities are qualitatively correct. It is seen that a LCST occurs only with a large negative value of s^E, whilst a UCST can occur with a small s^E of either sign. The former are much the less common and their interpretation in terms of the properties of molecules is less simple.

There are no thermodynamic restrictions on the signs or sizes of the excess heat capacity and excess volume at critical solution points, but it is most common for the former to be negative and the latter positive at a UCST, and vice versa at a LCST. If the negative excess heat capacity at a LCST keeps this sign as the temperature is raised, then h^E and s^E eventually become positive and the conditions for a UCST may be satisfied. Such a mixture shows a closed solubility loop, there being a range of temperature bounded above and below in which the components are only partially miscible. It is found that LCST are followed by UCST if the temperature is raised sufficiently, and if a gas–liquid critical point is not reached first. The sign of the excess volume (or, more strictly, of v_{2x}) determines the effect of pressure on a critical temperature. A displacement along a critical line in (p, T, x)-space is one for which g_{2x} remains equal to zero. Hence, for a small displacement δp, etc.,

$$\delta(g_{2x}) = g^c_{p2x}(\delta p) + g^c_{T2x}(\delta T) + g^c_{3x}(\delta x) = 0 \tag{4.146}$$

The third term is itself zero, and so

$$(dT^c/dp) = -g^c_{p2x}/g^c_{T2x} = v^c_{2x}/s^c_{2x} \tag{4.147}$$

The sign of v_{2x}^c is the opposite of that of v^E in the absence of a point of inflection on the v^E–x graph. The usual effect of an increase of pressure is, therefore, to raise critical solution temperatures. This is discussed in Chapter 6.

Much of the development of this section has been based on the assumptions, first, that the Taylor expansion (4.133) is valid near the critical point and, secondly, that g_{4x} is the first non-vanishing derivative at this point. Both assumptions are as inaccurate as were those of the continuity of a and its derivatives in the discussion of the gas–liquid critical point in Chapter 3. In particular, as is shown in the next section, the critical exponent γ is greater than unity, and δ is greater than 3, so that v_{2x}^c, s_{2x}^c and g_{4x}^c are all zero[26]. Equations such as (4.139) and (4.147) are therefore to be interpreted only in a limiting sense.

A further consequence of $\gamma^+ > 1$, for which there is some experimental evidence, is that coexistence curves calculated by extrapolation from values of g^E in the homogeneous mixture on the implicit assumption that $\gamma^+ = 1$ are too high at a UCST and too low at a LCST. It is common experience that this is so[26,27]. Scatchard and Wilson[27] show a closed solubility loop, calculated by such extrapolations, which has simultaneously too high a UCST and too low a LCST.

4.9 Exponents at the critical point of a binary mixture

The one-density function $g(p, T, x)$ for a binary system is the analogue of the one-density function $a(T, v)$ for a system of one component. This analogy leads to the identification of the critical exponents of a binary system shown in *Table 4.2*, which is derived from *Table 3.1*. (There is a similar set of entries for $g(p, x)$ at constant T.) The variables of the functions in this table are not shown but are $|T - T^c|$ for the first three lines and $|x - x^c|$ for the last, as follows from *Table 3.1*.

Classical thermodynamic arguments that parallel those of Section 3.2 show that the same inequalities hold between the various exponents, namely, (3.47), (3.48), (3.50) and (3.51). At first sight, it seems to be easier to measure the exponents for a binary liquid mixture than for the gas–liquid critical point of a one-component system, since the pressure can be held constant at or near atmospheric. Moreover, C_p is generally easier to

Table 4.2

Exponent	One component $a(T, v)$	Two components $g(T, x)$ p constant
α	c_v	$C_{p,x}$
β	v^l, v^g	x', x''
γ	$-(\partial p/\partial v)_T$	$(\partial^2 g/\partial x^2)_{p,T} = (\partial \mu_{21}/\partial x_2)_{p,T}$
δ	p	$-(\partial g/\partial x_2)_{p,T} = \mu_{21}$

measure than C_V. Gravitational effects can also be minimized by choosing components of similar mass density. Nevertheless, there are also disadvantages, of which one of the worst is the slow rate of diffusion in dense liquid mixtures, which makes thermal measurements difficult. The combined effects of gravity and diffusion are particularly complicated and have raised doubts about the interpretation of some measurements[28]. Measurements of C_p can be avoided by choosing instead the coefficient of thermal expansion which has the same exponent, as can be seen by writing down the entries in *Table 4.2* for $g(p, x)$ at constant T. The experimental difficulties and results (mainly from systems with UCST near room temperature) have been reviewed by Scott[23], Greer[23] and others[29] whose conclusions can be summarized as follows.

α The best values are from the coefficients of thermal expansions of 3-methylpentane + nitroethane[30], and i-butyric acid + water[31]; $\alpha = 0.10 \pm 0.06$.

β For many years the best values were around 0.35 but recent estimates[32,33] from orthobaric compositions within a few millikelvin of T^c are 0.32 ± 0.02. This shift parallels that for one-component systems, but there is evidence[33] that the temperature range over which the limiting behaviour can be seen is larger for mixtures.

γ Measurements of g_{2x} (from vapour pressures) are few, and all from one laboratory[23]; they yield $\gamma = 1.35 \pm 0.05$. Another method, which is experimentally easier but for which there are still some problems of interpretation, is the study of composition fluctuations by light scattering[23,34]. These measurements yield $\gamma = 1.25 \pm 0.05$, and are probably the more accurate.

δ This can also be determined from the vapour pressure via μ_{21}, but the experiment is difficult. The best results give[23] $\delta = 4.5 \pm 0.5$.

These results are so close to those of Chapter 3 that they need little further discussion. They suggest that the thermodynamic inequalities are, in fact, equalities, and that a scaling equation can do justice to all measurements. The scaling function t of equation (3.92) is $(|T - T^c|/T^c|x - x^c|^{1/\beta})$, and the parametric representations of the scaling equation are again the easiest to use.

This analysis leaves open the question of whether a different density might lead to a more symmetric representation of the data. It was noted in the last section that the volume fraction may sometimes be better than the mole fraction, but is difficult to use. The problem of choice becomes acute if the linearity of the orthobaric diameter is in question[23]. If we suppose that y' and y'' are the orthobaric compositions in the density of highest symmetry, y, then

$$y' = A + C(\delta T)^{1-\alpha} - B(\delta T)^{\beta}$$

$$y'' = A + C(\delta T)^{1-\alpha} + B(\delta T)^{\beta}$$

(4.148)

where $\delta T = |T - T^c|$. The mean $(y' + y'')/2$ is $A + C(\delta T)^{1-\alpha}$ and has the singularity in $(\delta T)^{1-\alpha}$ expected from the most probable behaviour of one-component systems. But if y is changed to a different composition variable, related to it by a smooth transformation, then the singularity in the diameter[27] is of order $(\delta T)^{2\beta}$. The same problem arises with a one-component system (Section 3.4) but is less discussed there since ρ is so far unchallenged as the most symmetric density, or order parameter. Attempts have been made to verify the $(1 - \alpha)$-singularity on nitromethane + carbon disulphide, and on cyclohexane + acetic anhydride[35], but the success of these claims cannot be decided until it is clear that the (2β)-problem has been overcome.

The addition of a third component leads to further complications. In a field space, we now have a surface of critical states which is cut at a single point by the two planes of constant p and T. Thus, if the components are chosen suitably, we can bring a three-component system to a critical point at (say) room pressure and temperature by changing the activities (or compositions) of the components. This experiment is usually represented in the familiar mole-fraction triangle (*Figure 4.14*) in which, for example, the addition of ethanol (c) to an immiscible mixture of water (a) and benzene (b) brings the system to a critical point. (The name *plait point* is sometimes used in this case but has little to recommend it.) The tongue-shaped curve resembles superficially the (T, x) curve for a two-component system, and the (T, v) or (T, ρ) curve for a one-component system. There is, however, an important distinction in that the coordinates in *Figure 4.14* are two densities (linear combinations of mole fractions), while in the one- and two-component systems the curves are drawn with coordinates of one density and one field. Hence the exponent β_x that determines the limiting shape of the curve in *Figure 4.14* is not the same as the β of the one- and two-component systems. It is, in fact, given by

$$\beta_x = \beta/(1 - \alpha_2^-) \tag{4.149}$$

and there are similar but not identical changes in other exponents[36]. These changes are examples of the *renormalization* of exponents, but this use of the

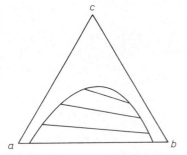

Figure 4.14 The phase diagram at fixed p, T for a ternary mixture, one pair of which is only partially miscible

word has no connection with its use in the *renormalization group* mentioned briefly in Section 3.5. The numerical change in β is small but the principle is important. Values of β_x of 0.38 and γ_x of 1.50 have been reported[37].

A more interesting set of complications arise in systems of three or more components if two critical end-points coincide in a field space. This is the subject of the next section.

4.10 Tricritical points

One-component systems rarely have three fluid phases; exceptions are helium and liquid crystals. Three fluid phases are common in binary systems, when two liquid phases coexist with a gas. As we have seen, the three phases can be reduced to two either by bringing the system to a critical point between the two liquid phases or to one between one of the liquids and the gas (*Figure 4.9*). Which point is reached first on raising the temperature depends on the similarity of the components. If they are reasonably similar, e.g., n-hexane + methanol (Section 5.8), then the system comes first to a liquid–liquid critical point (UCST = 34 °C). If the components are dissimilar, e.g., n-hexane + water (Section 6.5), then the system comes first to a critical point between the gas, i.e., the mixed vapours, and the liquid phase rich in hexane ($T^c = 210$ °C). *Figures 4.15*(a) and (b) are sketches of the two types of behaviour in a field space. There are three surfaces each of which represents a two-phase system. In *Figure 4.15*(a) there are, at low temperature, two liquid–gas surfaces, both of which are roughly parallel with the basal plane of the diagram, and a liquid–liquid surface which is almost vertical since pressure has little effect on liquid miscibility. The three surfaces meet at a triple-point line or three-phase line,

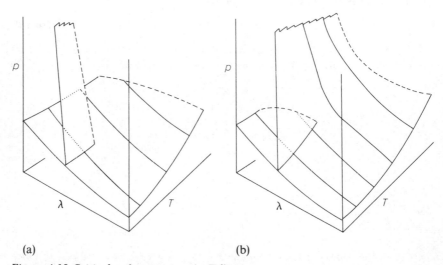

(a)   (b)

Figure 4.15 Critical end-points in a (p, T, λ) space

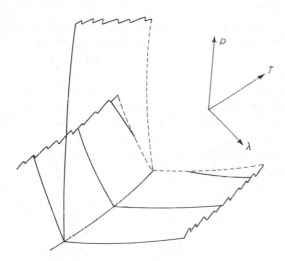

Figure 4.16 Tricritical point in a (p, T, λ) space. The two-phase surfaces in a line of triple points. Each surface ends at a critical line, and the three critical lines meet at a tricritical point

gas + liquid + liquid, which ends at the UCST. In *Figure 4.15*(b) it is one of the gas–liquid surfaces that comes to an end at a critical point. The points where a line of triple points meets a line of critical points is called a *critical end-point*, in these cases an upper critical end point or UCEP. Here, as is shown in *Figure 4.9*, there is a critical point between two phases in the presence of a third phase which plays no essential role in determining the critical behaviour of the system. Thus the UCEP, although clearly a special point in the phase diagram, does not differ in essentials from other critical points, in binary systems. We return to systems with critical end-points in Chapter 6 when we sketch also the more familiar (p, T, x) surfaces that correspond to *Figure 4.15*(a) and (b).

We now turn to the interesting question of what happens if the critical end-points coincide, that is, if the critical point between *phase'* and *phase''* is at the same p and T as that between *phase''* and *phase'''*. Clearly this will not happen in a binary system, unless induced by some chance symmetry[38,39], but may happen if we add a suitably chosen third component that raises the UCEP of either *Figure 4.15*(a) or (b), until it reaches the second critical line. The topology of the intersecting surfaces would then be that of *Figure 4.16*, which is a particular three-dimensional section of a four-dimensional field space, $(p, T, \lambda_2, \lambda_3)$. (We consider later a more precise geometry.) Here, a line of triple points ends at the meeting point of three critical lines. This is a *tricritical point*, at which, on making a suitable change in the fields, a system of three phases passes to a system of one (critical) phase, that is, in which both of the menisci in the sketch in *Figure 4.9* vanish simultaneously.

The phase rule gives the number of degrees of freedom of the different features—volumes, surfaces, lines and points—of diagrams such as *Figure*

4.15(a) and (b). In a binary system the two-phase surfaces are bivariant, the line of triple points is univariant and the critical end-points are invariant. In a ternary system there is a surface of triple points (bivariant) in a four-dimensional space, which ends in a line of critical end-points (univariant), which can end in a tricritical point (invariant). The addition of each further component raises the variance by one. In a four-component system the tricritical state is univariant and so can, in suitable cases, be found at an arbitrary value of one of the fields, e.g., at atmospheric pressure. In general, a critical point of order P, that is, where P phases become identical at one set of field variables, can occur only in a system of $(2P - 3)$ or more components[40].

The possibility of the occurrence of tricritical points was first suggested (under the name of 'critical point of higher order') by van der Waals and Kohnstamm[41], but none was discovered until 1962. It is of interest that two four-component systems which almost certainly have tricritical points above room temperature were studied earlier at room temperature only, and some of the properties of the surfaces that represent the systems in a density space were elucidated. These were the studies of $KI + HgI_2 + (C_2H_5)_2O + H_2O$ by Dunningham[42] in 1914, and $CH_3CN + C_6H_6 + n-C_7H_{16} + H_2O$ by Hartwig, Hood and Maycock[43] in 1955.

In ternary systems, when one of the phases is gas-like, the invariant tricritical point occurs at a pressure near that of the critical point of one of the pure components, say about 50 bar. We return to such systems in Chapter 6. Mertslin and Nikurashina[44] suggested that there might be a tricritical point (*triple critical point* in their terminology) in quaternary systems that had up to four fluid phases. They first discovered[45] a tricritical point in the system $(NH_4)_2SO_4 + C_6H_6 + C_2H_5OH + H_2O$, which has three liquid phases and a gas, and later[46] showed that there are two tricritical points in the system $C_6H_5OH + C_5H_5N + n-C_6H_{14} + H_2O$, and[47] one in $CH_3CN + C_6H_6 + n-C_6H_{14} + H_2O$. The first and third of their systems are the best studied because of the later work of Widom and his colleagues[48]. We take the first as our example.

Water and benzene are scarcely miscible but, at a fixed temperature (and atmospheric pressure), can be brought to a critical point by adding ethanol. A saturated solution of ammonium sulphate is scarcely miscible with ethanol but can be brought to a critical point by adding more water to the system, again at the same fixed temperature. If, therefore, the composition of the quaternary system is represented by a mole-fraction tetrahedron, then there will be critical points on two of the four triangular faces that represent the four ternary systems. These two faces have the binary system ethanol + water as their common line. When a little salt is added at the critical point of the first ternary system, the point moves into the interior of the tetrahedron. Similarly, when benzene is added to the second ternary system, the critical point can also be displaced into the interior. The fate of these two critical lines depends on the temperature, for above 49 °C they join to form a continuous curve through the tetrahedron from one face to another. Below 49 °C they do not meet but cut a three-phase region, as shown in *Figure 4.17*. One critical line MQ comes to a critical end-point at Q, where the critically identical phase Q is in equilibrium with a third phase at S. (We ignore any fourth or vapour phase.) Similarly the second critical line cuts the three-phase region at the critical end-point R, where the critical phase is in equilibrium with phase P. Between QS and RP there is a stack of three-phase triangles, of which two are drawn. Of the three liquid phases the lightest is richest in benzene and the heaviest richest in the salt. Indeed, the whole of the phase separation can be loosely considered as binary mixtures of ethanol and water from which a benzene layer has extracted some of the ethanol and the salt layer some of the water. As the temperature is raised towards 49 °C, the size of the stack of triangles diminishes until the three phases become identical at the tricritical point, $T^{tc} = 49\,°C$ and at mass fractions of ethanol 0.450, water 0.351, benzene 0.181 and salt 0.0175.

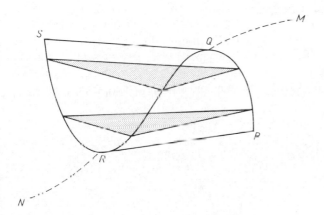

Figure 4.17 The three-phase equilibrium and two critical lines of a quaternary system shown in the interior of a composition tetrahedron

The phases can be represented more succinctly in an all-field space. We saw in Section 3.5 that the most general classical representation of an ordinary critical point is that of the Landau expansion in which a free energy is expanded in powers of an order-parameter s (a density) and then minimized with respect to s to obtain a function of the fields a_i. This result was obtained for a one-component system but can be carried over without essential change to ordinary critical points in multicomponent systems[49]. However, if there is to be a tricritical point the Landau expansion needs to be more elaborate[50]. The equations that parallel (3.66) are

$$\psi_s(a_i) = \min_s \psi(a_i, s)$$

$$\Psi(a_i, s) = a_1 s + a_2 s^2 + a_3 s^3 + a_4 s^4 + s^6 \qquad (4.150)$$

There is no need to include a term in s^5 since it could be removed by choosing a new zero for s, and stability requires that the coefficient of s^6 be positive. Even with these simplifications the structure of equations (4.150) is more complicated than that of (3.66). Whereas the ordinary critical point was located by the two conditions $a_1 = a_2 = 0$, the tricritical point generally requires $a_1 = a_2 = a_3 = a_4 = 0$, although symmetry sometimes introduces a welcome simplification.

The general surface $\psi_s(a_i)$ defined by equations (4.150) has been drawn by Griffiths in an all-field space and by Fox in a mixed space of fields and densities[50]. The former is best shown by three-dimensional sketches, each of which is a section at fixed a_4 of the four-dimensional (a_1, a_2, a_3, a_4)-space (*Figure 4.18*). If a_4 is the field that is closest to being a pure function of temperature then this figure is consistent with the results sketched in a density space in *Figure 4.17*. At low temperatures ($a_4 < 0$), there are two critical lines, M to the critical end-point (QS) and N to (PR). The line (QS) to (PR) is the junction of three sheets each representing two coexisting phases; it is therefore the triple-point line. This line shrinks as the temperature is raised and vanishes at $a_4 = 0$. For positive values of a_4 there is a continuous line of critical points crossing the whole of the accessible field space. Although this is not shown in the figure, the triple-point line is at $a_2 > 0$, if $a_4 < 0$, and falls to $a_2 = 0$ when $a_4 = 0$. The projection on the (a_1, a_3)-plane shows that the line always passes through $a_1 = a_3 = 0$. Hence the tricritical point is at the origin of the field space, $a_1 = a_2 = a_3 = a_4 = 0$.

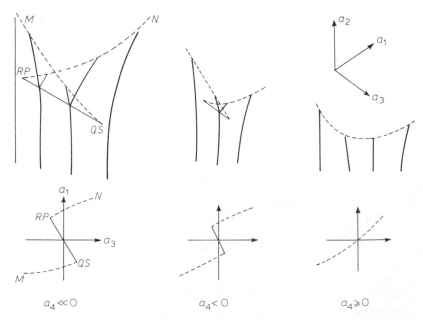

Figure 4.18 The phase equilibrium in an all-field space for a system that has a tricritical point at $a_4 = 0$. The upper figures are three-dimensional sketches, the lower the projection of the triple-point line and critical lines onto the $a_1 a_3$-plane

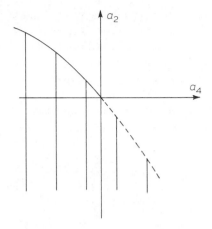

Figure 4.19 The phase equilibrium in a restricted field space $(a_1 = a_3 = 0)$ for a system with a tricritical point at $a_4 = 0$

Two further features of interest are, first, that as the stack of triangles, in the density representation *Figure 4.18*, shrinks it does so asymmetrically, so that the limiting shape of the stack is first a plane, then a line, and finally a point; the orientations of the plane and line single out particular directions in the density space. The second feature is that the triangles move with temperature in such a way that they finally shrink to a point (the tricritical composition) which lies outside any of the triangles.

An ordinary critical point is specified by the two conditions $a_1 = a_2 = 0$. If it is in a one-component system there are no further independent fields and so the point is invariant. If there are C components the variance is $(C - 1)$. Similarly, the variance of a tricritical point is generally $(C - 3)$, but there may be special symmetries that can reduce the number of constraints to less than four. This reduction can occur also at ordinary critical points but here these special cases are less important. Consider, for example, a one-component fluid in a sealed tube at $\rho = \rho^c$. On heating the system, it would pass from two phases to one phase at $a_2 = 0$, i.e., at $T = T^c$. In this restricted system there is no variable equivalent to a_1, for in all accessible states $a_1 = 0$. Similarly, a ferromagnet heated in zero external field, a_1, passes from the ordered state $(s \neq 0)$ to the disordered $(s = 0)$ at the critical or Curie point. Similarly restricted sets of variables suffice for the discussion of some tricritical points, although we must distinguish between cases where a variable is experimentally accessible but voluntarily suppressed (as chemical potential, $\mu - \mu^c$, or the field, H, in the two examples above) and the more important cases where it is physically inaccessible or non-existent.

Consider first a system with $a_1 = a_3 = 0$, so that the whole of the phase diagram can be represented in the (a_2, a_4)-plane, *Figure 4.19*. Here, a sheet of two-phase coexistence ends either at a line of critical points $(a_4 \geqslant 0)$, or at a line of triple points $(a_4 < 0)$. An example is the liquid mixture $^3He + {^4He}$ in which the superfluid component of ^4He plays the role of the third component[51]. The λ-point of ^{4}He is a critical point between the normal and superfluid components. Its temperature is lowered on adding ^{3}He until, at 0.83 K, it ends at the point where the two orthobaric lines of liquid–liquid immiscibility cut each other[52], *Figure 4.20*. If this diagram is drawn in the (T, λ)-plane rather than the (T, x), then the liquid–liquid area collapses to a line, and the system behaves as in *Figure 4.19*. Here, the field a_3 does not exist ($a_3 = 0$), and the field a_1 which is conjugate to the superfluid order-parameter can be said either to be non-existent, or at least to be experimentally inaccessible ($a_1 = 0$).

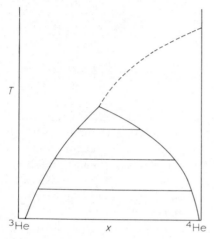

Figure 4.20 The phase diagram of ^{3}He + ^{4}He at fixed pressure. The dashed line is the λ-point line

A hypothetical ternary mixture in which g^E has the form

$$g^E = c_{12}x_1x_2 + c_{13}x_1x_3 + c_{23}x_2x_3 \tag{4.151}$$

with $c_{12} = c_{13} = c_{23} > 0$, was studied by Meijering[53]. The system has three tricritical points at one of which $x_1 = x_2 > x_3$, and similarly for the others by permutation. Because of the two-fold local symmetry of the phase diagram at each of these points, it follows that $a_3 = 0$. The remaining fields can be identified, in their asymptotic behaviour near the first tricritical point, as

$$a_1 = \lambda_1 - \lambda_2 \qquad a_2 = \lambda_3 - \lambda_3^{tc} \qquad a_4 = (\lambda_1 + \lambda_2) - (\lambda_1 + \lambda_2)^{tc} \tag{4.152}$$

If we were to study such a system under the restriction that $x_1 = x_2$, and so $\lambda_1 = \lambda_2$ or $a_1 = 0$, then it would behave as ^{3}He + ^{4}He. When component 3 is added to a mixture of 1 and 2 at their UCST, the critical point would be displaced into the interior of the mole-fraction triangle, until, after addition of sufficient of 3, the critical line would end at a tricritical point, and the addition of further 3 would produce three liquid phases. In general, however, the field a_1 is accessible in this system, but a_3 is not.

Meijering's assumption of the form of g^E was purely phenomenological, but there is a model with a well-defined hamiltonian which behaves in the same way, namely, the three-component penetrable-sphere model in the limit of infinite dimensionality[39]. Here, it is the symmetry of the hamiltonian that ensures that a_3 is identically zero.

There are therefore tricritical points which require the full set of four fields for their description (all real three- and four-component fluid systems), tricritical points near which $a_1 = a_3 = 0$ (^{3}He + ^{4}He and some magnetic systems[54]), and, at least in theory, points at which $a_3 = 0$ because of some symmetry in the hamiltonian. As far as is known, the rather unusual tricritical point[55] that would result from $a_1 = 0$, but with a_3 an accessible variable, does not exist.

Critical exponents can be defined for tricritical points. Thus β_1 describes the rate at which the base of the three-phase triangle vanishes as T approaches T^{tc}, and β_2 the rate at which its height vanishes. A third exponent β_3 describes how the third dimension of the stack of triangles vanishes. The classical Landau argument above gives $\beta_1 = \frac{1}{2}$, $\beta_2 = 1$, and $\beta_3 = \frac{3}{2}$. The best experimental results are those of Bocko and Widom[48] which yield $\beta_1 = 0.49 \pm 0.05$, $\beta_2 = 0.95 \pm 0.11$ and $\beta_3 = 1.50 \pm 0.05$. At the tricritical point of a ternary hydrocarbon system (*see* Section 6.6), Creek, Knobler and Scott[56] found values of β_3 between 1.4 and 1.5. Thus, in sharp contrast with ordinary critical points, it seems as if the classical description is adequate at least for a not-too probing description of behaviour at tricritical points. There are minor discrepancies[55], for example the critical line and the triple point line in ^{3}He + ^{4}He are not collinear, as shown in *Figure 4.19*, but appear to have a small difference of slope on the two sides of T^{tc}. The current theoretical view is that for ordinary critical points the transition from non-classical to classical behaviour would occur in a four dimensional world, but that for tricritical points this change occurs at a dimensionality of three.

References

1 MAITLAND, G. C., RIGBY, M., SMITH, E. B., and WAKEHAM, W. A., *Intermolecular Forces*, Oxford University Press, Oxford (1981)
2 BROWN, W. BYERS, *Phil. Trans.*, **A250**, 175 (1957)
3 YOUNG, T. F., and VOGEL, O. G., *J. Am. chem. Soc.*, **54**, 3025 (1932)
4 HILDEBRAND, J. H., and SCOTT, R. L., *Solubility of Nonelectrolytes*, 3rd Edn, 25–28, 239–246, 270–299, Reinhold, New York (1950); HILDEBRAND, J. H., PRAUSNITZ, J. M., and SCOTT, R. L., *Regular and Related Solutions*, Van Nostrand Reinhold, New York (1970)
5 ROWLINSON, J. S., *Handbuch der Physik*, Ed. Flügge, S., Vol. 12, Springer, Berlin (1958)
6 SCATCHARD, G., *Trans. Faraday Soc.*, **33**, 160 (1937); McGLASHAN, M. L., MORCOM, K. W., and WILLIAMSON, A. G., *Trans. Faraday Soc.*, **57**, 601 (1961); SCOTT, R. L., *J. phys. Chem., Ithaca*, **64**, 1241 (1960)
7 PRIGOGINE, I., and DEFAY, R., *Chemical Thermodynamics*, Transl. Everett, D. H., 350, Longman, London (1954)
8 FRANKS, F., and SMITH, H. T., *Trans. Faraday Soc.*, **64**, 2962 (1968)
9 VAN NESS, H. C., and MRAZEK, R. V., *A.I.Ch.E.Jl.*, **5**, 209 (1959)

10 MITCHELL, A. G., and WYNNE-JONES, W. F. K., *Discuss. Faraday Soc.*, **15**, 161 (1953); MALCOLM, G. N., and ROWLINSON, J. S., *Trans. Faraday Soc.*, **53**, 921 (1957)
11 GRIFFITHS, V. S., *J. chem. Soc.*, 860 (1954); SCHOTT, H., *J. chem. Engng Data*, **6**, 19 (1961)
12 SMYTH, C. P., and ENGEL, E. W., *J. Am. chem. Soc.*, **51**, 2660 (1929)
13 BEATTY, H. A., and CALINGAERT, G., *Ind. Engng Chem.*, **26**, 904 (1934)
14 COULSON, E. A., and HERINGTON, E. F. G., *Trans. Faraday Soc.*, **44**, 629 (1948); HERINGTON, E. F. G., *J. Inst. Petrol.*, **37**, 332 (1951); REDLICH, O., and KISTER, A. T., *Ind. Engng Chem.*, **40**, 341 (1948); IBL, N. V., and DODGE, B. F., *Chem. Engng Sci.*, **2**, 120 (1953); STEVENSON, F. D., and SATER, V. E., *A.I.Ch.E. Jl.*, **12**, 586 (1966)
15 HERINGTON, E. F. G., *J. appl. Chem.*, **18**, 285 (1968)
16 WADE, J., and MERRIMAN, R. W. J., *J. chem. Soc.*, **99**, 997 (1911); KLEINERT, T., *Angew. Chem.*, **46**, 18 (1933); OTSUKI, H., and WILLIAMS, F. C., *Chem. Engng Prog. Symp. Ser.*, **49**, No. 6, 55 (1953); OTHMER, D. F., and LEVY, S. L., *ibid.* 64; LARKIN, J. A., and PEMBERTON, R. C., *NPL Report*, Chem 43, National Physical Laboratory, Teddington (1976)
17 ROSCOE, H. E., and DITTMAR, W., *J. chem. Soc.*, **12**, 128 (1860); ROSCOE, H. E., *J. chem. Soc.*, **13**, 146 (1861); **15**, 270 (1862)
18 BONNER, W. D., and TITUS, A. C., *J. Am. chem. Soc.*, **52**, 633 (1930); BONNER, W. D., and WALLACE, R. E., *ibid.*, 1747
19 PRIGOGINE, I., and DEFAY, R., *Chemical Thermodynamics*, Transl. Everett, D. H., Chap. 28, Longman, London (1954); MALESINSKI, W., *Azeotropy and other Theoretical Problems of Vapour–Liquid Equilibrium*, Interscience, London (1965); HORSLEY, L. H., *Azeotropic Data III*, American Chemical Society, Washington, DC (1973)
20 SKOLNIK, H., *Ind. Engng Chem.*, **43**, 172 (1951)
21 GRIFFITHS, R. B., and WHEELER, J. C., *Phys. Rev.*, **A2**, 1047 (1970)
22 SCOTT, R. L., *Ber. Bunsenges. phys. Chem.*, **76**, 296 (1972)
23 SCOTT, R. L., in *Chemical Thermodynamics, Specialist Periodical Reports*, Ed. McGlashan, M. L., Vol. 2, Chap. 8., Chemical Society, London (1978); GREER, S. C., *Acc. chem. Res.*, 427 (1978)
24 HILDEBRAND, J. H., and SCOTT, R. L., *Regular Solutions*, Chap. 10, Prentice-Hall, Englewood Cliffs, NJ (1962); GILMOUR, J. B., ZWICKER, J. O., KATZ, J., and SCOTT, R. L., *J. phys. Chem., Ithaca*, **71**, 3259 (1967); SIMON, M., FANNIN, A. A., and KNOBLER, C. M., *Ber. Bunsenges. phys. Chem.*, **76**, 321 (1972); STEIN, A., and ALLEN, G. F., *J. phys. chem. ref. Data*, **2**, 443 (1974)
25 COPP, J. L., and EVERETT, D. H., *Discuss. Faraday Soc.*, **15**, 174 (1953)
26 ROWLINSON, J. S., in *Critical Phenomena*, Eds. Green, M. S., and Sengers, J. V., 9, National Bureau of Standards, Washington, DC (1966); DUNLAP, R. D., and FURROW, S. D., *J. phys. Chem., Ithaca*, **70**, 1331 (1966); MYERS, D. B., SMITH, R. A., KATZ, J. R., and SCOTT, R. L., *ibid.*, 3341; GAW, W. J., and SCOTT, R. L., *J. chem. Thermodyn.*, **3**, 335 (1971)
27 THORP, N., and SCOTT, R. L., *J. phys. Chem., Ithaca*, **60**, 670, 1441 (1956); CROLL, I. M., and SCOTT, R. L., *J. phys. Chem., Ithaca*, **62**, 954 (1958); **68**, 3853 (1964); SCATCHARD, G., and WILSON, G. M., *J. Am. chem. Soc.*, **86**, 133 (1964)
28 DICKINSON, E., KNOBLER, C. M., SCHUMACHER, V. N., and SCOTT, R. L., *Phys. Rev. Lett.*, **34**, 180 (1975); BLOCK, T. E., DICKINSON, E., KNOBLER, C. M., and SCOTT, R. L., *J. chem. Phys.*, **66**, 3786 (1977); KNOBLER, C. M., and SCOTT, R. L., *J. chem. Phys.*, **68**, 2017 (1978); GIGLIO, M., and VENDRAMINI, A., *J. chem. Phys.*, **68**, 2006 (1978); ALDER, B. J., ALLEY, W. E., BEAMS, J. W., and HILDEBRAND, J. H., *J. chem. Phys.*, **68**, 3009 (1978)
29 BEYSENS, D., TUFEU, R., and GARRABOS, Y., *J. Phys. (Fr.)*, **40**, L–623 (1979); BLOEMEN, E., THOEN, J., and VAN DAEL, W., *J. chem. Phys.*, **73**, 4628 (1980); SENGERS, J. V., in *Phase Transitions*, Eds. Levy, M., Le Guillou, J. C., and Zinn-Justin, J., Plenum, New York (1981)|
30 GREER, S. C., and HOCKEN, R., *J. chem. Phys.*, **63**, 5067 (1975)
31 KLEIN, H., and WOERMANN, D., *J. chem. Phys.*, **63**, 2913 (1975); MORRISON, G., and KNOBLER, C. M., *J. chem. Phys.*, **65**, 5507 (1976); WIDOM, B., and KHOSLA, M. P., *J. chem. Soc., Faraday Trans.1*, **76**, 2043 (1980)
32 WIMS, A. M., McINTYRE, D., and HYNNE, F., *J. chem. Phys.*, **50**, 616 (1969); REEDER, J., BLOCK, T. E., and KNOBLER, C. M., *J. chem. Thermodyn.*, **8**, 133 (1976)

33 GREER, S. C., *Phys. Rev.*, **14A**, 1770 (1976); *Ber. Bunsenges. phys. Chem.*, **81**, 1970 (1977)
34 CHU, B., SCHOENES, F. J., and KAO, W. P., *J. Am. chem. Soc.*, **90**, 3042 (1968); CHANG, R. F., BURSTYN, H., SENGERS, J. V., and BRAY, A. J., *Phys. Rev. Lett.*, **37**, 1481 (1976)
35 GOPAL, E. S. R., RAMACHANDRA, R., CHANDRA SEKHAR, P., GOVANDARAJAN, K., and SUBRAMANYAN, S. V., *Phys. Rev. Lett.*, **32**, 284 (1974); GOPAL, E. S. R., CHANDRA SEKHAR, P., ANANTHKRISHNA, G., RAMACHANDRA, R., and SUBRAMANYAN, S. V., *Proc. R. Soc.*, **A350**, 91 (1976)
36 ESSAM, J. W., and GARELICK, H., *Proc. phys. Soc.*, **92**, 136 (1967)
37 BAK, C. S., GOLDBURG, W. I., and PUSEY, P. N., *Phys. Rev. Lett.*, **25**, 1420 (1970); GOLDBURG, W. I., and PUSEY, P. N., *J. Phys. (Fr.)*, **33**, C1–105 (1972); ZOLLWEG, J. A., *J. chem. Phys.*, **55**, 1430 (1971)
38 SCOTT, R. L., *J. chem. Soc. Faraday Trans. 2*, **73**, 356 (1977); WHEELER, J. C., *J. chem. Phys.*, **73**, 5771 (1980)
39 GUERRERO, M. I., ROWLINSON, J. S., and MORRISON, G., *J. chem. Soc. Faraday Trans. 2*, **72**, 1970 (1976); DESROSIERS, N., GUERRERO, M. I., ROWLINSON, J. S., and STUBLEY, D., *J. chem. Soc. Faraday Trans. 2*, **73**, 1632 (1977)
40 ZERNIKE, J., *Recl Trav. chim. Pays-Bas Belg.*, **68**, 585 (1949)
41 VAN DER WAALS, J. D., and KOHNSTAMM, P., *Lehrbuch der Thermodynamik*, Vol. 2, 39, Barth, Leipzig (1912); KOHNSTAMM, P., in *Handbuch der Physik*, Eds. Geiger, H., and Scheel, K., Vol. 10, 223, Springer, Berlin (1926)
42 DUNNINGHAM, A. C., *J. chem. Soc.*, **105**, 368, 724, 2623 (1914)
43 HARTWIG, G. M., HOOD, G. C., and MAYCOCK, R. L., *J. phys. Chem.*, Ithaca, **59**, 52 (1955)
44 MERTSLIN, R. V., and NIKURASHINA, N. I., *J. gen. Chem. U.S.S.R.*, **29**, 2437 (1959)
45 RADYSHEVSKAYA, G. S., NIKURASHINA, N. I., and MERTSLIN, R. V., *J. gen. Chem. U.S.S.R.*, **32**, 673 (1962)
46 MYASNIKOVA, K. P., NIKURASHINA, N. I., and MERTSLIN, R. V., *Russ. J. phys. Chem.*, **43**, 223 (1969); NIKURASHINA, N. I., KHARITONOVA, G. I., and PICHIGINA, L. M., *Russ. J. phys. Chem.*, **45**, 444 (1971)
47 SINEGUBOVA, S. I., NIKURASHINA, N. I., *Russ. J. phys. Chem.*, **51**, 1230 (1977), SINEGUBOVA, S. I., NIKURASHINA, N. I., and TRIFONOVA, N. B., *Russ. J. phys. Chem.*, **52**, 1231 (1978)
48 WIDOM, B., *J. phys. Chem.*, Ithaca, **77**, 2196 (1973); LANG, J. C., and WIDOM, B., *Physica*, **81A**, 190 (1975); BOCKO, P. L., *Physica*, **103A**, 140 (1980)
49 FOX, J. R., *J. stat. Phys.*, **21**, 243 (1979)
50 GRIFFITHS, R. B., *J. chem. Phys.*, **60**, 195 (1974); FOX, J. R., *J. chem. Phys.*, **69**, 2231 (1978)
51 GRIFFITHS, R. B., *Phys. Rev. Lett.*, **24**, 715 (1970)
52 WALTERS, G. K., and FAIRBANKS, W. N., *Phys. Rev.*, **103**, 262 (1956); GRAFF, E. H., LEE, D. M., and REPPY, J. D., *Phys. Rev. Lett.*, **19**, 417 (1967)
53 MEIJERING, J. L., *Philips Res. Rep.*, **5**, 333 (1950); **6**, 183 (1951)
54 PARSONAGE, N. G., and STAVELEY, L. A. K., *Disorder in Crystals*, Oxford University Press, Oxford (1978)
55 ROWLINSON, J. S., in *High Pressure Science and Technology*, Eds. Vodar, B., and Marteau, P., Vol. 1, 20, Pergamon, Oxford (1980)
56 CREEK, J. L., KNOBLER, C. M., and SCOTT, R. L., *J. chem Phys.*, **67**, 366 (1977)

Chapter 5

Excess thermodynamic functions

5.1 Introduction

It was shown in the previous chapter that, at vapour pressures of less than a few bars, the extent to which real liquid mixtures deviate from ideality is best expressed through the use of thermodynamic excess functions. Nowadays, most experimentalists tabulate the results of their measurements on the thermodynamic properties of non-ideal mixtures in this form. Such data are used subsequently by a variety of physical scientists including chemical kineticists and spectroscopists interested in reactions occurring in solution and by chemical engineers engaged in the operation or design of chemical reactors, distillation columns or other types of separation device.

Another principal use of excess functions results from the fact that the statistical mechanical theories of liquid mixtures that are discussed in Chapter 8 lead naturally to the prediction of theoretical values of the excess functions. Precise experimental data on carefully-selected real mixtures are therefore desirable as test material against which the usefulness of competing theories can be judged.

This chapter begins with a discussion of the various experimental methods that are used for the measurement of excess functions, but the bulk of it is concerned with a review of the experimental excess functions of real mixtures. A comprehensive coverage is obviously impossible but all of the different classes of liquid mixture are included and mention is made of most of the numerous notable individual experimental achievements. Other reviews exist[1] and two extensive bibliographies have appeared recently[2,3]. One, a computer-tabulated compilation edited by Hicks[2] lists all the mixtures of non-electrolytes that have been studied up to 1974–5 and, as well as indicating the temperature and pressure range over which the principal excess functions have been measured, includes studies of high-pressure phase equilibria and critical properties. The other major bibliography, by Wisniak and Tamir[3], is more extensive and includes mixtures containing electrolytes, non-electrolytes and metals. It covers the period from 1900 to early 1977 and, although almost 6000 references are listed, users should note that the bibliography is not totally comprehensive.

Certain other bibliographies exist which, although they do not list excess functions explicitly, are most useful when used with the two previously-mentioned texts. The extensive bibliography of vapour–liquid equilibrium data by Wichterle, Linek and Hála[4] is, as far as can be judged, totally comprehensive while the more specialized compilation of Hiza, Kidnay and Miller[5] is restricted to mixtures of cryogenic interest. The annual Japanese publication edited by Sato, Hirata and Yoshimura[6] lists excess functions together with other thermodynamic and transport property data of interest to chemical engineers.

There are two sets of publications that list 'critically evaluated' experimental measurements on liquid mixtures. The *International Data Series* edited by Kehiaian[7] is concerned primarily with excess functions for mixtures of organic substances but also includes some tabulated and graphical data on phase equilibria. Volumes 1 and 6 of the extensive *Dechema Chemistry Data Series*[8] are concerned with vapour–liquid equilibrium data on organic and organic/aqueous mixtures and on cryoscopic mixtures respectively.

5.2 The experimental determination of excess functions

An apparatus in which the total vapour pressure and the compositions of both the liquid and vapour phases are measured at a known temperature is the recirculating equilibrium still. This is basically a single-plate distillation apparatus with provision for recycling the condensed vapour until true equilibrium is approached. Some arrangement is made to sample the bulk liquid and the condensed distillate and the composition of these two phases is usually determined by measurement of their densities or of their refractive indices. Any other suitable physical property can be used for this purpose and increasing use is being made of quantitative gas–liquid chromatography, especially for mixtures whose components are gaseous at room temperature.

Once p, T, x and y have been determined independently in this manner, it is then a trivial arithmetic exercise to obtain the activity coefficients and g^E by using equations (4.89) and (4.90). In practice, however, there are several experimental difficulties to be overcome before results of the highest precision can be obtained. The major problem is the proper establishment of thermodynamic equilibrium throughout an apparatus that is not thermostatted in the usual way. Modern practice is to incorporate stirrers in both the liquid boiler and in the condensed vapour trap to try and hasten the attainment and the maintenance of equilibrium[9]. Another minor problem results from the fact that the temperature in the still is controlled by a manostat. Because of fluctuations in atmospheric pressure, it is difficult to reproduce the same temperature over the length of time necessary to obtain a set of measurements covering the complete range of composition. The experimental measurements have to be corrected to truly isothermal conditions before g^E can be calculated.

Many of these difficulties can be avoided by using an apparatus in which only p and x or p and y are measured and the composition of the other phase is deduced by use of the Gibbs–Duhem equation. That is, it is sufficient to measure the total static vapour pressure over the liquid mixture as a function of either the liquid or vapour composition in order to obtain a complete knowledge of p, x, y and hence g^E at a single temperature. The former technique is the more popular and the total pressure may either be measured absolutely[10] or can be deduced by measuring the difference between the vapour pressure of the mixture and that of one or other of the two pure components[11]. The volume of the vapour is kept to a minimum so that the composition of the liquid mixture can be calculated from the total masses of the components in the apparatus, the apparatus volume, and the liquid mixture density, by using a simple iterative technique. In the past the overall composition of the mixture was determined by distilling into the apparatus known masses of the components from previously-weighed break-tip tubes[10,11] but recently precision injectors have been used to vary the composition of the mixture over a wide range[12]. This technique greatly increases the rate at which experimental results can be obtained.

Measurements of p and y can either be obtained by determining y directly using a buoyancy balance[13] or, more usually, by estimating the dew-point pressure of a vapour mixture of known composition by measuring the pressure as the volume is slowly varied[14]. This latter method has been used for certain mixtures where a static vapour pressure apparatus was found to be ineffective owing to condensation of the vapour on mercury surfaces leading to long-term cyclical fluctuations in the vapour pressure. A further variant[15] is to avoid accurate measurement of either x or y and to determine both the dew-point and the bubble-point pressures in a liquid–gas system. The subsequent analysis is simplified by the fact that y (dew) $= x$ (bubble).

The calculation of either y and g^E or of x and g^E from isothermal measurements of either p and x or of p and y requires solution of the Gibbs–Duhem differential equation. The most usual method is the algebraic one due to Barker[16] who assumed a particular form of g^E as a function of x with one or more adjustable parameters. The parameters are chosen by a least-squares method to minimize the errors in the total pressure. The function used by Barker is known variously as the Guggenheim, Scatchard or Redlich–Kister equation[17],

$$g^E = x_1 x_2 [\xi_1 + \xi_2(x_1 - x_2) + \xi_3(x_1 - x_2)^2 + \cdots] \qquad (5.1)$$

Such an equation satisfies the obvious requirement that g^E should be zero for both pure substances and, in principle, can be made to fit any set of experimental results by choosing sufficient terms. Differentiation of equation (5.1) gives

$$\mu_1^E = x_2^2 [\xi_1 + \xi_2(3 - 4x_2) + \xi_3(5 - 16x_2 + 12x_2^2) + \cdots] \qquad (5.2)$$

and similarly for component 2 but with a negative coefficient of ξ_2. It is seen that the lowest power of x_2 in the expansion for μ_1^E is the second. The absence of the first power is necessary in any expressions for μ_1^E and μ_2^E which satisfy the Gibbs–Duhem equation. Thus the simplest class of non-ideal mixtures is that for which ξ_2, and all higher coefficients in equations (5.1) and (5.2), vanish over a non-zero range of p and T. Such a system is often called a *regular mixture*, although many other meanings have been given to that over-worked adjective. They are called here *quadratic mixtures*, and for these we have

$$g^E = x_1 x_2 \xi_1 \qquad \mu_1^E = x_2^2 \xi_1 \qquad \mu_2^E = x_1^2 \xi_1 \tag{5.3}$$

$$h^E = x_1 x_2 [\xi_1 - T(\partial \xi_1 / \partial T)_p] \tag{5.4}$$

$$s^E = x_1 x_2 [-(\partial \xi_1 / \partial T)_p] \tag{5.5}$$

$$v^E = x_1 x_2 [(\partial \xi_1 / \partial p)_T] \tag{5.6}$$

If ξ_1 is assumed to be independent of temperature, then s^E is zero—a restriction which would bring the definition close to that of the regular mixture of Hildebrand[18]. There are many mixtures for which equations (5.3)–(5.5) are an adequate approximation, but in few of these is Ts_p^E (or Ts_V^E) small compared to h_p^E. Calculations of g^E from the total or partial pressures of many real mixtures show that the first term of equation (5.1) is usually much the most important at low vapour pressures. However, the first antisymmetric term $(x_1^2 x_2 - x_1 x_2^2)$ is often dominant in h^E and s^E.

Equation (5.1) can be easily extended to multicomponent mixtures[19]. The $X^E = f(x_1, x_2, x_3)$ surface for a ternary mixture can be represented by expressions of the type

$$\begin{aligned} X^E = &\ x_1 x_2 [\xi_{12} + \xi_{12}'(x_1 - x_2) + \cdots] \\ &+ x_2 x_3 [\xi_{23} + \xi_{23}'(x_2 - x_3) + \cdots] \\ &+ x_1 x_3 [\xi_{13} + \xi_{13}'(x_3 - x_1) + \cdots] + x_1 x_2 x_3 [\xi_{123} + \cdots] \end{aligned} \tag{5.7}$$

where X^E is any principal excess function and ξ_{12}, ξ_{13}, ξ_{12}', etc., are the experimental constants for the binary mixtures and ξ_{123}, etc., result from fitting ternary data.

A variant of the Barker method that has been used to calculate g^E from dew-point data is to fit the actual pressure deviations between the experimental total vapour pressure and the ideal total pressure given by equation (4.53) to a multi-parameter expression[20] similar in form to equation (5.1). A linear regression analysis then enables x to be calculated from the observed values of p, y and T.

Although equation (5.1) is adequate for mixtures where g^E is a reasonably symmetric function of x, it is unsatisfactory for the representation of highly skewed results. The form of equations (5.1) and (5.2) is such that unreliable values of μ^E at infinite dilution are obtained when four or more terms have to be used, owing to successive terms having alternating symmetry. Myers and Scott[21] have suggested the use of an alternative expression on such occasions,

$$g^{E} = x_1 x_2 \sum_{i=0}^{N} \xi_i (x_1 - x_2)^i / [1 + k(x_1 - x_2)] \qquad (5.8)$$

where k is a skewing factor with $-1 < k < +1$. The use of this equation enables some experimental results to be fitted with many fewer coefficients than would be required with equation (5.1).

Recently alternative expressions to equations (5.1) and (5.8) have been proposed[22] that are based on orthogonal polynomial series. They have the merit that the addition of extra terms to improve the fit with experiment has little effect on the magnitude of the earlier terms.

$$g^{E} = \sum_{i=2}^{N} \xi_i P_i (x_1, x_2) \qquad (5.9)$$

The first four polynomials, all of which contain $x_1 x_2$ as a common factor, are given below and all obey the orthogonality conditions $\int_0^1 P_i P_j \, dx_1 = 0$ and $\int_0^1 P_i P_i \, dx_1 \neq 0$.

$$P_2 = x_1 x_2 \qquad (5.10)$$

$$P_3 = x_1 x_2 (1 - 2x_1) \qquad (5.11)$$

$$P_4 = x_1 x_2 (1 - \tfrac{14}{3}x_1 + \tfrac{14}{3}x_1^2) \qquad (5.12)$$

$$P_5 = x_1 x_2 (1 - 8x_1 + 18x_1^2 - 12x_1^3) \qquad (5.13)$$

Additional terms can be obtained from the recurrence relationship

$$P_N = [(2N - 1)(1 - 2x_1)P_{N-1} - (N - 3)P_{N-2}]/(N + 2) \qquad (5.14)$$

where the first three polynomials are defined as 0, 0 and $x_1 x_2$.

Such an expression with temperature-dependent ξ_i has been used successfully to represent the highly asymmetric excess functions of ethanol + water mixtures[23]. Other expressions have been proposed[22] that differ only marginally from equations (5.9)–(5.14), which can be extended easily to multicomponent mixtures.

Many other relationships have been used to represent g^{E} and activity coefficients as functions of composition. One now widely used, especially by chemical engineers, is that due to Wilson[24],

$$g^{E}/RT = -C[x_1 \ln(x_1 + \tau_2 x_2) + x_2 \ln(x_2 + \tau_1 x_1)] \qquad (5.15)$$

In this equation $\tau_1 = (v_1/v_2)\exp(-G_1/RT)$ and $\tau_2 = (v_2/v_1)\exp(-G_2/RT)$, where v_1 and v_2 are the molar volumes of the two pure components and G_1, G_2 and C are adjustable parameters, the first two related to interaction energies. Originally this equation was used with $C = 1$ but in that form it was limited by its inability to predict partial miscibility in liquid phases. The equation can be used for multicomponent mixtures and modified forms have been suggested by Heil[25], by Renon[26] and by Abrams and Prausnitz[27].

Transpiration methods, similar to those that have been used for many decades to measure the vapour pressures of solids and of liquids of low volatility, have been used more recently to measure g^E. Several types of apparatus have been described[28] in which an 'inert' carrier gas, usually helium, is bubbled through a liquid mixture of known composition at constant temperature and total pressure. The resultant carrier gas–vapour mixture is fed to a gas chromatograph and the areas of the resultant peaks are directly related to the partial pressures of the components above the liquid mixture. The activity coefficients and hence g^E can be calculated directly and, although results of the highest precision have not yet been obtained, the method is quick and is capable of further refinement.

The development of gas–liquid chromatography has given us a method of measuring the activity coefficients of volatile components at high dilution in a solute of low volatility. Briefly, in an ideal column the activity coefficient is inversely proportional to the product of the vapour pressure and retention volume of the volatile component. This has been known for many years, but only recently have columns been built which are sufficiently flexible in their operating conditions to allow all the minor corrections to be measured precisely[29]. A by-product of this work is the measurement of the cross second virial coefficient, B_{13}, for the interaction of the volatile component and the carrier gas. Although the method is restricted in the type of mixture to which it is applicable and in the concentration range in which activity coefficients can be obtained, it is capable of high accuracy. The field has recently been reviewed by Letcher[30].

Isotropic light scattering has also been used to obtain g^E. The basic equation is due to Coumou and Mackor[31].

$$1 - (R_{id}/R_c) = \frac{x_1 x_2}{RT} \left(\frac{\partial^2 g^E}{\partial x_1 \partial x_2} \right)_T \tag{5.16}$$

For a quadratic mixture, equation (5.3), the right-hand side of equation (5.16) reduces to $2g^E/RT$. R_c, the Rayleigh ratio due to concentration fluctuations, can be obtained from the experimental isotropic Rayleigh ratio, the isothermal compressibility, β_T, n, the refractive index of the mixture, and $(\partial n/\partial p)_T$, the piezo-optic coefficient, while R_{id}, the 'ideal' Rayleigh ratio, requires experimental values of $(\partial n/\partial x)_T$. The need to measure these auxiliary functions with sufficient accuracy makes the method time-consuming and its use is only justified under exceptional circumstances[32]. Attempts to avoid measuring such functions as the piezo-optic coefficient by using approximate thermodynamic relationships lead to further loss of precision[33].

The molar heat of mixing, h^m (or h^E), is related to g^E by the Gibbs–Helmholtz equation but, in practice, this does not provide a satisfactory method of determining it. Measurements of vapour pressure, and so of g^E, rarely extend over a temperature range of more than 15 per cent of the absolute temperature. This limit is imposed partly by practical difficulties, for a given apparatus is rarely designed to measure pressure accurately over a range of greater than a factor of ten, and partly by the difficulty that has already been discussed of defining g^E at vapour pressures greater than 2–3 bar. If g^E is known to, say, 2 per cent over a temperature range of $1.0T$ to $1.1T$ and if Ts^E is about as large as g^E, then the probable errors in derived values of h^E and Ts^E, at $1.05T$, are 15 and 30 per cent, respectively. The accuracy of h^E and Ts^E obtained from the temperature derivative of g^E must always be an order of magnitude less than that of g^E, and some very bad estimates of h^E have been made by this method in the past.

Christensen *et al.*[34] have attempted to reverse this procedure by using an equation such as (5.15) for g^E and then obtaining an expression for h^E by differentiation using the Gibbs–Helmholtz equation assuming that the interaction parameters G_1 and G_2 are independent of temperature. The resultant equation is fitted to experimental excess enthalpies to obtain explicit values of the interaction parameters that are then used to obtain g^E. Such analyses are basically unsound for they depend on the original equation possessing the correct temperature dependence and have been rightly criticized by Vonka *et al.*[34]

The only fully satisfactory method of obtaining h^E is a direct, calorimetric one and the pioneering work of Hirobe[35] is an impressive example of how scientific ingenuity was used to outweigh the limitations of the primitive apparatus available at the time and yield worthwhile results albeit at the expense of considerable labour. Hirobe's work, although published in 1925, was actually carried out some 15 years previously and describes the use of an isothermal calorimeter to measure h^E for 51 binary mixtures and, in many cases, his experimental results are within a few per cent of the best modern measurements.

Most modern calorimeters avoid the commonest design fault inherent in earlier models by eliminating vapour spaces and thus the possible vaporization of the mixture under investigation during the mixing process. They also try to ensure complete mixing of the components. Three types of calorimeter are currently employed: batch, isothermal-dilution, and flow.

The type of adiabatic batch calorimeter designed originally by McGlashan[36] is one of the most successful and has now been used to obtain results over a wide range of temperature. When used to study mixtures with positive h^E, the calorimeter is operated in a near-isothermal manner by employing electrical energy to maintain a constant temperature during the mixing and thus reducing errors due to heat loss. Such calorimeters use mercury to separate the two pure liquids prior to mixing and obviously cannot be operated at low temperatures. Under these conditions, the liquids can be separated either by a thin metal diaphragm that is punctured to initiate mixing[37] or, as in the most recent design of Staveley[38], by a large-bore metal valve. This latter calorimeter has enabled values of h^E to be obtained on mixtures of liquified gases at cryoscopic temperatures with a precision equal to that attainable for normal liquids at room temperature, even though a vapour space is present and a consequent correction is required.

The isothermal dilution calorimeter due originally to Van Ness[39] has been developed extensively by Marsh[40] and his colleagues and is now capable of producing results of the highest precision and accuracy. Mixtures of various compositions are made by injecting known quantities of the pure components into the calorimeter from thermostatted calibrated syringes. Individual results are thus obtained at a much faster rate than is possible with a batch calorimeter and a measurement of h^E at a single temperature over the complete range of composition is possible within one day. Isothermal conditions are maintained either by direct electrical compensation in the case of positive excess enthalpies, or by the use of Peltier effect thermoelectric cooling devices when exothermal processes are encountered[41].

Flow calorimeters, in which the pure components are injected continuously at known rates into the calorimeter vessel, are also capable of producing high quality data and several designs have been described[42]. One disadvantage of both flow and isothermal-dilution calorimeters is that, unlike batch calorimeters, relatively large quantities, typically one mole, of each of the pure components are required to produce a complete set of results.

The flow calorimeter due originally to Picker[43], and now available commercially, is particularly interesting in that the flow rates can be varied in a continuous, known manner so that a complete h^E–composition curve can be produced quickly with a precision of around 1 per cent. A variant of the Picker calorimeter enables the heat capacities of the pure components and of mixtures to be determined[44] and thus c_p^E, the excess heat capacity, can be estimated. Few such direct measurements of c_p^E are currently available and this function is usually obtained from h^E measurements made at different temperatures, $c_p^E = (\partial h^E/\partial T)_p$.

Most of the early determinations of the excess volume, v^E, were calculated from measurements of the density of the mixture and of the pure components. Highly accurate densities are required if v^E is to be estimated with acceptable precision and many measurements obtained in this manner are grossly in error. Battino[45] has produced a valuable review. Excess volumes at cryoscopic temperatures are still obtained by this method[46], while the use of commercially-available vibrating-tube densimeters[47] that enable liquid densities to be obtained routinely to better than 1 part in 10^5 has revived interest in the indirect determination of v^E at room temperature and above. Liquid densities can also be measured by the use of a variety of magnetic float techniques[48] which have also been used at cryoscopic temperatures.

The most common method of determining v^E is the direct one, using either a batch or a continuous-dilution dilatometer, which measures the actual volume change that occurs on mixing. Like the equivalent calorimeters, the batch dilatometer[49] is simple to use and yields precise results with only small amounts of material. The more complex continuous-dilution designs are capable of higher precision but use much larger quantities of the pure liquids. Many variants of the continuous-dilution design have been described[50] and those of Bottomley and Scott and of Kumaran and McGlashan are worthy of special mention for, unlike most of the earlier models, these eliminate completely the passage of either of the liquids or of mercury through greased taps.

Another indirect method that is applicable only to mixtures of small, non-polar molecules results from the observation that, for such mixtures, the Clausius–Mossotti function, $\mathcal{M}(\varepsilon - 1)/v(\varepsilon + 2)$, is a linear function of the liquid mole fraction, i.e., the excess Clausius–Mossotti function is zero[51]. With this assumption, measurement of ε, the dielectric constant of a mixture, enables v, the molar volume of the mixture, to be calculated, and hence v^E. The advantage of this technique is that ε can be measured quite easily over a wide range of p and T using simple apparatus, whereas the direct measurement of a liquid density at high p or T is difficult.

5.3 Mixtures of condensed gases

The excess thermodynamic properties of mixtures of non-polar liquids have long been the testing ground of statistical theories of solutions. The most favoured were the three binary systems formed from benzene, cyclohexane and carbon tetrachloride, although these are by no means the most suitable for the purpose. The constituent molecules are of different sizes and shapes. None is spherical, and little is known of the variation of the intermolecular forces with orientation. It is better, therefore, to examine first the experimental results for more simple systems, choosing, as far as possible, those containing monatomic and diatomic molecules.

We consider, first, mixtures of the noble gases. Unfortunately, their critical and triple points are so far apart that it is difficult to choose mixtures in which the components have liquid ranges that overlap. Only the pairs argon + krypton and krypton + xenon satisfy this criterion, and even here the vapour pressure of the first component is inconveniently high at the triple point of the second, e.g., 9.5 bar for argon at the triple point of krypton. Measurements of vapour pressure and of heats of dilution can be made usefully even at temperatures below the triple point of one of the components, but the range of composition becomes increasingly restricted as the temperature is lowered. Useful discussion of such results can be made only insofar as the liquid vapour pressure of the second component can be extrapolated below its triple point.

Values of g^E, etc., in the remainder of this chapter are those for the equimolar mixture, unless otherwise defined, and it is implied that this value is also close to the maximum.

Argon + Krypton

There is substantial agreement between the measurements of g^E and of v^E by Staveley and his co-workers [52,53,54] and by Chui and Canfield[55]. The early German work can now be discarded[56]. Staveley's results are shown in the table below and also included are two values of g^E at 138 K and at 143 K calculated from the vapour–liquid equilibrium measurements of Schouten *et al.*[57]. Although these latter measurements extend up to 193 K, it is impractical to attempt to calculate g^E at high temperatures for, even at 143 K, many auxiliary data, such as the third virial coefficients[58], are required. The excess enthalpy has only been measured at a single temperature[54] for, at higher temperatures where h^E is likely to become negative, the difference in the vapour pressures of the two pure components is sufficient to make the operation of the batch calorimeter extremely difficult. Flow calorimetry is probably the only option if the measurements are to be extended.

T/K	$g^E/\text{J mol}^{-1}$	$h^E/\text{J mol}^{-1}$	$v^E/\text{cm}^3\,\text{mol}^{-1}$	*References*
103.94	82.5			53
115.77	83.9		-0.518	55
116.9		43		54
138.0	92.3			57
143.0	107.9			57

Krypton + Xenon

Both Calado and Staveley[59] and Chui and Canfield[55] have measured g^E and v^E at 161.38 K, the triple point of xenon. Not only is there a discrepancy of 10 per cent in g^E, 114.5 and 103.3 J mol^{-1}, respectively, but there is an even larger variation in v^E. Calado and Staveley's figure is -0.695 cm^3 mol^{-1} compared to Chui and Canfield's value of -0.459 cm^3 mol^{-1} and the two v^E–composition curves are skewed in the opposite sense.

Argon + Xenon

This mixture can only be studied in the liquid state at low xenon concentrations. Yunker and Halsey[60] have measured the activity coefficient of xenon at 84–88 K and Chui and Canfield[55] obtained a value of v^E of -0.248 cm^3 mol^{-1} at an argon mole fraction of 0.875 and at a temperature of 115.77 K.

Argon + Nitrogen

Recent measurements[38,61] on both g^E and h^E are in accord with the earlier work of Pool *et al.*[62] and of Sprow and Prausnitz[63] and all are listed below. Massengill and Miller[64] have measured v^E at six temperatures up to 113 K and some of their results, after correction to zero pressure are shown below. The large negative value of v^E at the highest temperature is typical of many mixtures where one of the components is at a much higher reduced temperature than the other. Increase of pressure greatly reduces these large negative excess volumes and such measurements have been reported recently by Singh and Miller[51].

T/K	$g^E/J\,mol^{-1}$	$h^E/J\,mol^{-1}$	$v^E/cm^3\,mol^{-1}$	*References*
83.82	33.9, 34.4	51	−0.18	63, 62
84.5	34.7	53.2		38
89.89			−0.224	64
112.0	35.6			61
113.0			−1.385	64

Argon + Oxygen

There are many measurements of g^E for this mixture[62,65] at temperatures from 83.8 to 110 K. All are reasonably consistent and are well represented by the expression

$$g^E/J\,mol^{-1} = 61.6 - 0.292(T/K)$$

A recent calorimetric determination[66] of h^E at 86.6 K of 61.1 J mol^{-1} is in good agreement with an earlier value[62] of 60 J mol^{-1} at 83.8 K. The results of Knobler et al.[67] at 86 K, although they are in fair agreement with these results at equimolar concentrations, must be discarded because of their unreasonable temperature dependence. The excess volume[62] appears to be almost independent of temperature with a value of 0.136 cm^3 mol^{-1} at 83.8 K.

Argon + Carbon monoxide

Only g^E and v^E at 83.82 K have been measured[68] with equimolar values of 56.7 J mol^{-1} and 0.094 cm^3 mol^{-1}, respectively.

Argon + Methane

Kidnay et al.[69] have made a critical assessment of the numerous measurements[61,63,70] of g^E for this mixture and find that they are well represented by the expression

$$g^E/J\,mol^{-1} = 197.87 + 1.053(T/K)\ln(T/K) - 6.094(T/K)$$

This equation was derived under the assumption that h^E is a linear function of temperature[66] and, while this is reasonable at temperatures up to 110 K, the recent measurements of Mosedale and Wormald[71] made at moderate pressures and at temperatures up to 145 K show that h^E decreases rapidly at higher temperatures and is already strongly negative at 140 K.

Early measurements of v^E at low temperatures[70] have been confirmed by the more extensive work of Lui and Miller[72]. Their values range from $+0.18\,cm^3\,mol^{-1}$ at 91 K to $-0.045\,cm^3\,mol^{-1}$ at 120 K. The recent indirect measurements of Singh and Miller[51] extend these data to high pressures of up to 500 bar. Under such conditions, v^E becomes positive at all the experimental temperatures.

Argon + Ethane

The excess Gibbs function, enthalpy and volume have all been measured and vapour–liquid equilibria studied[73] in the temperature range $81.44 < T/K < 113.52$. The mixture is predicted to exhibit a UCST below 81 K but this has not yet been confirmed experimentally.

T/K	$g^E/J\,mol^{-1}$	$h^E/J\,mol^{-1}$	$v^E/cm^3\,mol^{-1}$	References
90.16	374.4	239.1		38
108			-0.889	51

In the final chapter it will be shown how statistical theories of liquid mixtures have been extended to include binary systems, one or both of whose components is non-spherical either by virtue of its shape or by its possession of a permanent electrostatic multipole, dipole, quadrupole, octopole, etc. Inspired by the need to produce precise experimental data on the simplest of such mixtures, groups at Oxford under Staveley and in Lisbon under Calado have published a large body of results in recent years which are summarized in *Table 5.1*.

Table 5.1 Simple mixtures containing multipolar molecules

Mixture	$g^E/J\,mol^{-1}$	$h^E/J\,mol^{-1}$	$v^E/cm^3\,mol^{-1}$	T/K	Ref.
$Kr + C_2H_4$	240.4	314.9 (117.7 K)	$+0.213$	115.76	74
$Kr + NO$	389.9		0.451	115.76	75
$C_2H_4 + NO$	337.8			115.76	76
$Xe + C_2H_4$	145.4		0.346	161.39	77
$Xe + HCl$	654.6	813.2 (182.3 K)	0.664	195.42	78
$Xe + HBr$	573.8		0.120	195.42	79
$HCl + HBr$	91.9		0.072	195.42	79
$Xe + CH_3Cl$	533.8		-0.625	182.32	80
$Xe + N_2O$	448.2	901.8 (184 K)	0.654	182.32	81
$Xe + CF_4$	644.9	959.3 (163 K)	1.937	159.01	82
$Xe + C_2F_6$	621	1079 (175 K)	2.35	173.11	83
$Ar + C_2H_6$	317.1		-0.379	90.69	76
$Kr + C_2H_6$	79.8	49.1 (117 K)	-0.218	115.76	76
$Xe + C_2H_6$	-29.2	-51.7 (163 K)	-0.115	161.39	76
$N_2O + C_2H_4$	168.9	2709 (184 K)	-0.066	182.32	84
$HCl + CF_4$	Partially miscible up to at least 173 K				85

Some of these mixtures will be discussed in more detail in Chapter 8 and all that need be noted here is the great diversity of molecular types that are represented. HCl, HBr and CH_3Cl are dipolar (and also possess quadrupole moments), while C_2H_4, C_2H_6, C_2F_6 and N_2O are all quadrupolar. The dipole moment of N_2O is small and its solution properties are governed by its quadrupole moment which has the opposite sign to that of C_2H_4. CF_4 possesses a strong permanent octopole and NO is known to form dimers in the liquid state. The most remarkable feature of the data listed in *Table 5.1* is the great similarity in the signs and even in the magnitudes of the excess Gibbs functions and excess enthalpies for all mixtures of the type spherical + non-spherical. Both g^E and h^E are large and positive for all such systems with the one exception of $Xe + C_2H_6$. The excess volume is strongly dependent on the ratio of the molar volumes of the two pure components and thus the experimental values are less directly related to molecular type.

Nitrogen + Oxygen

The earlier discrepancy that existed between the g^E measurements at various temperatures and the calorimetrically determined h^E has been almost resolved by the recent calorimetric work of Lewis and Staveley. Uncertainty still remains in v^E.

T/K	g^E/J mol^{-1}	h^E/J mol^{-1}	v^E/cm^3 mol^{-1}	References
63.1	45.5			68
70.0	44			86
77.5	40			86
78.0			−0.21	87
80.5		61.1, 57.9		38, 66
83.8			−0.31	62

Nitrogen + Carbon monoxide

No recent measurements have been reported and only g^E and v^E are known.

T/K	g^E/J mol^{-1}	v^E/cm^3 mol^{-1}	References
68.1	26.7		68
83.8	23.0	+0.127	62
83.8	22.7		63

Carbon monoxide + Methane

All the principal excess functions are known and, although there is reasonable agreement in the magnitude of the equimolar values, the uncertainty in the degree of asymmetry in h^E and hence in s^E remains to be resolved.

T/K	$g^E/J\,mol^{-1}$	$h^E/J\,mol^{-1}$	$v^E/cm^3\,mol^{-1}$	*References*
90.7	110	105	−0.32	88, 89
90.7	121			63
91.2		106	−0.34	37

5.4 Liquified natural-gas mixtures

The principal components of natural gas are the lower paraffinic hydrocarbons, C_1 to C_4, N_2 and smaller amounts of CO_2, H_2O, H_2S and higher hydrocarbons. Much effort has gone into the measurement of the thermodynamic properties of binary and ternary mixtures composed of these commercially-important substances and a selected group of mixtures will be considered in this section. Many more measurements have been made on the volumetric behaviour of these mixtures than of g^E or h^F, for the ability to calculate the density of a real multicomponent natural-gas mixture from a knowledge of its composition, temperature and pressure remains a desirable commercial objective.

Nitrogen + Methane

McClure et al.[90] have measured h^E at 91.5 and 105.0 K, obtaining values of 138.1 and 100.6 J mol^{-1}, respectively. They have undertaken a critical evaluation of all measurements of g^E with the assumption that h^E is a linear function of temperature up to 110 K and have shown that all[61,90-93] except two early determinations[63,70] are self-consistent. The excess Gibbs function is large, reaching 192.6 J mol^{-1} at 112 K. Recent h^E results[94] show that, like argon + methane mixtures discussed previously[71], a strongly non-linear temperature dependence develops above 120 K.

Many measurements of v^E over wide ranges of T and p appear in the literature[64,70,72,95,96] and a selection is shown in *Figure 5.1*. Such behaviour, with large negative values of v^E at high temperatures that rapidly become less negative with increase of pressure, is typical of binary mixtures whose components have widely different critical temperatures. The highest experimental temperature shown in *Figure 5.1*, 120 K, at which $v^E/v \simeq 0.2$ is only some 6 K below the gas–liquid critical temperature of pure nitrogen.

Nitrogen + Ethane and + Propane

Although no excess functions have been calculated, vapour–liquid equilibrium measurements have been made on both systems[91,96] and the presence of partial miscibility confirmed at low temperatures in each case. The ternary phase diagram has been measured[96] for nitrogen + methane + propane at temperatures between 114 and 122 K where the area of the region of partial miscibility decreases rapidly with increase of temperature.

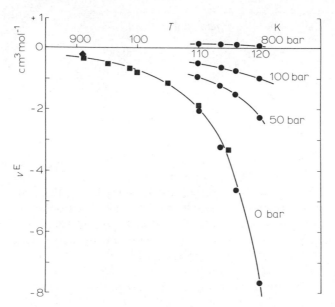

Figure 5.1 The equimolar excess volume for nitrogen + methane as a function of temperature and pressure. ■ *the data of Liu and Miller*[72], *corrected by Massengill and Miller*[64], ● *Nunes da Ponte et al.*[96], *and Hiza et al.*[95], ◆ *Fuks and Bellemans*[70]

Nitrogen + Methane + Argon

This ternary system has been intensively studied by Miller and his co-workers[51,61,64]. Both g^E and v^E are available over the whole range of composition, the former at 112 K, the latter at temperatures ranging from 90 to 115 K and at pressures up to 500 bar. No new effects are observed, the experimental ternary excess functions being closely approximated by the mole fraction average of the appropriate binary systems.

Methane + Ethane

The results of the numerous vapour–liquid equilibrium studies that have been made on this important mixture have been critically reviewed by Parrish and Hiza[97] and many of the existing measurements are shown to be

T/K	$g^E/J\,mol^{-1}$	$h^E/J\,mol^{-1}$	$v^E/cm^3\,mol^{-1}$	References
91.5		86.6		100
90.69	117.2			76
103.99	120.5		−0.453	76
112		68.6	−0.586	100, 51
115.77	124			100
125			−0.820	48
135			−0.085	48

mutually inconsistent. The results of Lu[91] must be discarded as being too high, and, of the remainder, only the early work of Bloomer *et al.*[98] and the more recent studies of Wichterle and Kobayashi[99] and of Miller and Staveley[100] show any semblance of consistency. The position with respect to v^E is much better, with substantial agreement being obtained for the data of four different groups[48,51,93,101], only the work of Klosek and McKinley[102] and of Jensen and Kurata[103] is discordant. h^E has also been determined[100].

Methane + Propane

There is fair agreement between the g^E data reported by several groups of research workers; only those of Poon and Lu[96] seem anomalous. The recent calorimetric determinations of h^E show that the earlier values that were derived from vapour pressure measurements were gross underestimates[106]. The excess volume has also been measured over quite a wide temperature range. All the principal excess functions, including $c_p^{E\,100,106}$, are strongly skewed towards mixtures rich with respect to methane.

T/K	$g^E/J\,mol^{-1}$	$h^E/J\,mol^{-1}$	$v^E/cm^3\,mol^{-1}$	*References*
90.68	187.1, 180.7		−0.510	104, 105, 95
91.5		154.1		100
112.0		129.2	−0.845	100, 95
115.77	209.5			105
130.0			−1.336	95
134.83	225.4			105

Methane + Butane

Parrish and Hiza[97] have evaluated approximate values of g^E from vapour pressures and find that those of Kahre[107] and of Elliot *et al.*[108] are consistent whereas those of Wang and McKetta[109] are not. The excess volume is known between 120 and 130 K[95] with an equimolar value of $-1.30\,cm^3\,mol^{-1}$ at 120 K.

Methane + 2-methylpropane

Only v^E is available[51,95] with values comparable with those of methane + propane at the same temperature and thus considerably less negative than those of the previous mixture.

Ethane + Propane

Recent determinations of v^E using a reliable magnetic float technique[95] show that, in contrast to some early results[101], v^E is small ($-0.04\,cm^3\,mol^{-1}$)

and virtually independent of temperature from 100 to 125 K. An approximate value of h^E is also small[100], 5 J mol^{-1} at 100 K.

Other mixtures

Hiza et al.[95] have measured v^E for the binary mixtures ethane + butane, propane + butane, propane + 2-methylpropane and butane + 2-methylbutane in the temperature range $110 < T/K < 130$, although not at equimolar concentrations.

The excess volume for the following ternary mixtures of natural gas components has been determined over the listed temperature range.

Mixture	T/K	References
Methane + ethane + propane	91–115	95, 101
Nitrogen + methane + ethane	91–115	95
Nitrogen + methane + propane	91–108	95
Methane + ethane + argon	91–115	51

Methane + Ethene

All the available thermodynamic measurements on this industrially-important and much-studied mixture have been evaluated critically by Lobo et al.[110]. Some values of the excess functions are listed below. Lobo et al. have used these to produce a nine-parameter expression that represents g^E (and h^E) from 100 to 200 K. The vapour–liquid equilibrium results of Miller et al.[111] and of Volova[112] that extend to the highest temperatures are in good agreement with the derived expression whereas the excess Gibbs functions of Hsi and Lu[113] are too low and were not included in the analysis.

T/K	g^E/J mol^{-1}	h^E/J mol^{-1}	v^E/cm^3 mol^{-1}	References
103.99	216.14		−0.184	114
115.76	209.2	254.8		114, 110
140.12		217.5		110
150.00		174.0		110

Ethane + Ethene

Calado et al.[76] have studied this mixture and their results are appended.

T/K	g^E/J mol^{-1}	h^E/J mol^{-1}	v^E/cm^3 mol^{-1}
161.39	98.9	192.7	0.156

5.5 Mixtures of the higher hydrocarbons

Linear paraffinic hydrocarbons are non-dipolar and their mixtures with one another form a particularly simple class of mixture whose excess functions at and above room temperature were studied extensively between 1950 and 1970. Much of this experimental work was inspired by a desire to test the empirical 'principle of congruence' put forward by Brønsted and Koefoed[115] in 1946 but which has since been more closely integrated with statistical mechanical theory and so is dealt with in Section 8.7.

The experimental excess functions of n-alkane mixtures follow a common pattern whose general features are:

(a) The excess Gibbs free energy and excess volumes are negative and approximately quadratic. They increase in magnitude as the temperature increases.
(b) The excess heat is positive and quadratic at low temperatures but becomes negative at high temperatures. It crosses the axis $h^E = 0$ as a sigmoid curve which has $h^E < 0$ for mixtures weak, and $h^E > 0$ for those strong in the longer component. The temperature[56] at which $h^E = 0$ at $x = \frac{1}{2}$ is 338 K for $n\text{-}C_6H_{14} + n\text{-}C_{16}H_{34}$, 353 K for $n\text{-}C_6H_{14} + n\text{-}C_{24}H_{50}$, and 368 K for $n\text{-}C_8H_{18} + n\text{-}C_{24}H_{50}$.
(c) The magnitudes of g^E, h^E and v^E increase rapidly with the differences in chain length.

The most accurate results are those of McGlashan and his co-workers[116] and measurements from Amsterdam[117]. Recently, precise values for all the principal excess functions have been published[118] for $n\text{-}C_6H_{14} + n\text{-}C_{10}H_{22}$ and $+ n\text{-}C_{11}H_{24}$. Results for a typical mixture $n\text{-}C_6H_{14} + n\text{-}C_{16}H_{34}$ are shown in *Table 5.2*.

Since 1970 a large body of high-quality experimental work has been undertaken on mixtures of pseudo-spherical hydrocarbons by Marsh and his co-workers in Armidale[119] and by a Swiss group[120]. The main objective has been to produce data against which statistical theories of mixtures of molecules with spherically-symmetric intermolecular potentials can be tested. Much of the experimental work concerned mixtures of the cycloalkanes, from C_5 to C_8 and also mixtures of the cycloalkanes with the small globular hydrocarbon 2,3-dimethylbutane. Results for cyclopentane + cyclohexane are given in *Table 5.2* and the general trends can be summarized thus:

(a) The excess Gibbs function is very small for cycloalkane mixtures and exhibits no discernible trend with size difference. The excess enthalpy and volume are both small and positive for cyclopentane + cyclohexane and become negative for cyclopentane + cyclo-octane.
(b) Both g^E and h^E are small for mixtures of 2,3-dimethylbutane with cyclopentane but both rapidly increase as the size of the cycloalkane is itself increased. The excess volume is negative for all such mixtures and becomes increasingly so as the temperature is raised.

Table 5.2 The excess functions of some mixtures containing hydrocarbons

System	T/K	$g^E/J\,mol^{-1}$	$h^E/J\,mol^{-1}$	$c_p^E/J\,mol^{-1}\,K^{-1}$	$v^E/cm^3\,mol^{-1}$	References
n-C_6H_{14} + n-$C_{16}H_{34}$	293	-65.3	129	-3.6	-0.495	10, 116
c-C_6H_{12} + n-$C_{16}H_{34}$	298	-177	502	-5.6	$+0.632$	129, 130
C_6H_6 + n-$C_{16}H_{34}$	298	91	1256	-5.5	$+1.116$	135, 136
CCl_4 + n-$C_{16}H_{34}$	298	-118	577		0.67	135, 150, 151
c-C_5H_{10} + c-C_6H_{12}	298	-4	27.8	-0.76	0.0412	118
c-C_5H_{10} + OMCTS	298	-208	212		0.046	121
c-C_6H_{12} + TKEBS	298	-494	167		0.25	122
c-C_6H_{12} + C_6H_6	298	397 (313 K)	799.3	-2.92	0.654	14, 40, 41, 50

Because cycloalkanes with more than eight carbon atoms can no longer be considered to be pseudo-spherical, the Armidale group has sought to study binary mixtures with large size differences between the two components species by investigating mixtures of small, compact pseudospherical molecules with molar volumes $\simeq 100\,cm^3$ with either octa-methylcyclotetrasiloxane (OMCTS)[121] or tetrakis-(2-ethylbutoxy)silane (TKEBS)[122] whose molar volumes at room temperature are 312 and $487\,cm^3$, respectively. Both these large molecules, although not strictly hydrocarbons, present hydrocarbon-like surfaces to neighbouring molecules and are wholly miscible with the cycloalkanes. Typical results are listed in *Table 5.2* where the most interesting feature is the unexpected positive sign of v^E. As will be seen in Chapter 8, most theories predict large negative excess volumes for binary mixtures with such disparate molar volumes and so it is likely that these two large molecules, both of which possess considerable degrees of intramolecular vibration and rotation, cannot be properly described by a simple centro-symmetric intermolecular potential.

Another set of pseudo-spherical molecules that are not strictly hydrocarbons but may be amenable to theoretical treatment are the tetramethyl derivatives of carbon, silicon and tin. The excess functions for the three possible binary pairs and for two other closely related mixtures are noted below.

System	T/K	$g^E/J\,mol^{-1}$	$h^E/J\,mol^{-1}$	$v^E/J\,mol^{-1}$	References
$CMe_4 + SiMe_4$	283	7.0 (280 K)	22.3	+0.051	123
$SiMe_4 + SnMe_4$	283	70.7 (280 K)	67.6	−0.394	123
$CMe_4 + SnMe_4$	283	95.9 (280 K)	99.1	−0.312	123
$CMe_4 + c\text{-}C_6H_{12}$	298		97.7	−1.441	124
$SiMe_4 + c\text{-}C_6H_{12}$	298		191.1	−1.095	124

Although it is possible to discern trends in the excess functions when both components in a binary mixture are of the same type, either both linear or both pseudo-spherical, it is much more difficult to generalize about mixtures containing molecules of differing shape. Patterson, Delmas and their co-workers at McGill[125–128] have accumulated many experimental results over the past few years in an attempt to understand the various relevant effects. Most of their measurements have been of h^E and g^E (and hence s^E), but v^E and c_p^E have also been determined for a great variety of predominantly hydrocarbon mixtures containing linear, branched and cyclic molecules. Most of the results can be accounted for in a semi-empirical way by postulating two effects.

(a) Destruction of short-range orientational order in the longer n-alkanes by the addition of a 'structure-breaker' consisting of a highly branched or globular second component. This gives rise to positive contributions to both h^E and s^E. These orientational correlations can be detected by other techniques, such as depolarized Rayleigh light scattering, and decrease rapidly with increase in temperature so that c_p^E is strongly negative. The effect is clearly illustrated when the excess enthalpies of both n-hexane and cyclohexane with n-hexadecane, 129 and 502 J mol^{-1} respectively, are compared with the values for the same two small hydrocarbons mixed with a highly-branched C_{16} compound, 2,2,4,4,6,8,8-heptamethylnonane, -68 and 129 J mol^{-1}. Patterson *et al.*[125-127] assume that there are no significant orientational correlations in this latter isomeric globular molecule in the pure state and so the differences between the pairs of excess enthalpies, 197 and 373 J mol^{-1}, give a measure of the effectiveness of n-hexane and of cyclohexane in destroying orientational order. Not surprisingly the pseudo-spherical cyclohexane proves to be the more efficient. Although this basic idea is fully established, a few anomalies remain. The role of orientational effects on v^E is still unclear and it is curious for example that, although both cyclohexane and benzene are efficient 'structure-breakers', tetra-hydronaphthalene is not.

(b) A 'condensation effect' was introduced by Patterson[128] to account for the remarkable differences that certain highly-branched molecules display in their apparent ability to destroy orientational order. Some relevant results at 298 K are listed in the table below where it can be seen that large differences are apparent between both h^E and c_p^E for mixtures of n-hexadecane, n-octane and, more surprisingly, cyclopentane with the two highly-branched nonanes. Patterson *et al.* interpret these by postulating that, because 2,2,5-trimethylhexane is an open molecule with the possibility of almost unrestricted internal rotation whereas 3,3-diethylpentane possesses a more compact structure and is sterically hindered, the freely-rotating segments of the two n-alkane molecules 'condense' in the presence of the latter nonane with a consequent exothermal contribution to h^E and a corresponding positive addition to c_p^E. This effect is absent in mixtures with cyclohexane and also with the higher cycloalkanes but is present with cyclopentane. It is assumed that the presence of the hindered nonane causes restricted intermolecular rotation of the small anisotropic cycloalkane. This rather speculative idea is supported by measurements on mixtures of cycloalkanes with anisotropic hydrocarbons such as methyl and dimethylcyclohexanes and with *cis-* and *trans-*decalin. In every case, mixtures with cyclopentane give large negative h^E's and positive c_p^E's, whereas mixtures of these molecules with the larger more spherically symmetric cycloalkanes give small positive h^E's and negative c_p^E's. Once again few results are available for v^E and even fewer for g^E on relevant mixtures.

Several other groups have studied the thermodynamic properties of hydrocarbon mixtures of the type pseudo-spherical + non-spherical, usually represented by a cycloalkane mixed with a linear n-alkane[129-134]. The most

	$h^E/\text{J mol}^{-1}$	$c_p^E/\text{J mol}^{-1}\,\text{K}^{-1}$
2,2,5-trimethylhexane + n-C_{16}	315	−5.4
+ n-C_8	95	−0.99
+ c-C_6	192	−0.38
+ c-C_5	88	+0.11
3,3-diethylpentane + n-C_{16}	18	−3.18
+ n-C_8	−83	+0.15
+ c-C_6	179	−0.04
+ c-C_5	−106	+1.11

recent work of Scott and his associates[133] has shown that, when a series of isomeric hydrocarbons with differing degrees of branching is mixed with a common second component, regardless of whether this component is linear (n-$C_{16}H_{34}$) or globular (c-C_6H_{12}), the differences in h^E and v^E on going from one branched hydrocarbon to another are very comparable and do not seem to depend strongly on the absolute magnitude of the excess function. Some of these differences are small and the explanation of such subtle effects is not possible using current statistical theory.

Mixtures formed from paraffinic hydrocarbons and aromatic hydrocarbons[135,136] generally have larger positive excess free energies and enthalpies than those, of similar molar mass ratio, formed solely from paraffins. A comparison of both n-C_6 + n-C_{16} and c-C_6 + n-C_{16} with C_6H_6 + n-C_{16} is given in the *Table 5.2*, where an increase in all the excess functions, including v^E, for the last system is displayed. If the aromatic molecule contains an aliphatic side group and thus acquires an intermediate status, this effect is much reduced; for example, h^E for toluene + n-C_{16}[136] is only 735 J mol^{-1}, and for mixtures of n-C_{16} with p-xylene and with mesitylene h^E is further reduced to 444 and 427 J mol^{-1}, respectively[137]. The equimolar excess enthalpies, excess volumes and (dv^E/dT) increase slowly along the series benzene + n-alkane in much the same manner as in the series branched alkane or cyclohexane + n-alkane. In contrast, the negative excess heat capacities become increasingly negative and the scanty results for the excess Gibbs function[135] also show a decrease with increasing n-alkane chain length.

There have been few studies of hydrocarbon mixtures that contain an unsaturated hydrocarbon and the recent calorimetric measurements of Wóycicki[138] are by far the most comprehensive. n-Alkane + n-alkene systems containing the same number of carbon atoms invariably possess positive h^E with substantial variations in the absolute magnitude depending on the configuration of the alkene. Some of Wóycicki's results at 298 K are given below and it is evident that the differences in h^E when two isomers are mixed with a common second component greatly exceed h^E when the two isomers are mixed directly, a similar situation to that of mixtures of *cis*- and *trans*-decalin with each other[139] and when mixed separately with cyclohexane or cyclopentane[140].

The scanty results for n-alkane + n-alkyne mixtures[141] indicate that h^E is quite large and positive due to the fact that alkyne–alkyne interactions are considerably stronger than their counterparts in alkene mixtures.

Mixture	$h^E/J\,mol^{-1}$	References
n-hexane + *cis*-2-hexene	74.5	138
n-hexane + *trans*-2-hexene	54.0	138
cis-2-hexane + *trans*-2-hexene	− 3.4	138
n-hexane + 3-hexyne	409	141

The simplest aromatic + alicyclic mixture, benzene + cyclohexane, is probably the most-studied of all liquid mixtures for it is commonly used as a test system for new designs of calorimeter and dilatometer. Its excess functions are therefore well established with the equimolar h^E at 298 K[40] being close to 799.3 J mol^{-1} falling to 568 J mol^{-1} at 393 K [42]. The excess volume is almost independent of temperature[142] and its value at 298 K is 0.650 cm^3 mol^{-1}. The excess Gibbs function is also moderately large and positive[14,143] with a value of 297.3 J mol^{-1} at 313 K.

The addition of one methyl group to either molecule leads to a small reduction in both g^{E}[144] and h^{E}[145] but the addition of longer aliphatic side groups to the cyclohexane nucleus leads to a reversal of this trend. The excess enthalpy for cyclohexane + ethylbenzene has fallen to 540 J mol^{-1}[146] whereas the figure for benzene + ethylcyclohexane has risen to 810 J mol^{-1}[147].

Over the past decade, Benson and his co-workers[148] have produced many results of high-quality on binary mixtures in which both species are aromatic hydrocarbons. Wóycicki[147] and Hsu and Clever[145] have also contributed to these studies. Excess enthalpies, volumes and heat capacities are now known for almost all the possible binary pairs containing benzene, toluene, the three xylenes, ethyl benzene and the three isomeric trimethylbenzenes. The earlier measurements of Rastogi[149] have been shown to be seriously in error. It is impossible to generalize about the magnitudes of the excess functions except to say that both h^E and v^E tend to be small and positive although occasionally negative functions are encountered, sometimes over only a portion of the concentration range, the excess function thus being 'cubic' in form. The excess heat capacity is, almost without exception, negative and often exhibits quite large variations in magnitude when different aromatic isomers are mixed with a common second component. Few measurements of g^E are available.

5.6 Mixtures containing halides

Most organic halides and some of the covalent inorganic ones are liquids of low polarity which mix readily with each other and with hydrocarbons.

Mixtures containing such molecules have been widely studied and some of the more important results are collected in this section.

The excess functions of the series carbon tetrachloride + n-alkane are very similar to those of the series cyclohexane + n-alkane in sign, absolute magnitude, and variation with alkane chain length[135,150–152]. Results for $CCl_4 + n$-$C_{16}H_{34}$ are in *Table 5.2*. Other haloalkane + n-alkane series show similar trends. Delmas and Purves[153] have measured both h^E and g^E (the latter by light scattering) for 1,2-dibromoethane with a large number of linear and branched hydrocarbons. Both sets of excess functions are much larger than those for mixtures containing CCl_4, with h^E for $C_2H_4Br_2 + n$-$C_{16}H_{34}$ exceeding $2 \, kJ \, mol^{-1}$ at 298 K.

There have been many measurements of the excess functions of mixtures of the type CCl_4 + pseudo-spherical hydrocarbon because such systems are more amenable to quantitative theoretical treatment than mixtures containing an n-alkane. CCl_4 + cyclopentane has values of the equimolar Gibbs function, enthalpy and volume of $+34.1 \, J \, mol^{-1}$, $+77.1 \, J \, mol^{-1}$ and $-0.0351 \, cm^3 \, mol^{-1}$, respectively[154]. The excess Gibbs function and enthalpy for CCl_4 + cyclohexane are both twice as large but the excess volume is positive and numerically five times larger[155]. Mixtures of CCl_4 + neopentane are interesting in that $g^E \simeq h^E \simeq 320 \, J \, mol^{-1}$ at 273 K with an excess volume of $-0.5 \, cm^3 \, mol^{-1}$. In contrast, mixtures of CCl_4 with the extremely large pseudo-spherical molecule TKEBS are characterized by large negative values of both g^E and h^E, -633 and $-228 \, J \, mol^{-1}$ respectively[122], and v^E is of the same sign, -0.241 $cm^3 \, mol^{-1}$. The principal excess functions of CCl_4 + both TKEBS and OMCTS are all considerably more negative compared to the functions for the same two large molecules when mixed with benzene[121,156].

The six possible binary mixtures that can be formed from the four tetrachlorides of carbon, silicon, tin and titanium have been studied by Sackmann and his co-workers[157]. The excess functions at 323 K (v^E at 293 K) are tabulated below. These data should be useful for testing statistical theories of mixtures composed of tetrahedral octopolar molecules.

Mixture	$g^E/J \, mol^{-1}$	$h^E/J \, mol^{-1}$	$Ts^E/J \, mol^{-1}$	$v^E/cm^3 \, mol^{-1}$
$CCl_4 + SiCl_4$	88	147	59	0.05
$CCl_4 + TiCl_4$	74	190	116	0.08
$CCl_4 + SnCl_4$	122	285	163	0.46
$SiCl_4 + TiCl_4$	205	153	-52	-0.38
$SiCl_4 + SnCl_4$	173	293	120	0.13
$TiCl_4 + SnCl_4$	80	222	142	0.08

Partially chlorinated alkanes such as chloroform and methylene dichloride are weakly polar and therefore possess larger principal excess functions with hydrocarbons than does carbon tetrachloride. A typical example is the series n-hexane + CCl_4, $+CHCl_3$ and $+CH_2Cl_2$, where

Table 5.3

System	$g^E/\text{J mol}^{-1}$	$h^E/\text{J mol}^{-1}$	$c_p^E/\text{J mol}^{-1}\,\text{K}^{-1}$	$v^E/\text{cm}^3\,\text{mol}^{-1}$	References
CCl_4 + benzene	+82	115.5	0.96	+0.0054	158, 160
CCl_4 + toluene	+5 (293 K)	−18.5	1.39	−0.0435	158, 160
CCl_4 + o-xylene	−84	−23.2	0.77	−0.0045	158, 159, 161
CCl_4 + m-xylene	−60	4.1	0.64	+0.1125	158, 159, 161
CCl_4 + p-xylene	−94	−75.9	1.26	−0.0113	158, 159, 161
CCl_4 + mesitylene		107.7	−0.47	+0.3754	158, 161

the equimolar enthalpies are respectively 315, 756 and 1319 J mol^{-1} and the excess volumes, $+0.044$, $+0.335$ and $+0.724$ cm^3 mol^{-1} [152]. All these figures are increased as the size of the n-alkane is itself increased.

There is good spectroscopic evidence that mixtures of CCl$_4$ with aromatic hydrocarbons form weak intermolecular complexes in the liquid and solid states and the presence of these unusual interactions in such mixtures has initiated several precise thermodynamic investigations of their excess functions. Some results at 298 K are given in *Table 5.3*.

One interesting feature of these systems is that three of them show double maxima in certain excess functions at certain temperatures, e.g., h^E for CCl$_4$ + toluene at around 313 K [158], h^E for CCl$_4$ + m-xylene at around 293 K [160] and, perhaps most surprisingly, v^E for CCl$_4$ + benzene exhibits a highly-skewed double maxima in an extremely limited temperature range between 295.4 and 296.2 K [162]. A simple theoretical model of complex formation in which the equilibrium constant increases in the sequence CCl$_4$ + benzene < + toluene < + p-xylene < + mesitylene but the magnitude of the (negative) enthalpy of complex formation decreases in the same order is sufficient to represent satisfactorily these experimental results (including the double maxima in the excess enthalpies). This apparent complexing interaction between the two species is enhanced when weakly polar, partially chlorinated hydrocarbons are mixed with aromatics. Both CHCl$_3$ [163] and CH$_2$Cl$_2$ [164] mix exothermally with benzene and the excess volumes of the polar 1,1-dichloroethane with a series of aromatics are everywhere much smaller than the excess volumes of the isomeric non-polar 1,2-dichloroethane and the same aromatic hydrocarbons [165].

Mixtures of CCl$_4$ with unsaturated hydrocarbons have excess functions that range from small and positive to large and negative in the case of CCl$_4$ + 3-hexyne, where there are strong indications of incipient compound formation between the two components. Some results of Wóycicki[138] at 298 K are appended.

Mixture	h^E/J mol^{-1}
CCl$_4$ + cis-2-hexene	-33
CCl$_4$ + trans-2-hexene	$+41$
CCl$_4$ + 1,5-hexadiene	-54
CCl$_4$ + 3-hexyne	-514

The excess enthalpies and volumes of chlorobenzene and of o-, m- and p-dichlorobenzene mixed with aromatic hydrocarbons[166] are all quite large and negative except for mixtures of the chlorinated aromatics with benzene itself. No simple pattern can be discerned among the measured excess functions due no doubt to the complicated nature of the intermolecular potentials in such mixtures that includes contributions from π–π interactions as well as the more usual angle-dependent multipolar terms.

A large number of measurements exist on mixtures of two fully- or partly-halogenated aliphatic hydrocarbons[167-170]. The components range from moderately polar (1,1,1-trichloroethane) to non-polar (C_2Cl_4 and 1,1,2,2-tetrachloroethane). The excess functions for mixtures of isomers are quite large[168], the excess enthalpy of *cis*-1,2-dichloroethene + *trans*-1,2-dichloroethene is 150.7 J mol^{-1} and the excess volume +0.025 cm^3 mol^{-1}. McGlashan[171] has measured the enthalpies of mixing of seven mixtures of the type $CCl_nMe_m + CCl_xMe_y$ and of three similar mixtures with Si instead of C as the central atom. All the excess enthalpies are small and positive at 298 K.

Carbon disulphide has no permanent dipole and is only weakly quadrupolar. Its solution properties with hydrocarbons are very similar to those of carbon tetrachloride and some results at 298 K are noted in the table[172].

Mixture	g^E/J mol^{-1}	h^E/J mol^{-1}	v^E/cm^3 mol^{-1}
$CS_2 + C_6H_6$	255	550	0.526
$CS_2 + n\text{-}C_6H_{14}$		588	0.248

5.7 Mixtures containing fluorocarbons

It will be seen in the next section that most liquid mixtures that have large positive excess functions and are but partially miscible at some temperatures contain at least one polar component. A striking exception is the large class of mixtures in which one of the components is a fully- or near-fully-fluorinated non-polar compound of carbon, silicon or sulphur and the second an aliphatic or alicyclic hydrocarbon or another organic halide with small or zero polarity. The thermodynamic properties of these and other fluorocarbon-containing mixtures have been reviewed recently[173].

The first three mixtures listed in *Table 5.4* below conform to the above specifications and it can be seen that the excess Gibbs function is $\frac{1}{3}RT$ to $\frac{1}{2}RT$ and the excess enthalpy is twice as large. All such mixtures exhibit a positive excess volume representing a volume expansion of some 2 to 10 per cent and will separate into two liquid phases at temperatures below an upper critical solution temperature unless prevented from doing so by the formation of a solid phase. The actual UCSTs for these three mixtures are, respectively, 94.5, 296 and 319 K. Although the excess entropies are also large and positive when measured at constant pressure, the large positive excess volumes cause the calculated constant-volume excess entropies to be close to zero or even small and negative.

The magnitudes of the excess functions (and hence the tendency to phase separation) are reduced when a fully fluorinated aliphatic or alicyclic fluorocarbon is mixed with a partially fluorinated hydrocarbon and a system composed of two perfluorocarbons forms a near-ideal mixture. Mixtures of partially chlorinated fluorocarbons also exhibit a decreased

159

Table 5.4

Mixture	T/K	$g^{E}/\mathrm{kJ\,mol^{-1}}$	$h^{E}/\mathrm{kJ\,mol^{-1}}$	$v^{E}/\mathrm{cm^{3}\,mol^{-1}}$	References
$CF_4 + CH_4$	110	0.36	0.537 (93 K)	0.88	174, 38
$n\text{-}C_6F_{14} + n\text{-}C_6H_{14}$	308	1.329	2.157	5.38	175
$c\text{-}CF_3.C_6F_{11} + c\text{-}CH_3.C_6H_{11}$	338	1.39	2.80	7.5	176
$C_6F_6 + c\text{-}C_6H_{12}$	313	0.776	1.517	2.57	179–181
$C_6F_6 + C_6H_6$	313	−0.044	−0.433	0.801	179–181
$C_6F_6 + 1,4\text{-}(CH_3)_2.C_6H_4$	313	−0.395	−1.66	0.086	179–181
$C_6F_6 + (C_2H_5)_3N$	283	0.453	0.65	1.37	182–184

tendency to phase separation. For example, no UCST is found for either $CF_4 + CF_3Cl$ or $CH_4 + CF_3Cl$[177].

A satisfactory quantitative explanation of these large thermodynamic effects is still lacking even though the basic experimental pattern was established by Scott and his group from the 1950s onwards[178]. The large difference in molar volume between the perfluorocarbon and the parent hydrocarbon is most definitely not a major factor as was once thought to be the case and it is likely that a fuller understanding of the forces involved will only come about when the quite recent experimental measurements described earlier on such simple systems as $Xe + CF_4$, $Xe + C_2F_6$, $CF_4 + CH_4$, etc., are interpreted using perturbation theory.

It can be seen from the data on $C_6F_6 + c\text{-}C_6H_{12}$ in *Table 5.4* that, when an aliphatic or alicyclic perfluorocarbon is replaced by C_6F_6, the principal excess functions are still positive but are reduced by about 50 per cent in magnitude. This trend is further continued when both components are aromatic. Not only are g^E, h^E and Ts^E now quite large and negative and v^E much reduced but, with only one known exception ($C_6F_6 + C_6Me_6$), such mixtures form congruently-melting equimolar compounds, in many cases melting at a higher temperature than either of the two pure components. The equimolar $C_6F_6:C_6H_6$ compound melts at $297.2\,K$[185], almost $20\,K$ higher than either pure C_6F_6 or C_6H_6. Such behaviour, in complete contrast to the partial liquid–liquid miscibility noted earlier, is a manifestation of enhanced unlike interactions, and the large volume of research effort that has been undertaken in this area since 1960 when these effects were first detected has been stimulated by a desire to understand the molecular reasons for these gross differences in thermodynamic behaviour[186].

Onc unusual feature of $C_6H_6 + C_6F_6$ mixtures is the occurrence of both positive and negative azeotropy at one temperature. Some results of Gaw and Swinton[180] at 343 K are listed below,

x_F	0.0000	0.1847	0.7852	1.0000
p/bar	0.7341	0.7432	0.7129	0.7185

and the presence of a positive azeotrope at a C_6F_6 mole fraction of about 0.2 and of a negative one at around 0.8 is clearly seen. The presence of double azeotropy in this mixture has been confirmed by subsequent measurements using equilibrium stills[187] but, to date, no other mixture has been found to possess this interesting behaviour. Recent measurements[308] have shown that this mixture is non-azeotropic above $460\,K$ and indicate that it is likely that, as the temperature is raised above $343\,K$, the positive and negative azeotropes disappear together at a horizontal point of inflection that occurs at a near-equimolar composition around $443\,K$.

It can be seen from a comparison of the fifth with the sixth entry in *Table 5.4* that additional methyl groups on the aromatic hydrocarbon give a further negative contribution to the excess functions and even the excess volume for C_6F_6 + 1,3,5-trimethylbenzene becomes negative with a value of $-0.357 \, cm^3 \, mol^{-1}$ at $313 \, K^{179}$. This negative trend is also apparent if methyl groups are added to the cyclohexane nucleus of the C_6F_6 + c-C_6H_{12} system and measurements such as these have led to the abandonment of the early explanations for the unusual thermodynamic properties of these mixtures that postulated the presence of charge transfer interactions between the dissimilar molecular species. It now seems likely that a combination of anisotropic overlap and multipolar forces combined with normal dispersion forces are sufficient to account for the measured properties.

Most of the early measurements on systems containing C_6F_6 were performed by Scott at UCLA[188] and by Swinton at Strathclyde[179, 181, 185]. Over the past decade, Stubley and co-workers[189] have measured the excess enthalpies and volumes of many mixtures of the type C_6F_5H + alkyl benzene and C_6F_5H + alkylcyclohexane.

The trend in the excess functions with degree of substitution of the hydrocarbon closely parallels the trends found previously for mixtures containing C_6F_6. Both the excess enthalpy and the volume decrease regularly as the extent of methyl substitution of the aromatic or alicyclic hydrocarbon is increased. If, however, a series of mono-substituted alkyl benzenes as one component is considered, h^E and v^E increase as the chain length of the alkyl side group lengthens. This trend is due to the hydrocarbon changing its overall character from aromatic to aliphatic with a resultant positive contribution to the excess functions. The group led by Fenby[190] has considered mixtures of the type C_6H_6 + C_6F_5X and c-C_6H_{12} + C_6F_5X, where X is H, F, Cl, Br, or I. Parallel trends are observed for both series with negative contributions to both the excess enthalpy and volume as the X substituent becomes larger and more polarizable.

It has been indicated earlier that it is likely that the thermodynamic properties of most hydrocarbon + fluorocarbon mixtures are interpretable in terms of the presence of 'normal' dispersion and multipolar intermolecular forces but this is not so when C_6F_6 and similar molecules are mixed with amines, ethers and heterocyclic molecules. Many such mixtures of fluorocarbons + n-donors exhibit strong charge-transfer bands in the visible and uv region and their thermodynamic properties can usually be rationalized using a version of the 'ideal associated solution' theory where a bimolecular complex is postulated[191] and its equilibrium constant can be obtained both from spectroscopic measurements and from the thermodynamic excess functions. In this latter case, it is usual to choose a reference mixture that is closely similar to the real system under investigation but in which no complexing reaction is believed to be present. Armitage *et al.*[192] have studied C_6F_6 + N,N-dimethylaniline, and + N,N- dimethyltoluidine using C_6F_6 + isopropylcyclohexane as the reference system. They have also investigated C_6F_5CN + N,N-dimethylaniline[193]. In all cases almost exact agreement was found for the enthalpy of complex formation as derived by the two completely different methods.

Other studies of similar systems include the measurement of the excess enthalpy of C_6F_6 with pyridine and with five substituted pyridines[183,194] and the measurement of all the principal excess functions for C_6F_6 + triethylamine[182–184], a typical n-donor that also possesses a considerable permanent dipole. Results for this last mixture are listed in *Table 5.4* where it is seen that, although all the excess functions are positive, they are only half the magnitude of the values for the C_6F_6 + c-C_6H_{12} system. There are many comparable studies[195] on mixtures of C_6F_6 with other n-donors, particularly aliphatic and cyclic ethers where mixtures with the appropriate alkanes and cycloalkanes make suitable 'reference systems' that enable a crude estimate to be made of the thermodynamic parameters of any complexing interaction.

Although mixtures of perfluorocarbons tend to be nearly ideal, such behaviour is not found for mixtures of inorganic fluorides. Staveley *et al.*[196] have measured the excess Gibbs function at 158.95 K for the two mixtures $NF_3 + BF_3$ and $NF_3 + HCl$. NF_3 is believed to be a weak base and might be expected to associate with either a typical Lewis acid like BF_3 or a typical Brönsted acid such as HCl. This expectation is not fulfilled for the experimental excess Gibbs function for the former system is large and positive with an equimolar value of $632\,J\,mol^{-1}$ while the $NF_3 + HCl$ system is characterized by even larger positive deviations from ideality leading to the formation of two partially-miscible liquid phases at the experimental temperature. The actual UCST was not determined.

5.8 Mixtures containing polar liquids

A polar substance is usually defined as one whose molecules have permanent dipole moments. However, this simple criterion is not very useful in discussing the effect of polarity on the properties of liquids or mixtures.

It will be shown in Chapter 7 that the effective polarity of a molecule increases with increasing size of the ratio $(\mu^2/\sigma^3 kT)$, where μ is the permanent dipole moment and σ the collision diameter. Even this expression is unable to show the increase in effective polarity that follows from an unsymmetrical disposition of the polar group within the whole molecule, as, for example, occurs with alcohols and primary amines. However, it is adequate to explain why mixtures of, say, benzene with substances of moderate dipole moment and reasonable symmetry, such as toluene ($\mu = 0.4\,D$) or even chlorobenzene ($\mu = 1.5\,D$), may reasonably be classed as non-polar mixtures. This section is given to mixtures in which the polarity of one or both of the components plays a decisive part in determining the excess properties. There are many hundred such mixtures and only a representative selection can be discussed here. They are placed in order of increasing complexity. Aqueous mixtures are discussed in the next section.

The more conventional polar organic molecules can be divided into two classes: first, those which cannot act as proton donors in a hydrogen bond, such as nitriles, ketones and nitro compounds, and, secondly, proton donors such as the primary and secondary amines and the alcohols. We consider them in this order.

The aliphatic nitriles exhibit partial miscibility with n-alkanes, the UCST for ethanenitrile (acetonitrile) + n-C_5 and + n-C_{18} being 341.2 and 426.2 K respectively[197]. The UCST falls as the aliphatic portion of the nitrile increases in length and for mixtures with n-hexane the temperatures are: ethanenitrile (350.2 K), propanenitrile (287.2 K) and n-butanenitrile (244.2 K). The excess functions have been measured for the lower n-alkanes + propanenitrile and + n-butanenitrile. All are large and positive, as is to be expected for a mixture near a UCST. The equimolar excess enthalpy and volume for mixtures containing n-hexane are propanenitrile (1428 J mol^{-1}, 0.44 cm^3 mol^{-1}) and n-butanenitrile (1280 J mol^{-1}, 0.24 cm^3 mol^{-1})[197]. These values increase quite rapidly as the carbon number of the hydrocarbon is raised.

In contrast, the lower nitriles are wholly miscible with CCl_4 and aromatic hydrocarbons with resultant reduced values of the excess functions. Some equimolar data at 318 K for mixtures containing ethanenitrile are given below[198-200]. Note that s^E is negative. The excess volume is small and negative with CCl_4 and 'cubic' with benzene.

	CCl_4	C_6H_6	$CH_3.C_6H_5$
g^E	+1190	+670	+780
h^E	+920	+490	+500

The excess enthalpy is large and positive for mixtures of the aromatic benzonitrile + cyclohexane (1390 J mol^{-1}) but is much reduced for mixtures with aromatic hydrocarbons[201]. Some experimental curves are shown in *Figure 5.2* and illustrate the complicated concentration dependence. The curve for benzonitrile + toluene is one of the best examples in the literature of an excess function with three extrema.

Mixtures of aliphatic and aromatic ketones with hydrocarbons form another series of systems that can be classified as polar + non-polar. Mixtures containing 2-propanone (acetone) have been studied for many years and, recently, Benson and his co-workers in Ottawa[208-210] have undertaken a major project by determining the principal excess functions for many other ketone + hydrocarbon mixtures. Some typical results are listed overleaf.

The trends are comparable to those for systems containing nitriles. The excess functions are large and positive for mixtures with n-alkanes where the major effect is the destruction of the dipole–dipole interactions in the pure liquid ketones. The excess enthalpy with a common hydrocarbon falls in the order 2-propanone > 2-butanone > 2-pentanone and the same trend

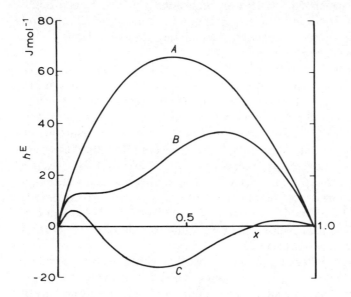

Figure 5.2 The excess enthalpy[201] at 298 K for the mixtures benzonitrile + chlorobenzene, curve A; benzonitrile + benzene, curve B; benzonitrile + toluene, curve C, as functions of x, the mole fraction of benzonitrile

System	T/K	$g^E/J\,mol^{-1}$	$h^E/J\,mol^{-1}$	$v^E/cm^3\,mol^{-1}$	References
2-propanone + $n\text{-}C_6H_{14}$	298		1548	0.24	202–204
2-propanone + C_6H_6	298	292 (323 K)	144	−0.07	205–207
2-butanone + C_6H_6	298	121 (323 K)	−50	−0.116	205, 208, 209
2-pentanone + C_6H_6	298	40 (323 K)	−75	−0.110	205, 208, 209

is evident in mixtures with benzene and carbon tetrachloride. This effect is primarily due to the fact that, although the actual dipole moments of all the ketones listed above are approximately constant, the effective dipole decreases as the length of the aliphatic chain increases and thus weakens the ketoxy interactions. The excess enthalpies of mixtures of ketones with benzene and other aromatics and with carbon tetrachloride are often negative or 'cubic' in shape and reflect the presence of strong dipole-induced dipole interactions between the two components, either μ–π in the case of ketone + aromatic or n–π in ketone + CCl$_4$ mixtures.

The excess enthalpies for mixtures of n-hexane, benzene or carbon tetrachloride with aromatic ketones lie in the order phenylacetone >

acetophenone > benzylacetone[210]. Although all the excess enthalpies are positive for these nine mixtures, the enthalpies for systems containing benzene and carbon tetrachloride are much smaller and are evidence of π–π and n–π interactions between these two components.

Nitroalkyls differ from ketones and nitriles in that their excess functions for mixtures with a given hydrocarbon tend to be much more positive. This is mainly because of the larger effective dipole moments of the nitroalkyls. Nitromethane is only partially miscible with aliphatic hydrocarbons at room temperature[211,212] but the UCSTs with aromatics are reduced to below 273 K. The excess functions tend to diminish as the molar volume of the nitroalkyl increases; the excess enthalpies with benzene at 298 K are: nitromethane $(790 \, \text{J mol}^{-1})$, nitroethane $(277 \, \text{J mol}^{-1})$, 1-nitropropane $(63 \, \text{J mol}^{-1})$, 1-nitrobutane $(-27 \, \text{J mol}^{-1})$, 1-nitropentane $(-110 \, \text{J mol}^{-1})$ and 1-nitrohexane $(-123 \, \text{J mol}^{-1})$[213].

Scott, Knobler and their colleagues[214] have recently drawn attention to the fact that the degree of branching of hydrocarbons can influence the excess volumes of polar + hydrocarbon mixtures to quite marked extents although the excess enthalpy is little affected. They have studied both nitroethane and 1-chloropentane mixed with isomeric hexanes and noted similar large variations in the excess volume. These effects may well be closely related to the 'orientation' and 'condensation' factors introduced by Patterson[125–128] and discussed in Section 5.5.

Mixtures containing cyclic or linear ethers mixed with hydrocarbons constitute yet another large class of polar + non-polar systems whose excess properties have been studied in considerable detail, especially over the last decade[215–223]. The pattern is so similar to that of nitriles, ketones and nitroalkyls + hydrocarbons, with large positive excess functions for ether + aliphatic or alicyclic hydrocarbon and small positive or negative values for ether + aromatic hydrocarbon, that no values need be quoted. Mention must be made of the comprehensive studies on mixtures containing a cyclic ether by Andrews and Morcom[215] and by Grolier et al.[216] on systems containing an aliphatic or aromatic ester.

One set of mixtures where a considerable difference exists between the ethers and other non-hydrogen bond-forming polar molecules is that of mixtures with chlorinated hydrocarbons where rather large negative values of both the excess enthalpy and volume are found, indicating the presence of strong forces between the dissimilar molecular species. Some results are listed below where it is seen that the values for mixtures containing carbon tetrachloride are much more negative than comparable results noted earlier in this section. The even larger negative figures for chloroform- and dichloromethane-containing mixtures are probably a consequence of dipole–dipole interactions although both molecules are quite weakly polar.

Mixtures with diethers show similar effects. The excess enthalpy for equimolar mixtures of diethoxymethane $+ CCl_4$ and $+ CHCl_3$ are -459 and $-2434 \, \text{J mol}^{-1}$, respectively[224].

Mixture	T/K	$h^E/J\,mol^{-1}$	$v^E/cm^3\,mol^{-1}$	References
Diethylether + CCl_4	298	-487	-0.71	220, 219
Diethylether + CH_2Cl_2	298	-977	-0.75	220, 222
Diethylether + $CHCl_3$	298	-2646	-1.34	224

The three binary mixtures that can be formed from acetone, ethanenitrile and nitromethane all display small and irregular deviations from ideal behaviour, indeed ethanenitrile + nitromethane obeys Raoult's law within experimental error at 333 K [225]. Such small deviations are to be expected for mixtures composed of similarly-sized molecules with comparable effective polarities.

Binary mixtures containing an amine are usually more difficult to interpret in simple qualitative terms than mixtures containing other types of polar molecule such as those described in the previous sections. This is because primary and secondary amines, as well as exerting the usual intermolecular forces associated with polar molecules, can also form hydrogen bonds either in self-associated complexes or with the appropriate groups on dissimilar molecular species. We begin by considering tertiary amines where this possibility is absent.

Some relevant results are listed below. The principal excess functions for triethylamine + benzene are all positive and not too different from the functions for mixtures with n-alkanes. The enormous negative values for chloroform mixtures are a consequence of the formation of strong inter-component H-bonds for which there is abundant spectroscopic evidence. It is quite reasonable to use the ideal associated mixture theory to interpret the experimental data for such a system where, to a large extent, it is the 1–2 interactions that determine the magnitude of the excess functions. Reasonably concordant results are obtained from analysis of the total excess functions, the functions at infinite dilution, and of the spectroscopic measurements with an equilibrium constant of 3.5 ± 1 and an enthalpy of complex formation of $-15 \pm 2\,kJ\,mol^{-1}$ [227]. This is a reasonable value for the strength of a typical H-bond. Attempts have been made to use more sophisticated 'non-ideal associated mixture' treatments to analyse the experimental data [233] but they offer little advantage.

Data for pyridine + benzene and + chloroform are also listed in the table,

	T/K	$g^E/J\,mol^{-1}$	$h^E/J\,mol^{-1}$	$v^E/cm^3\,mol^{-1}$	Ref.
$(C_2H_5)_3N + C_6H_6$	298	122 (333 K)	329	0.005	226
$(C_2H_5)_3N + CHCl_3$	283	-1070	-4570	-1.93	227
$C_5H_5N + C_6H_6$	298	125	8	-0.201	228, 229
$C_5H_5N + CHCl_3$	308	-700 (323 K)	-1909	-0.132	230
$C_6H_5 \cdot NH_2 + C_6H_6$	298	393	720	-0.24	231, 232

where the same trend is displayed, although the H-bond strength in pyridine + chloroform is obviously weaker.

Data for aniline + benzene are listed at the bottom of the table and are typical of mixtures of primary amines (both aliphatic and aromatic) with hydrocarbons. The excess Gibbs function, enthalpy and entropy are all moderately large and positive and v^E is small and negative. The positive sign and magnitudes are indicative of strong H-bonded self-association in the pure amine and can be contrasted with the much smaller positive values for benzene + triethylamine and + pyridine. The same trend can be seen when h^E for the series of mixtures of benzene with aniline (720 J mol^{-1}), with *N*-ethylaniline (321 J mol^{-1}) and with *N,N*-diethylaniline (167 J mol^{-1}) are compared[232]. A very similar decrease in self-association is detected if methylene groups are interspersed between the NH_2 and the phenyl ring. The h^F's at 303 K are: C_6H_6 + benzylamine (502 J mol^{-1}), + 2-phenyl-ethylamine (343 J mol^{-1}), and + 3-phenylpropylamine (296 J mol^{-1})[232].

Aniline is only partially miscible with aliphatic hydrocarbons at room temperature. The UCSTs vary with chain length and branching, and with the number of rings[234]. This temperature is therefore a useful property for characterizing petroleum fractions. The UCSTs with normal paraffins of increasing chain length pass through a minimum of 342.3 K at n-hexane. The temperatures are sensitive to impurities in either component. The activity coefficients[235] and excess partial molar heats[236] are positive on both sides of the miscibility gap in aniline + n-hexane, and in aniline + cyclohexane, as is to be expected. However, the excess volume is positive in mixtures rich in hexane but negative in those rich in aniline[237].

Large positive excess functions are observed for mixtures containing phenols where there is the possibility of cyclic H-bonded structures being present in the pure liquid. The excess enthalpy for benzene + phenol is 860 J mol^{-1} at 313 K[238].

The two lowest alcohols, methanol and ethanol, differ in their behaviour towards hydrocarbons. Methanol is the more polar and is not completely miscible with the paraffins at room temperature. The UCSTs rise from 287 K with n-pentane to 307 K with n-hexane and 339 K with n-octane, whilst that with cyclohexane is 319 K[239]. Ethanol is miscible with the lower paraffins at room temperature and so are both alcohols with carbon tetrachloride and the aromatic hydrocarbons. Mixtures of methanol with paraffins have received less attention than the other mixtures discussed below, but positive heats of mixing have been measured[239] above and below the UCST in mixtures with n-hexane, n-heptane and cyclohexane. The excess volumes are positive[240].

Methanol + benzene was studied in detail by Scatchard and his colleagues[241], and by many others since[242]. The excess volume is 'cubic', being positive at 298 K only if the mole fraction of methanol is less than 0.5.

Methanol + carbon tetrachloride is, perhaps, the simplest example of a hydrogen-bonded liquid in an inert diluent. *Figure 5.3* shows the results of Scatchard and his colleagues[241]. Later measurements, mainly by Missen

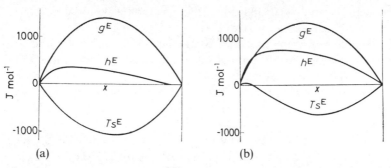

Figure 5.3 The excess functions[241] at 308 K for the mixtures (a) *methanol + carbon tetrachloride, and* (b) *methanol + benzene, as functions of x, the mole fraction of methanol*

and his co-workers[243], have shown that the excess heat falls with decreasing temperature but retains its asymmetry. It has assumed a 'cubic' form by 273 K. n-Propanol behaves similarly.

The excess Gibbs function and enthalpy of ethanol with the lower paraffins are both large and positive and the excess entropy is negative. The excess Gibbs function increases with the length of the alkane chain until a UCST is reached. There have been some recent notable experimental investigations of these mixtures[244].

The phase diagram for ethanol + n-hexadecane is highly skewed with the UCST occurring at 325 K and at an ethanol mole fraction of 0.80[245]. All the principal excess functions, g^E, h^E, c_p^E and v^E, have been measured for this system and the results have been successfully interpreted using an elaborate alcohol-association model in which the alcohol is assumed to form both H-bonded chains of varying length and also cyclic structures. A physical interaction term is included to allow for non-polar interactions.

Stokes and his co-workers[245,246] have utilized the full capability of the new techniques of continuous dilution calorimetry and dilatometry to investigate the marked asymmetry of h^E and of v^E for alcohol + diluent mixtures by making measurements in extremely dilute solution. The experimental functions are unexceptional in mixtures rich with respect to alcohol but, in dilute alcohol solutions, the enthalpy of dilution, h^E/x_{alc}, increases rapidly, and invariably exhibits a point of inflection. A typical experimental curve is shown in *Figure 5.4*(a) where it is seen that, although the limiting value in infinitely dilute alcohol solution is only weakly dependent on temperature, the steepness of the function is extremely temperature-sensitive, implying that effects due to interactions involving more than two alcohol molecules are present even at these extremely low concentrations. The excess volume for the same system is shown, plotted in a similar manner, in *Figure 5.4*(b) and is seen to be comparable in shape except that the limiting intercept, in this instance, increases markedly with increase in temperature. The fact that both $h^E/x(1-x)$ and $v^E/x(1-x)$ are only weakly concentration-dependent at high alcohol concentrations is evident from the

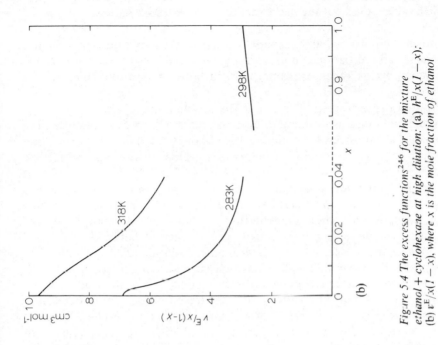

(b)

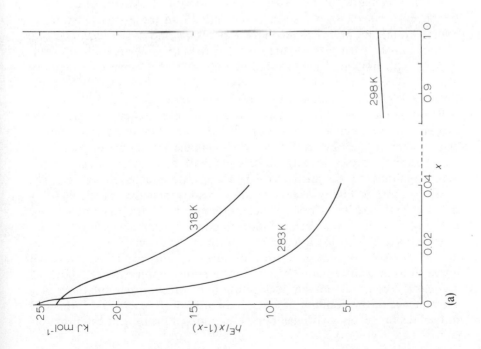

(a)

Figure 5.4 The excess functions[246] for the mixture ethanol + cyclohexane at high dilution: (a) $h^E/x(1-x)$; (b) $v^E/x(1-x)$, where x is the mole fraction of ethanol

right-hand sides of *Figures 5.4*(a) and *5.4*(b). Note that the same vertical scale has been used but the mole fraction scale has been reduced by a factor of five.

Stokes and his research group[246] have studied mixtures of primary, secondary and tertiary alcohols with a great variety of second components, n-alkanes, cycloalkanes, aromatic hydrocarbons, carbon tetrachloride and carbon disulphide and find very little variation in the complicated concentration dependence of h^E and v^E. The absolute magnitudes vary over a considerable range as do the partial molar quantities. A full understanding of these fascinating results is still lacking although the non-ideal association model mentioned earlier goes a long way towards providing a satisfactory interpretation.

Alcohol + aromatic hydrocarbon mixtures have more positive excess heats and entropies and lower excess free energies than those of alcohols with paraffins or carbon tetrachloride[247]. Barker[248] attributes this behaviour to a more favourable energy of interaction between a hydroxy group and the π-electrons of an aromatic molecule than with the less polarizable electrons of a saturated molecule. Such interaction, although not as strong as a conventional hydrogen bond, is less demanding in its geometrical requirements, leads to the breaking of more hydroxy bonds and so, paradoxically, to a slightly larger excess heat and a considerably larger excess entropy. The greater miscibility of all highly polar liquids with aromatic than with aliphatic hydrocarbons has already been noted and is also evidence for such interaction.

Among the many notable experimental papers in this area mention must be made of that of Brown, Fock and Smith[249] on the excess functions of many normal and branched alcohols in n-hexane and in benzene, and that of Ramalho and Ruel[244] on the excess enthalpies of 1-alkanols up to 1-decanol with n-alkanes up to n-tetradecane and also on mixtures of two alcohols, These latter mixtures all have negative values of c_p^E where the opposite was found for all alcohol + n-alkane mixtures.

Benson and his group have made a comprehensive study of cycloalkane + cycloalkanol mixtures[250] with rings containing from five to eight carbon atoms. The results for cycloheptane are shown in *Figure 5.5* and are seen to bear a striking similarity in both concentration dependence and magnitude to the methanol + benzene data illustrated earlier. The excess volume is highly skewed and is weakly negative in alcohol-rich mixtures and positive at other concentrations, reaching a maximum value of $+0.14\,cm^3\,mol^{-1}$ at an alcohol mole fraction of 0.22.

There have been many attempts to interpret the thermodynamic properties of alcohol + diluent mixtures in terms of the probable molecular interactions in such systems[248,249,251]. At a purely empirical level, the small and nearly constant enthalpy of dilution in alcohol-rich mixtures is a consequence of the addition of a small amount of the inert diluent leading to the breaking of few hydrogen bonds, for most of the diluent can probably

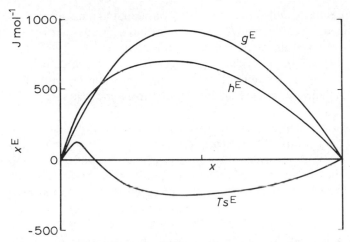

Figure 5.5 The excess functions[250] for the mixture cycloheptanol + cycloheptane at 298 K as functions of x, the mole fraction of cycloheptanol

be accommodated interstitially within the matrix of bonded alcohol molecules. The enthalpy of dilution is much larger for alcohol-weak mixtures where the presence of a large excess of diluent must break up most of the H-bonded clusters, although it is obvious from the recent measurements of Stokes *et al.*[246] that some self-association is still apparent even in extremely dilute solution, $x_{alc} \ll 0.001$.

The excess entropy is positive in alcohol-weak mixtures and this is almost certainly due to the loss of orientational order following the breaking up of the alcohol–alcohol H-bonds, but the entropy quickly becomes positive as the concentration of alcohol and thus the number of hydrogen-bonded, highly-structured clusters increases. The asymmetry of the excess entropy is therefore similar to that of the excess enthalpy and the resultant excess Gibbs function is almost symmetrical in mole fraction.

Thermodynamic properties alone cannot be used to give explicit structural information concerning condensed phases, but infrared spectroscopy has been of great assistance in this respect. Freymann, in 1931[252], was the first to observe that —OH stretching modes in alcohol + diluent mixtures appeared as two absorption bands in the near infrared, one, with a relatively sharp profile, due to free, uncomplexed alcohol molecules and another much broader absorption that resulted from the overlapping of several bands due to H-bonded vibrations in various small associated clusters. The variation in the intensity of these bands with concentration and with temperature can be used to obtain values of the equilibrium constants and enthalpies of formation of the different associated species believed to be present. The initial analyses were extremely speculative as there was no acceptable method of deciding on the nature of the associated clusters, whether they were dimers, trimers or higher polymers and whether they were linear or cyclic in structure. More recently the use of matrix isolation techniques has enabled the spectra of individual associated entities to be obtained unequivocally[253] and these studies have greatly assisted the interpretation of the results on the neat liquid mixtures[254]. A full correlation of the thermodynamic data with the detailed information now available from infrared measurements is still outstanding although Prigogine and his colleagues made an initial attempt some years ago with limited success[255].

It will be shown in Chapter 8 that statistical mechanics has made recent significant advances towards the construction of a rigorous theory of liquid mixtures the components of which have anisotropic intermolecular potentials either by virtue of their non-spherical shape or by their possession of a permanent electrostatic multipole. The range of applicability of such theories is restricted at present to relatively simple molecules containing few atoms, and strongly energetic and directional interactions such as hydrogen-bonding are excluded from consideration. Over the years more approximate theories have been developed that are, in principle, applicable to mixtures containing all types of molecular liquid. Most of the theories adopt a group contribution approach and a necessarily brief account of their development and use is now given.

Langmuir[256], over a half a century ago, was the first to suggest the estimation of the thermodynamic properties of mixtures by considering the interactions between the different functional groups of which the individual molecules were composed. Proper development of the theory had to await the advent of electronic computers, for quite considerable amounts of both computation and data storage are required for the successful operation of any useful scheme.

Many semi-empirical group-contribution schemes have been developed over the past decade. Two of the most useful are UNIFAC (Universal quasi-chemical Functional group Activity Coefficients) due to Prausnitz, Abrams, Fredenslund and others[257,258], and ASOG (Analytical Solution of Groups) introduced by Wilson, Deal, Derr, Tochigi and Kopina[259,260]. Both systems have as their ultimate objective the prediction of the activity coefficients and hence the excess Gibbs function and, if present, the region of partial liquid miscibility, for all possible multicomponent liquid mixtures.

The activity coefficient is divided arbitrarily into two parts, the first a combinatorial term due to differences in the sizes and shapes of the component molecules and the second, residual term, related directly to energy differences

$$\ln \gamma_i = \ln \gamma_i^{comb} + \ln \gamma_i^{res}$$

Each molecule is divided into its component functional groups, $-CH_3$, $-CH_2-$, $-OH$, $-CHO$, etc., and each pair of groups is assigned a value of the group interaction parameter. These are computed from experimental vapour–liquid equilibria data on a large number of carefully selected 'key' binary mixtures. Each group is also assigned a volume and a surface area.

The ASOG schemes used the athermal Flory–Huggins equation to compute $\ln \gamma_i^{comb}$ and $\ln \gamma_i^{res}$ is represented by the Wilson equation (5.15). The group interaction parameters are simple functions of temperature. The UNIFAC procedure is slightly more complex in that both group size and surface area are required for the calculation of $\ln \gamma_i^{comb}$. The residual activity coefficients are represented by part of the UNIQUAC equation[257] which is based on Guggenheim's quasi-chemical approximation for non-ideal mixtures but temperature-independent group interaction parameters are used.

Computer programs are available for the operation of both schemes and it is claimed that the thermodynamic properties of 70 per cent of all known binary and multicomponent mixtures can be predicted with acceptable accuracy[258]. The principal users (and originators) of such methods are chemical engineers who require to be able to predict the thermodynamic properties of a wide variety of highly non-ideal mixtures for many purposes, typically the design of suitable distillation columns.

More detailed theories have also been developed that attempt to correlate both g^E and h^E and are based on the lattice theory of liquid mixtures first developed by Barker for mixtures containing an alcohol[248]. These theories use a more elaborate selection of functional groups to try and account for the subtle intramolecular effects caused by the interaction of pairs of functional groups situated close to one another in the same molecule. A comprehensive review by Kehiaian is available[261].

There is no large body of systematic work on mixtures of two polar components, except for that on aqueous solutions which is discussed in the next section, and the interpretation of such results as there are is difficult. In a few cases it is clear that there is a strong and specific interaction between the unlike molecules which is not present between those of either of the pure components.

The classical example of this behaviour is the system chloroform + acetone[262]. Here, the excess free energy, heat and entropy are all negative and also the excess volume except for mixtures weak in chloroform. The specific interaction is here a hydrogen bond between the hydrogen atom of a chloroform molecule and the carbonyl group of an acetone molecule. Hydrogen bonding cannot occur in either of the pure components.

Chloroform can interact similarly with the oxygen atom of ethers[263] and the π-electrons of benzene[264]. A weak hydrogen bond between chloroform and aromatic and olefinic systems has been detected by the 'chemical shift' of the proton magnetic resonance frequency[265]. Such a bond is of a type between the conventional localized hydrogen bond and the formation of an electron-transfer complex. There are, of course, several examples of mixtures where the negative contribution to the excess functions resulting from a very strong inter-component hydrogen bond is sufficient to outweigh the weaker positive contributions caused by hydrogen-bonded self-association in both pure components. A good example is methanol + piperidine where all the principal excess functions, including v^E, are large and negative[266].

It is important to know the relative strengths of the hydrogen bonds formed from a hydroxy group in an alcohol or in water towards proton acceptors such as ethers, ketones, nitriles, amines and other alcohols. Gordy and Stanford[267] classified such bonds by studying the perturbation of the OD fundamental frequency in solutions of CH_3OD in a large number of polar solvents. They took the frequency of the band in dilute solution in benzene as that of the unperturbed monomer and found the shifts to lower frequencies to increase in order: nitriles, ketones, ethers, amines.

Thus the nitrile and carbonyl groups are not good proton acceptors in spite of their high dipole moments. Ethers and amines—primary, secondary, tertiary and aromatic—are better acceptors because the electrons around the oxygen and nitrogen atoms have less *s* and more *p* character than those in ketones and nitriles. This order of increasing affinity for protons is found also from a study of the thermodynamic properties[268]. The excess heat of isopropanol in equimolar mixtures of the following liquids at 298 K is[269]: propionitrile $(+1720 \, J \, mol^{-1})$, acetone $(+1710 \, J \, mol^{-1})$, isopropanol $(0 \, J \, mol^{-1})$, isopropylamine $(-2500 \, J \, mol^{-1})$. This sequence confirms the order chosen by Gordy and Stanford from their spectroscopic results and is of considerable use in interpreting the thermodynamic results for aqueous systems. Hussein and Millen[270] have used both spectroscopic and thermo-dynamic techniques to investigate the strengths of the hydrogen bonds formed between methanol and a series of aliphatic amines. They find that the bond strength increases in the order $NH_3 < MeNH_2 < EtNH_2 < Me_2NH < Me_3N \simeq Et_2NH < Et_3N$, a logical sequence in terms of electron-donating ability.

5.9 Aqueous mixtures

No one has yet proposed a quantitative theory of aqueous solutions of non-electrolytes, and such solutions will probably be the last to be understood fully. Nevertheless, they may be usefully classified by the strength and number of the proton-accepting groups in a molecule of the second component. The mixtures are in many ways analogous to those of highly polar substances in an inert solvent, as, for example, of methanol in carbon tetrachloride. But in aqueous solutions the organic liquid, although nec-essarily polar, plays the role of the inert diluent, as water is one of the most strongly and most regularly hydrogen-bonded liquids.

The addition to water of any molecules containing inert groups such as CH_3 reduces the total number of hydrogen bonds, even if the inert group is itself attached to a highly polar group such as OH. Extremely dilute solutions in water are exceptions where the presence of so-called 'hy-drophobic interactions' between molecules of slightly soluble species, such as the lower aliphatic and aromatic hydrocarbons, is believed to be due to an enhancement of the degree of hydrogen-bonding between the solvent water molecules above that in pure water[271]. The presence of hydrophobic interactions is deduced from the experimentally-observed decrease in the solubility of these simple non-polar substances in aqueous solution as the temperature is raised[272] and by the direct measurement of exothermal enthalpies of solution by Arnett[273]. Such interactions play a major role in determining the configurations adopted by proteins and other biologically-active molecules in solution and in the action of long-chain surfactants in forming micellar structures[274].

Aqueous solutions are discussed in this section in the order of the increasing strength of the second component as a proton acceptor. This order is that established above, namely, nitriles, ketones, alcohols and amines. It is interesting to note that, in contrast to most other classes of binary mixture, there have been few recent measurements of the excess functions of aqueous mixtures over the last decade. This apparent neglect is probably due to the lack of an adequate rigorous theory of such systems to provide a quantitative interpretation of the experimental data. Wisely, experimental work has been concentrated on the determination of the phase equilibria of such mixtures at high temperatures and pressures and is considered in the next chapter.

The equimolar excess free energy of the system water + ethanenitrile is about $+1100\,\mathrm{J\,mol^{-1}}$ at 308 K, and the excess enthalpy at 298 K has a maximum of $+840\,\mathrm{J\,mol^{-1}}$ at a mole fraction of water of 0.38 and apparently becomes slightly negative at mole fractions greater than 0.96[275]. The excess entropy is, therefore, everywhere negative. Qualitatively the system resembles methanol + carbon tetrachloride (*Figure 5.3(a)*) with water taking the part of methanol. The system forms a positive azeotrope but is miscible at all temperatures.

Acetone[276] behaves very similarly to ethanenitrile but forms an azeotrope only above 358 K.

Methylethylketone, acetylacetone and propanenitrile are immiscible with water and have upper critical solution temperatures[234] above 373 K.

These two classes of proton acceptors, nitriles and ketones, have their excess properties in the following order: $g^E > h^E > 0 > Ts^E$ and have a UCST if g^E is sufficiently large. Such substances rarely, if ever, show LCST in equilibrium with the saturated vapour, but methyl ethyl ketone has one at high pressures and would clearly have one at the vapour pressure if the system did not solidify first (*see* Chapter 6). Nevertheless, the typical phase behaviour of these systems is the formation of a UCST only.

If the solute is capable of accepting a small number (per molecule) of reasonably strong hydrogen bonds from water, then there is added at room temperature to an excess heat of the type described a negative and symmetrical term. The sign and symmetry follow from the fact that the term arises from a direct interaction between molecules of opposite species. The total excess heat is now generally positive in mixtures rich, and negative in those weak, in solute. The effect on the excess entropy is less spectacular but must be a decrease. 1,4-Dioxan[277] is a good example of such a solute

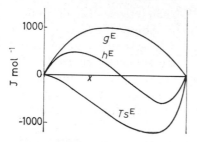

Figure 5.6 The excess functions[277] for water + 1,4-dioxan at 298 K as a function of x, the mole fraction of water

(*Figure 5.6*) and many other di- and tri-ethers, both cyclic and linear, behave similarly[278] as do ethanol (above room temperature, see below), n-propanol and isopropanol[279].

Such systems have negative excess volumes, caused mainly by the solute–water bond, and positive excess heat capacities which arise from the increased orientational freedom of both components as the temperature is raised. The solute–water bond is strong enough with many alcohols and ethers for them to be miscible at all temperatures, although some come close to the formation of a closed solubility loop. 1,4-Dioxan is one of these, for it has been shown[277] that g^E is too small by only 10 per cent over a range of at least 125 K for the formation of a closed loop. n-Butanol[280] shows only a UCST of 398 K, isobutanol[281] has a UCST of 406 K and is tending towards a LCST at 273 K; secondary butanol forms a closed loop at high pressures (*see* Chapter 6), and tertiary butanol is completely miscible at all temperatures. Diethyl ether + water is clearly tending towards a LCST as the mutual solubilities decrease with increasing temperatures[280]. Tetrahydrofuran + water is one mixture where closed loop behaviour has been established with a LCST at 345 K, a few degrees above the normal boiling point of the cyclic ether, and a UCST some 65 K higher. The equimolar g^E is already 1640 J mol^{-1} at 343 K and v^E is strongly negative. The excess enthalpy at 298 K is negative in aqueous-rich mixtures and positive over a small concentration region at high ether mole fractions[278,282]. The positive c_p^E means that h^E will probably be everywhere positive at temperatures approaching that of the UCST.

Hydrogen bonds are weakened at higher temperatures and the resultant changes in the excess functions can be quite complicated in their concentration and temperature dependence, due to the different types of interaction that are present. Mixtures of water + ethanol have been studied by Pemberton and Larkin at the National Physical Laboratory, Teddington, and their results at three temperatures are shown in *Figure 5.7*[283]. The excess Gibbs function, enthalpy and entropy all become more positive at the higher temperatures. The complex behaviour of h^E is particularly apparent and, in addition, a double maximum develops in this function at temperatures around 323 K. The excess heat capacity is positive

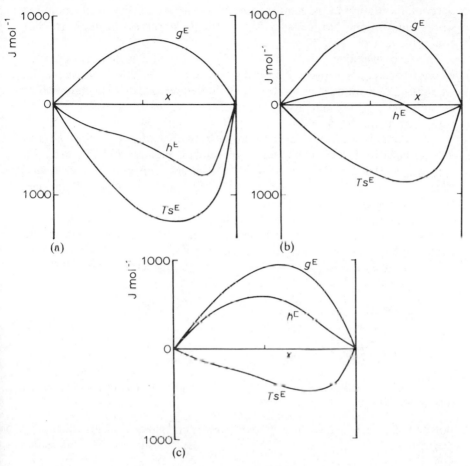

Figure 5.7 *The excess functions[284] for water + ethanol at three temperatures as a function of x, the mole fraction of water:* (a) 298 K; (b) 343 K; (c) 383 K

at all the experimental temperatures and is highly skewed towards water-rich mixtures at low temperatures but becomes more symmetrical in its composition-dependence as the temperature is increased.

The excess volume of water + ethanol has been studied recently by Marsh and Richards[284] over a wide temperature range using a continuous-dilution dilatometer and some unusual effects were detected at low alcohol concentrations. Although the partial molar excess volume of ethanol at infinite dilution in water is extremely temperature sensitive and becomes more negative with increase in temperature, this sensitivity is greatly reduced at slightly higher mole fractions until, at $x_{alc} = 0.038$, the partial molar excess volume becomes independent of temperature. The excess volume itself exhibits similar behaviour at $x_{alc} = 0.082$ with a constant value of $-0.42\,\text{cm}^3\,\text{mol}^{-1}$. Marsh and Richards interpret these effects in terms of a particular variation of the temperature of maximum density of the mixtures with composition. No unusual behaviour is observed at high alcohol concentrations.

The large negative values of both h^E and Ts^E at low temperatures for alcohol + water mixtures are indicative of a considerable increase in the strength and/or number of solute–water interactions. Other substances that exhibit similar behaviour in aqueous solution at or below room temperature are methanol[285], glycol ethers[286], polymethylene glycols[287], glycerol[288], aliphatic amines[289] and pyridine bases[290]. Although both h^E and Ts^E are negative for all such systems, g^E remains positive and this situation can lead to the exhibition of a lower critical solution point; indeed, such points are almost entirely confined to the list above. As c_p^E is still positive, the systems often show a UCST as well.

Finally, there are mixtures in which the solute–water bonds are so strong or so numerous that h^E is more negative than Ts^E. These mixtures are thermodynamically most stable, as g^E is negative, and never separate into two liquid phases. Substances in this category are generally polyalcohols or

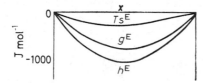

Figure 5.8 The excess functions[291] for water + hydrogen peroxide at 298 K as a function of x, the mole fraction of water

polyamines, such as hydrogen peroxide[291] (*Figure 5.8*), hydrazine[292], diethylamine[293], ethylene diamine[294], glucose[295], sucrose[295], and agar–agar[296].

This classification of aqueous solutions of non-electrolytes by the strength of the hydrogen bonds from the hydroxy group in water is not very precise but does bring some order into this confused field. It is summarized in *Table 5.5*.

Table 5.5 Aqueous solutions of non-electrolytes in order of increasing strength and number of hydrogen bonds

H^E	G^E	TS^E	V^E	$(C_p)^E$	Examples	Typical phase behaviour
+	+	−	+	− ?	nitriles, ketones	UCST
(+, −)	+	−	−	+	ethers, alcohols	UCST but tending towards closed loop
(+, −)	+	−	−	+	polyethers, amines	Closed loop
−	−	−	−	+	polyalcohols, polyamines	Completely miscible

5.10 Mixtures of quantum liquids

Quantum liquids are outside the scope of this book, but it would be pedantic to omit all discussion of some remarkable phase equilibria and UCSTs which have been found in mixtures of these liquids.

At high temperatures, isotopic mixtures of heavy molecules are perhaps the best examples we have of ideal mixtures, and remain so at temperatures down to the melting curves. h^E and v^E for benzene + perdeuterobenzene are only $0.65 \, \text{J mol}^{-1}$ and $0.0004 \, \text{cm}^3 \, \text{mol}^{-1}$, respectively, at 298 K and the values for cyclohexane + perdeuterocyclohexane are some four times larger[297]. However, much larger relative deviations are found for the system $^3\text{He} + {}^4\text{He}$ which does not solidify at atmospheric pressure. Some years ago, Prigogine and his colleagues[298] predicted that this mixture would have a small positive excess free energy and hence, since neither component is solid at zero temperature, that there must be a UCST. They estimated this at 0.7 K. Shortly afterwards the critical point was observed by Walters and Fairbank[299] and by Zinoveva and Peshkov[300] at 0.83 K. The phase-boundary curve has its maximum close to the equimolar composition and coincides with the λ-point curve (as a function of composition) for mole fractions of ^3He above 0.7. The junction is not a simple UCST, but a tricritical point, and has been discussed in Section 4.11. The complete phase diagram is one of the most complicated known for a binary system, since we have not only liquid–liquid equilibrium but also azeotropy[301], a λ-surface[302] which is a function of p, T and x, a tricritical point, and freezing of both components at higher pressures to more than one solid phase[302].

Perhaps the most remarkable feature of the system is that at zero temperature the compositions of the phases do not approach those of the pure components[303]. The limiting mole fractions of ^3He are 1.000 and $0.0637 + 0.0005$. At first sight, this result appears to be in conflict with the third law of thermodynamics, but theoretical analysis shows that this is not so[304]. The excess functions have been measured and, at 0.9 K, just above the tricritical point, g^E and h^E are 3.2 and $2.5 \, \text{J mol}^{-1}$, respectively[305]. The fact that the two isotopes mix endothermally is the basis of the helium dilution refrigerator. This device is now available commercially and enables temperatures of a few mK to be obtained and maintained routinely.

Phase separation in isotopic mixtures is caused primarily by differences in mass and not in quantal statistics[298]. It is, therefore, not surprising that immiscibility is found also in liquid mixtures of neon with the different forms of hydrogen. The following thermodynamic properties have been reported:

Mixture	T/K	$g^E/\text{J mol}^{-1}$	$h^E/\text{J mol}^{-1}$	$UCST/\text{K}$	References
$\text{Ne} + n\text{-}\text{H}_2$	30	121	208	28.93	306
$\text{Ne} + n\text{-}\text{D}_2$	27	104	165	25.71	307

References

1 KEHIAIAN, H., *Thermodynamics of Organic Mixtures, MTP International Review of Science*, Vol. 10: *Thermochemistry and Thermodynamics*, Butterworths, London (1972); STOKES, R. H., and MARSH, K. N., *A. Rev. phys. Chem.*, **23**, 65 (1972); SWINTON, F. L., *A. Rev. phys. Chem.*, **27**, 153 (1976)

2 HICKS, C. P., in *Chemical Thermodynamics, Specialist Periodical Reports*, Ed. McGlashan, M. L., Vol. 2, 275, Chemical Society, London (1978)

3 WISNIAK, J., and TAMIR, A., *Mixing and Excess Thermodynamic Properties*, Elsevier, Amsterdam (1978)

4 WICHTERLE, I., LINEK, J., and HÁLA, E., *Vapour–Liquid Equilibrium Data Bibliography*, Elsevier, Amsterdam (1973); *Supplement 1* (1976); *Supplement 2* (1979)

5 HIZA, M. J., KIDNAY, A. J., and MILLER, R. C., *Equilibrium Properties of Fluid Mixtures: A Bibliography of Data on Fluids of Cryogenic Interest*, IFI/Plenum, New York (1975)

6 SATO, K., HIRATA, M., and YOSHIMURA, S., *Physico-Chemical Properties for Chemical Engineering*, Maruzen, Tokyo, Vol. 1 (1977); Vol. 2 (1978); Vol. 3 (1979)

7 KEHIAIAN, H., (Ed.) *International DATA Series: Selected Data on Mixtures*, Thermodynamics Research Center, College Station, Texas (1973)

8 *Dechema Chemistry Data Series*, Vol. 1 (1977), GMEHLING, J., ONKEN, U., and ANLT, W., *Vapour–Liquid Equilibrium Data Collection*; Vol. 6 (1979), DÖRING, R., KNAPP, H., OELLRICH, L. R., PLÖCKER, U. J., and PRAUSNITZ, J. M., *Vapour–Liquid Equilibria for Mixtures of Low Boiling Substances*, Deutsche Gesellschaft für Chemisches Apparatewesen eV, Frankfurt

9 BOUBLIK, T., and BENSON, G. C., *Can. J. Chem.*, **47**, 539 (1969); MARIPURI, V. O., and RATCLIFF, G. A., *J. chem. Engng Data*, **17**, 366 (1972)

10 McGLASHAN, M. L., and WILLIAMSON, A. G., *Trans. Faraday Soc.*, **57**, 588 (1961)

11 GAW, W. J., and SWINTON, F. L., *Trans. Faraday Soc.*, **64**, 637 (1968)

12 GIBBS, R. E., and VAN NESS, H. C., *Ind. & Eng. Chem., Fundam.*, **11**, 410 (1972); TOMLINS, R. P., and MARSH, K. N., *J. chem. Thermodyn.*, **8**, 1185 (1976)

13 CHRISTIAN, S. D., NEPARKO, E., and AFFSPRUNG, H. E., *J. phys. Chem., Ithaca*, **64**, 442 (1960)

14 BREWSTER, E. R., and McGLASHAN, M. L., *J. chem. Soc., Faraday Trans. 1*, **69**, 2046 (1973); *J. chem. Thermodyn.*, **9**, 1095 (1977); DIXON, D. T., and HEWITT, F. A., *J. chem. Soc., Faraday Trans. 1*, **75**, 1940 (1979)

15 DIXON, D. T., and McGLASHAN, M. L., *Nature, Lond.*, **206**, 710 (1965)

16 BARKER, J. A., *Aust. J. Chem.*, **6**, 207 (1953)

17 GUGGENHEIM, E. A., *Trans. Faraday Soc.*, **33**, 151 (1937); SCATCHARD, G., *Chem. Rev.*, **44**, 7 (1949); REDLICH, O., and KISTER, A. T., *A.I.Ch.E.Jl.*, **40**, 341 (1948)

18 HILDEBRAND, J. H., and SCOTT, R. L., *Solubility of Nonelectrolytes*, 3rd Edn, Reinhold, New York (1950); *Regular Solutions*, Prentice-Hall, Englewood Cliffs, NJ (1962); HILDEBRAND, J. H., PRAUSNITZ, J. M., and SCOTT, R. L., *Regular and Related Solutions*, Van Nostrand Reinhold, New York (1970)

19 MORRIS, J. W., MULVEY, P. J., ABBOTT, M. M., and VAN NESS, H. C., *J. chem. Engng Data*, **20**, 403 (1975); SHATAS, J. P., ABBOTT, M. M., and VAN NESS, H. C., *ibid.*, 406; DIELSI, D. P., PATEL, R. B., ABBOTT, M. M., and VAN NESS, H. C., *J. chem. Engng Data*, **23**, 242 (1978); WILSON, S. R., PATEL, R. B., ABBOTT, M. M., and VAN NESS, H. C., *J. chem. Engng Data*, **24**, 130, 133 (1979); ANDERSON, D., HILL, R. J., and SWINTON, F. L., *J. chem. Thermodyn.*, **12**, 483 (1980)

20 VAN NESS, H. C., BYER, S. M., and GIBBS, R. E., *A.I.Ch.E.Jl.*, **19**, 238 (1973)

21 MYERS, D. M., and SCOTT, R. L., *Ind. Engng Chem.*, **55**, 43 (1963); MARSH, K. N., *J. chem. Thermodyn.*, **9**, 719 (1977)

22 CHRISTIANSEN, L. J., and FREDENSLUND, A., *A.I.Ch.E.Jl.*, **21**, 49 (1975); KLAUS, R. L., and VAN NESS, H. C., *Chem, Engng Prog. Symp. Ser.*, **63**(81), 88 (1967)

23 LARKIN, J. A., and PEMBERTON, R. C., *NPL Report*, Chem 43, National Physical Laboratory, Teddington (1976)

24 WILSON, G. M., *J. Am. chem. Soc.*, **86**, 127 (1964)

25 HEIL, J. F., and PRAUSNITZ, J. M., *A.I.Ch.E.Jl.*, **12**, 678 (1966)

26 RENON, H., and PRAUSNITZ, J. M., *A.I.Ch.E.Jl.*, **14**, 135 (1968)

27 ABRAMS, D. S., and PRAUSNITZ, J. M., *A.I.Ch.E.Jl.*, **21**, 116 (1975)

28 WICHTERLE, I., and HÁLA, E., *Ind. & Eng. Chem., Fundam.*, **2**, 155 (1963); ARNIKAR, H. J., RAO, T. S., and BODHE, A. A., *J. chem. Educ.*, **47**, 826 (1970); RAO, T. S., BODHE, A. A., and GANDHE, B. R., *J. Chromat.*, **59**, 151 (1971); HALL, D. J., MASH, C. J., and PEMBERTON, R. C., *NPL Report*, Chem, 95, National Physical Laboratory, Teddington (1979); PEMBERTON, R. C., and MASH, C. J., *J. chem. Thermodyn.*, **9**, 867 (1979)
29 EVERETT, D. H., and STODDART, C. T. H., *Trans. Faraday Soc.*, **57**, 746 (1961), EVERETT, D. H., *Trans. Faraday Soc.*, **61**, 1637 (1965); CRUICKSHANK, A. J. B., GAINEY, B. W., and YOUNG, C. L., *Trans. Faraday Soc.*, **64**, 337 (1968); LUNGER, S. H., and PURNELL, J. H., *J. phys. Chem., Ithaca*, **67**, 263 (1963); CONDOR, J. R., and PURNELL, J. H., *Trans. Faraday Soc.*, **64**, 1505 (1968)
30 LETCHER, T. M., in *Chemical Thermodynamics, Specialist Periodical Reports*, Ed. McGlashan, M. L., Vol. 2, 46, Chemical Society, London (1978)
31 COUMOU, D. J., MACKOR, E. L., and HIJMANS, J., *Trans. Faraday Soc.*, **60**, 1539 (1964)
32 BROWN, N. M. D., MAGUIRE, J. F., and SWINTON, F. L., *J. chem. Thermodyn.*, **9**, 855 (1978)
33 MYERS, R. S., and CLEVER, H. L., *J. chem. Thermodyn.*, **2**, 53 (1970); DELMAS, G., and PURVES, P., *J. chem. Soc. Faraday Trans. 2*, **73**, 1838 (1977)
34 HANKS, R. W., GUPTA, A. C., and CHRISTENSEN, J. J., *Ind. & Eng. Chem., Fundam.*, **10**, 504 (1971); HANKS, R. W., TAN, R. L., and CHRISTENSEN, J. J., *Thermochim. Acta*, **23**, 41 (1978); VONKA, P., NOVAK, J. P., SUSKA, J., and PICK, J., *Chem. Eng. Commun.*, **2**, 51 (1975)
35 HIROBE, H., *J. Fac. Sci. Tokyo Univ.*, **1**, 155 (1925)
36 LARKIN, J. A., and McGLASHAN, M. L., *J. chem. Soc.*, 3425 (1961); ARMITAGE, D. A., and MORCOM, K. W., *Trans. Faraday Soc.*, **65**, 688 (1969); HILL, R. J., and SWINTON, F. L., *J. chem. Thermodyn.*, **11**, 383, 489 (1980)
37 LAMBERT, M., and SIMON, M., *Physica*, **28**, 1191 (1962); JEENER, J., *Rev. Scient. Instrum.*, **28**, 263 (1957)
38 LEWIS, K. L., and STAVELEY, L. A. K., *J. chem. Thermodyn.*, **7**, 855 (1975)
39 MRAZEK, R. V., and VAN NESS, H. C., *A I Ch E.Jl.*, **7**, 190 (1961), SAVINI, C. G., WINTERHALTER, D. R., KOVACH, L. H., and VAN NESS, H. C., *J.chem. Engng Data*, **11**, 40 (1966)
40 STOKES, R. H., MARSH, K. N., and TOMLINS, R. P., *J. chem. Thermodyn.*, **1**, 211 (1969); EWING, M. B., MARSH, K. N., STOKES, R. H., and TUXFORD, C. W., *J. chem. Thermodyn.*, **2**, 751 (1970)
41 WINTERHALTER, D. R., and VAN NESS, H. C., *J. Chem. Engng Data*, **11**, 189 (1969); MURAKAMI, S., and BENSON, G. C., *J. chem. Thermodyn.*, **1**, 559 (1969); COSTIGAN, M. J., HODGES, L. J., MARSH, K. N., STOKES, R. H., and TUXFORD, C. W., *Aust. J. Chem.*, **33**, 2103 (1980)
42 McGLASHAN, M. L., and STOECKLI, H. F., *J. chem. Thermodyn.*, **1**, 589 (1969); GOODWIN, S. R., and NEWSHAM, D. M. T., *J. chem. Thermodyn.*, **3**, 325 (1971); HARSTED, B. S., and THOMSEN, E. S., *J. chem. Thermodyn.*, **6**, 549 (1974); ELLIOT, K., and WORMALD, C. J., *J. chem. Thermodyn.*, **8**, 881 (1976)
43 PICKER, P., JOLICOURR, J., and DESNOYERS, J. E., *J. chem. Thermodyn.*, **1**, 469 (1969)
44 FORTIER, J.-L., and BENSON, G. C., *J. chem. Thermodyn.*, **8**, 411 (1976)
45 BATTINO, R., *Chem. Rev.*, **71**, 5 (1971)
46 TERRY, M. J., LYNCH, J. T., BUNCLARK, M., MANSELL, K. R., and STAVELEY, L. A. K., *J. chem. Thermodyn.*, **1**, 413 (1969); MASSENGILL, D. R., and MILLER, R. C., *J. chem. Thermodyn.*, **5**, 207 (1973); CALADO, J. C. G., and SOARES, V. A. M., *J. chem. Thermodyn.*, **9**, 911 (1977)
47 KIYOHARA, O., and BENSON, G. C., *Can. J. Chem.*, **51**, 2489 (1973); GROLIER, J.-P. E., BENSON, G. C., and PICKER, P., *A.I.Ch.E.Jl.*, **20**, 243 (1975); RADOJKOVIĆ, N., TASIĆ, A., DJORDJEVIĆ, B., and GROZDANIĆ, D., *J. chem. Thermodyn.*, **8**, 111 (1976)
48 HALES, J. L., *J. Phys. (Fr.)*, **E3**, 855 (1970); WEEKS, I. A., and BENSON, G. C., *J. chem. Thermodyn.*, **5**, 107 (1973); HAYNES, W. M., HIZA, M. J., and FREDERICK, N. V., *Rev. Scient. Instrum.*, **47**, 1237 (1976); **48**, 39 (1977); HIZA, M. J., and HAYNES, W. M., *Adv. cryogen. Engng*, **23**, 594 (1978)
49 DUNCAN, W. A., SHERIDAN, J. P., and SWINTON, F. L., *Trans. Faraday Soc.*, **62**, 1090 (1966); STOOKEY, D. J., SALLAK, H. M., and SMITH, B. D., *J. chem. Thermodyn.*, **5**, 741 (1973)

50 BEATH, L. A., O'NEILL, S. P., and WILLIAMSON, A. G., *J. chem. Thermodyn.*, **1**, 293 (1969); PFLUG, H. D., and BENSON, G. C., *Can. J. Chem.*, **46**, 287 (1968); STOKES, R. H., LEVIEN, B. J., and MARSH, K. N., *J. chem. Thermodyn.*, **2**, 43 (1970); BOTTOMLEY, G. A., and SCOTT, R. L., *J. chem. Thermodyn.*, **6**, 973 (1974); KUMARAN, M. K., and McGLASHAN, M. L., *J. chem. Thermodyn.*, **9**, 259 (1977);
51 PAN, W. P., MADY, M. H., and MILLER, R. C., *A.I.Ch.E.Jl.*, **21**, 283 (1975); SINGH, S. P., and MILLER, R. C., *J. chem. Thermodyn.*, **10**, 747 (1978); **11**, 395 (1979)
52 DUNCAN, A. G., DAVIES, R. H., BYRNE, M. A., and STAVELEY, L. A. K., *Nature, Lond.*, **209**, 1236 (1966)
53 DAVIES, R. H., DUNCAN, A. G., SAVILLE, G., and STAVELEY, L. A. K., *Trans. Faraday Soc.*, **63**, 855 (1967)
54 LEWIS, K. L., LOBO, L. Q., and STAVELEY, L. A. K., *J. chem. Thermodyn.*, **10**, 351 (1978)
55 CHUI, C. H., and CANFIELD, F. B., *Trans. Faraday Soc.*, **67**, 2933 (1971)
56 SCHMIDT, H. *Z. phys. Chem. Frankf. Ausg.*, **20**, 363 (1959); **24**, 265 (1960); WILHELM, G., and SCHNEIDER, G., *Z. phy. Chem. Frankf. Ausg.*, **32**, 62 (1962)
57 SCHOUTEN, J. A., DEERENBERG, A., and TRAPPENIERS, N. J., *Physica*, **81A**, 151 (1975)
58 CALIGARIS, R. E., and HENDERSON, D., *Molec. Phys.*, **30**, 1853 (1975)
59 CALADO, J. C. G., and STAVELEY, L. A. K., *Trans. Faraday Soc.*, **67**, 289 (1971)
60 YUNKER, W. H., and HALSEY, G. D., *J. phys. Chem., Ithaca*, **64**, 484 (1960)
61 MILLER, R. C., KIDNAY, A. J., and HIZA, M. J., *A.I.Ch.E. Jl.*, **19**, 145 (1973)
62 POOL, R. A. H., SAVILLE, G., HERINGTON, T. M., SHIELDS, B. D. C., and STAVELEY, L. A. K., *Trans. Faraday Soc.*, **58**, 1692 (1962)
63 SPROW, F. B., and PRAUSNITZ, J. M., *A.I.Ch.E.Jl.*, **12**, 780 (1966)
64 MASSENGILL, D. R., and MILLER, R. C., *J. chem. Thermodyn.*, **5**, 207 (1973)
65 BURN, I., and DIN. F., *Trans. Faraday Soc.*, **58**, 1341 (1962); NARINSKII, G. B., *Z. phys. Chem. Frankf. Ausg.*, **34**, 1778 (1960); *Russ. J. phys. Chem.*, **40**, 1093 (1966)
66 LEWIS, K. L., SAVILLE, G., and STAVELEY, L. A. K., *J. chem. Thermodyn.*, **7**, 389 (1975)
67 KNOBLER, C. M., VAN HEIJNINGEN, R. J. J., and BEENAKKER, J. M. M., *Physica*, **27**, 309 (1961)
68 DUNCAN, A. G., and STAVELEY, L. A. K., *Trans. Faraday Soc.*, **62**, 548 (1966)
69 KIDNAY, A. J., LEWIS, K. L., CALADO, J. C. G., and STAVELEY, L. A. K., *J. chem. Thermodyn.*, **7**, 847 (1975)
70 MATHOT, V., *Nuovo Cim.*, **9**, Supp. 1, 356 (1958); FUKS, S., and BELLEMANS, A., *Bull. Soc. chim. Belg.*, **76**, 290 (1967); CALADO, J. C. G., and STAVELEY, L. A. K., *J. chem. Phys.*, **56**, 4718 (1972); GIRAVELLE, D., and LU, B. C-Y., *Can. J. chem. Engng*, **49**, 144 (1971)
71 MOSEDALE, S. E., and WORMALD, C. J., *J. chem. Thermodyn.*, **9**, 483 (1977)
72 LIU, Y.-P., and MILLER, R. C., *J. chem. Thermodyn.*, **4**, 85 (1972)
73 ECKERT, C. A., and PRAUSNITZ, J. M., *A.I.Ch.E.Jl.*, **11**, 886 (1965)
74 CALADO, J. C. G., NUNES DA PONTE, M., SOARES, V. A. M., and STAVELEY, L. A. K., *J. chem. Thermodyn.*, **10**, 35 (1978)
75 CALADO, J. C. G., and STAVELEY, L. A. K., *Fluid Phase Equilib.*, **3**, 153 (1979)
76 CALADO, J. C. G., unpublished measurements
77 CALADO, J. C. G., and SOARES, V. A. M., *J. chem. Thermodyn.*, **9**, 911 (1977)
78 CALADO, J. C. G., KOZDON, A. F., MORRIS, P. J., DE PONTE, M. N., STAVELEY, L. A. K., and WOOLF, L. A., *J. Chem. Soc., Faraday Trans. 1*, **71**, 1372 (1975); LOBO, L. Q., STAVELEY, L. A. K., CLANCY, P., and GUBBINS, K. E., *J. chem. Soc., Faraday Trans. 1*, **76**, 174 (1980)
79 CALADO, J. C. G., GRAY, C. G., GUBBINS, K. E., PALAVRA, A. M. F., SOARES, V. A. M., STAVELEY, L. A. K., and TWU, C.-H., *J. chem. Soc. Faraday Trans. 1*, **74**, 893 (1978)
80 CALADO, J. C. G., GOMES DE AZEVEDO, E. J. S., and SOARES, V. A. M., unpublished measurements
81 MACHADO, J. R. S., GUBBINS, K. E., LOBO, L. Q., and STAVELEY, L. A. K., *J. chem. Soc., Faraday Trans. 1*, **76**, 2496 (1980)
82 CLANCY, P., GUBBINS, K. E., and GRAY, C. G., *Chem. Soc. Faraday Discuss.*, **66**, 116 (1979); LOBO, L. Q., McCLURE, D. W., STAVELEY, L. A. K., CLANCY, P., GUBBINS, K. E., and GRAY, C. G., *J. chem. Soc., Faraday Trans. 2*, **77**, 425 (1981)

83 ALDERSLEY, S. C., LOBO, L. Q., and STAVELEY, L. A. K., *J. chem. Thermodyn.*, **11**, 597 (1979)
84 LOBO, L. Q., STAVELEY, L. A. K., and GUBBINS, K. E., unpublished measurements
85 LOBO, L. Q., STAVELEY, L. A. K., and GUBBINS, K. E., unpublished measurements
86 ARMSTRONG, G. T., GOLDSTEIN, J. M., and ROBERTS, D. E., *J. Res. natn. Bur. Stand.*, **55**, 265 (1955)
87 KNAAP, H. F. P., KNOESTER, M., and BEENAKKER, J. J. M., *Physica*, **27**, 309 (1961)
88 MATHOT, V., STAVELEY, L. A. K., YOUNG, J. A., and PARSONAGE, N. G., *Trans. Faraday Soc.*, **52**, 1488 (1956)
89 POOL, R. A. H., and STAVELEY, L. A. K., *Trans. Faraday Soc.*, **53**, 1186 (1957)
90 McCLURE, D. W., LEWIS, K. L., MILLER, R. C., and STAVELEY, L. A. K., *J. chem. Thermodyn.*, **8**, 785 (1976)
91 CHANG, S.-D., and LU, B. C.-Y., *Chem. Engng Progr. Symp. Ser.*, **63**, 18 (1967)
92 PARRISH, W. R., and HIZA, M. J., *Adv. cryogen. Engng*, **19**, 300 (1974)
93 STRYJEK, R., CHAPPELEAR, P. S., and KOBAYASHI, R., *J. chem. Engng Data*, **19**, 334 (1974)
94 LEWIS, K. L., MOSEDALE, S. E., and WORMALD, C. J., *J. chem. Thermodyn.*, **9**, 121 (1977)
95 HIZA, M. J., HAYNES, W. M., and PARRISH, W. R., *J. chem. Thermodyn.*, **9**, 873 (1977); RODOSEVICH, J. B., and MILLER, R. C., *A. I. Ch. E. Jl.*, **19**, 729 (1973)
96 NUNES DA PONTE, M., STREETT, W. B., and STAVELEY, L. A. K., *J. chem. Thermodyn.*, **10**, 151 (1978), POON, D. P. L., and LU, B. C.-Y., *Adv. cryogen. Engng*, **19**, 292 (1974)
97 PARRISH, W. R., and HIZA, M. J., *Adv. cryogen. Engng*, **21**, 485 (1976)
98 BLOOMER, O. T., GAMI, D. C., and PARENT, J. D., *Res. Bull. Inst. Gas Technol.*, **22**, 1 (1953)
99 WICHTERLE, I., and KOBAYASHI, R., *J. chem. Engng Data*, **17**, 9 (1972)
100 MILLER, R. C. and STAVELEY, L. A. K., *Adv. cryogen. Engng*, **21**, 493 (1976)
101 SHANA'A, M. Y., and CANFIELD, F. B., *Trans. Faraday Soc.*, **64**, 2281 (1968)
102 KLOSEK, J., and McKINLEY, C., in *Proc. 1st Int. Conf. on LNG*, Eds. White, J. W., and Neumann, A. E. S., Institute of Gas Technology, Chicago (1968)
103 JENSEN, R. H., and KURATA, F., *J. Petrol. Technol.*, **21**, 683 (1969)
104 STOECKLI, H. F., and STAVELEY, L. A. K., *Helv. chim. Acta.*, **53**, 1961 (1970)
105 CALADO, J. C. G., GARCIA, G. A., and STAVELEY, L. A. K., *J. chem. Soc., Faraday Trans. 1*, **70**, 1445 (1974)
106 CUTLER, J. B., and MORRISON, J. A., *Trans. Faraday Soc.*, **61**, 429 (1965)
107 KAHRE, L. C., *J. chem. Engng Data*, **19**, 67 (1974)
108 ELLIOT, D. G., CHEN, R. J. J., CHAPPELEAR, P. S., and KOBAYASHI, R., *J. chem. Engng Data*, **19**, 71 (1974)
109 WANG, R. H., and McKETTA, J. J., *J. chem. Engng Data*, **9**, 30 (1964)
110 LOBO, L. Q., CALADO, J. C. G., and STAVELEY, L. A. K., *J. chem. Thermodyn.*, **12**, 419 (1980)
111 MILLER, R. C., KIDNAY, A. J., and HIZA, M. J., *J. chem. Thermodyn.*, **9**, 167 (1977)
112 VOLOVA, L. M., *Zh. fiz. Khim.*, **14**, 268 (1940)
113 HSI, C., and LU, B. C.-Y., *Can. J. chem. Engng*, **49**, 140 (1971)
114 CALADO, J. C. G., and SOARES, V. A. M., *J. chem. Soc. Faraday, Trans. 1*, **73**, 1271 (1977)
115 BRØNSTED, J. H., and KOEFOED, J., *K. danske Vidensk. Selsk. (Mat. Fys. Skr.)*, **22**, No. 17 (1946)
116 McGLASHAN, M. L., and WILLIAMSON, A. G., *Trans. Faraday Soc.*, **57**, 588, 601 (1961); McGLASHAN, M. L., and MORCOM, K. W., *ibid.*, 581
117 HOLLEMAN, T., *Physica*, **29**, 585 (1963); **31**, 49 (1965); *PhD Thesis*, Univ. Amsterdam (1964); DESMYTER, A., and VAN DER WAALS, J. H., *Recl. Trav. chim. Pays-Bas Belg.*, **77**, 53 (1958); HIJMANS, J., *Molec. Phys.*, **1**, 307 (1958)
118 MARSH, K. N., OTT, J. B., and COSTIGAN, M. J., *J. chem. Thermodyn.*, **12**, 343 (1980); MARSH, K. N., OTT, J. B., and RICHARDS, A. E., *ibid.*, 897
119 EWING, M. B., LEVIEN, B. J., MARSH, K. N., and STOKES, R. H., *J. chem. Thermodyn.*, **2**, 689 (1970); LEVIEN, B. J., *J. chem. Thermodyn.*, **5**, 679 (1973); EWING, M. B., and MARSH, K. N., *J. chem. Thermodyn.*, **5**, 651, 659 (1973); **6**, 35, 43, 395, 1087 (1974); **9**, 357, 863 (1977)
120 D'ALMEIDA, M. D., GARCIA, J. G. F., and BOISSONNAS, C. G., *Helv. chim. Acta*, **53**,

1389 (1970); LUONG, M. D., ABDELIAH, Z., and STOECKLI, H. F., *Helv. chim. Acta*, **56**, 2513 (1973)

121 MARSH, K. N., *J. chem. Thermodyn.*, **2**, 359 (1970); **3**, 355 (1971); LEVIEN, B. J., and MARSH, K. N., *J. chem. Thermodyn.*, **2**, 227 (1970); TOMLINS, R. P., and MARSH, K. N., *J. chem. Thermodyn.*, **8**, 1185 (1976); **9**, 651 (1977)

122 TOMLINS, R. P., and ADAMSON, M., *J. chem. Thermodyn.*, **6**, 757 (1974); TOMLINS, R. P., and BURFITT, C., *J. chem. Thermodyn.*, **6**, 659 (1974); TOMLINS, R. P., *J. chem. Thermodyn.*, **8**, 1195 (1976)

123 AHMED, A., DIXON, D. T., and McGLASHAN, M. L., *J. chem. Thermodyn.*, **9**, 1087 (1977); BREWSTER, E., and McGLASHAN, M. L., *ibid.*, 1095

124 DIXON, D. T., and HEWITT, F. A., *J. chem. Thermodyn.*, **10**, 501 (1978)

125 LAM, V. T., PICKER, P., PATTERSON, D., and TANCRÈDE, P., *J. chem. Soc., Faraday Trans. 2*, **70**, 1465, 1479 (1974)

126 TURRELL, S., and DELMAS, G., *J. chem. Soc., Faraday Trans. 1*, **70**, 572 (1974); DELMAS, G., and THANH, N. T., *J. chem. Soc., Faraday Trans. 1*, **71**, 1172 (1975); DELMAS, G., DE ST. ROMAIN, P., and PURVES, P., *ibid.*, 1181

127 TANCRÈDE, P., PATTERSON, D., and LAM, V. T., *J. chem. Soc., Faraday Trans. 2*, **71**, 985 (1975); TANCRÈDE, P., BOTHOREL, P., DE ST. ROMAIN, P., and PATTERSON, D., *J. chem. Soc., Faraday Trans. 2*, **73**, 15 (1977); TANCRÈDE, P., PATTERSON, D., and BOTHOREL, P., *ibid.*, 29; PATTERSON, D., *Pure appl. Chem.*, **47**, 305 (1976)

128 DE ST. ROMAIN, P., VAN, H. T., and PATTERSON, D., *J. chem. Soc., Faraday Trans. 1*, **75**, 1700, 1708 (1979)

129 GÓMEZ-IBÁÑEZ, J. D., and SHIEH, J. J. C., *J. phys. Chem., Ithaca*, **69**, 1660 (1965); GÓMEZ-IBÁÑEZ, J. D., and WANG, F. T., *J. chem. Thermodyn.*, **3**, 811 (1971); YANG, S. K., and GÓMEZ-IBÁÑEZ, J. D., *J. chem. Thermodyn.*, **8**, 209 (1976)

130 SANCHEZ-PAJARES, R. G., and DELGADO, J. N., *J. chem. Thermodyn.*, **11**, 815 (1979); ARENOSA, R. L., MENDUIÑA, C., TARDAJOS, G., and DIAZ PEÑA, M., *ibid.*, 159

131 BRENNAN, J. S., HILL, R. J., and SWINTON, F. L., *J. chem. Thermodyn.*, **10**, 169 (1978); MAIRS, T. E., and SWINTON, F. L., *J. chem. Thermodyn.*, **12**, 573 (1980); HILL, R. J., MAIRS, T. E., and SWINTON, F. L., *ibid.*, 581

132 GOATES, J. R., OTT, J. B., and GRIGG, R. B., *J. chem. Thermodyn.*, **11**, 497 (1979); OTT, J. B., GRIGG, R. B., and GOATES, J. R., *Aust. J. Chem.*, **33**, 1921 (1980); MARTIN, M. L., and YOUINGS, J. C., *ibid.*, 2103

133 REEDER, J., KNOBLER, C. M., and SCOTT, R. L., *J. chem. Thermodyn.*, **7**, 345 (1975); HANDA, Y. P., REEDER, J., KNOBLER, C. M., and SCOTT, R. L., *J. chem. Engng Data*, **22**, 218 (1977); FENBY, D. V., KHURMA, J. R., KOONER, Z. S., BLOCK, T. E., KNOBLER, C. M., REEDER, J., and SCOTT, R. L., *Aust. J. Chem.*, **33**, 1927 (1980)

134 HARSTED, B. S., and THOMSEN, E. S., *J. chem. Thermodyn.*, **7**, 369 (1975)

135 JAIN, D. V. S., GUPTA, V. K., and LARK, B. S., *J. chem. Thermodyn.*, **5**, 451 (1973); JAIN, D. V. S., and LARK, B. S., *ibid.*, 455

136 DIAZ PEÑA, M., and MENDUIÑA, C., *J. chem. Thermodyn.*, **6**, 387, 1097 (1974); DIAZ PEÑA, M., and NUÑEZ DELGADO, J., *J. chem. Thermodyn.*, **7**, 201 (1975); ARENOSA, R. L., MENDUIÑA, C., TARDAJOS, G., and DIAZ PEÑA, M., *J. chem. Thermodyn.*, **11**, 825 (1979); DIAZ PEÑA, M., and DELGADO, J. N., *An. Quim.*, **70**, 678 (1974)

137 PICQUENARD, E., KEHIAIAN, H. V., ABELLO, L., and PANNETIER, G., *Bull. Soc. chim. Fr.*, 120 (1972)

138 WÓYCICKI, W., *J. chem. Thermodyn.*, **7**, 77, 1007 (1975)

139 LAL, M., and SWINTON, F. L., *J. phys. Chem., Ithaca*, **73**, 2883 (1969)

140 BENSON, G. C., MURAKAMI, S., LAM. V. T., and SINGH, J., *Can. J. Chem.*, **48**, 211 (1970); JONES, D. E. G., WEEKS, I. A., and BENSON, G. C., *Can. J. Chem.*, **49**, 2481 (1971)

141 WÓYCICKI, W., and RHENSIUS, P., *J. chem. Thermodyn.*, **11**, 153 (1979)

142 POWELL, R. J., and SWINTON, F. L., *J. chem. Engng Data*, **13**, 260 (1968); STOOKEY, D. J., SALLAK, H. M., and SMITH, B. D., *J. chem. Thermodyn.*, **5**, 741 (1973)

143 YOUNG, K. L., MENTZER, R. A., GREENKONN, R. A., and CHAO, K. C., *J. chem. Thermodyn.*, **9**, 979 (1977)

144 DIAZ PEÑA, M., COMPOSTIZO, A., and CRESPO COLIN, A., *J. chem. Thermodyn.*, **11**, 447 (1979)

145 HSU, K. Y., and CLEVER, H. L., *J. chem. Thermodyn.*, **7**, 435 (1975)

146 MURAKAMI, S., and FUJISHIRO, R., *Bull. chem. Soc. Japan*, **40**, 1784 (1967)

147 WÓYCICKI, W., *J. chem. Thermodyn.*, **4**, 1 (1972)

148 SINGH, J., PFLUG, H. D., and BENSON, G. C., *J. phys. Chem., Ithaca*, **72**, 1939 (1968); MURAKAMI, S., LAM, V. T., and BENSON, G. C., *J. chem. Thermodyn.*, **1**, 397 (1969); LAM, V. T., MURAKAMI, S., and BENSON, G. C., *J. chem. Thermodyn.*, **2**, 17 (1970); TANAKA, R., and BENSON, G. C., *J. chem. Thermodyn.*, **8**, 259 (1976); FORTIER, J. L., and BENSON, G. C., *J. chem. Engng Data*, **24**, 34 (1979)
149 RASTOGI, R. P., NATH, J., and MISRA, J., *J. phys. Chem., Ithaca*, **71**, 1277, 2524 (1967)
150 HARSTED, B. S., and THOMSEN, E. S., *J. chem. Thermodyn.*, **6**, 549, 557 (1974)
151 JAIN, D. V. S., LARK, B. S., CHAMAK, S. S., and CHANDLER, P., *Indian J. Chem.*, **8**, 66 (1970); JAIN, D. V. S., YADAV, O. P., and GILL, S. S., *Indian J. Chem.*, **9**, 339 (1971)
152 BISSELL, T. G., OKAFOR, G. E., and WILLIAMSON, A. G., *J. chem. Thermodyn.*, **3**, 393 (1971); BISSELL, T. G., and WILLIAMSON, A. G., *J. chem. Thermodyn.*, **7**, 131, 815 (1975)
153 DELMAS, G., and PURVES, P., *J. chem. Soc., Faraday Trans. 2*, **73**, 1829, 1838 (1977)
154 BOUBLIK, T., LAM, V. T., MURAKAMI, S., and BENSON, G. C., *J. phys. Chem., Ithaca*, **73**, 2356 (1969)
155 SANNI, S. A., and HUTCHINSON, P., *J. chem. Engng Data*, **18**, 317 (1973)
156 MATHOT, V., and DESMYTER, A., *J. chem. Phys.*, **21**, 782 (1953)
157 SACKMANN, H., and ARNOLD, H., *Z. Elektrochem.*, **63**, 565 (1959); KOLBE, A., and SACKMANN, H., *Z. phys. Chem. Frankf. Ausg.*, **31**, 281 (1962); WEIWAD, F., KEHLEN, H., KUSCHEL, F., and SACKMANN, H., *Z. phys. Chem.*, **253**, 114 (1973)
158 McGLASHAN, M. L., STUBLEY, D., and WATTS, H., *J. chem. Soc.*, A, 673 (1969); HOWELL, P. J., and STUBLEY, D., *ibid.*, 2489; HOWELL, P. J., SKILLERNE DE BRISTOWE, B. J., and STUBLEY, D., *J. chem. Soc.*, A, 397 (1971)
159 JAIN, D. V. S., and WADI, R. K., *J. chem. Thermodyn.*, **8**, 493 (1976)
160 SCATCHARD, G., WOOD, S. E., and MOCHEL, J. M., *J. Am. chem. Soc.*, **62**, 712 (1940); KIND, R., KAHNT, G., SCHMIDT, D., SCHUMANN, J., and BITTRICH, H. J., *Z. phys. Chem.*, **238**, 277 (1968)
161 KIYOHARA, O., and BENSON, G. C., *J. chem. Thermodyn.*, **9**, 691 (1977); FORTIER, J.-L., and BENSON, G. C., *ibid.*, 1181
162 BOTTOMLEY, G. A., and SCOTT, R. L., *J. chem. Thermodyn.*, **6**, 973 (1974)
163 BENNETT, J. E., and BENSON, G. C., *Can. J. Chem.*, **43**, 1912 (1965)
164 MURAKAMI, S., and BENSON, G. C., *J. chem. Thermodyn.*, **1**, 559 (1969)
165 DHILLON, M. S., *J. chem. Thermodyn.*, **6**, 915, 1107 (1974)
166 TANAKA, R., and BENSON, G. C., *J. chem. Engng Data*, **21**, 320 (1976); **22**, 291 (1977); **23**, 75 (1978); **24**, 37 (1979)
167 MEIJER, E. L., BROUWER, N., and MILTENBURG, J. C., *J. chem. Thermodyn.*, **8**, 703 (1976); MILTENBURG, J. C., OBBINK, J. H., and MEIJER, E. L., *J. chem. Thermodyn.*, **11**, 37 (1979)
168 TANAKA, R., MURAKAMI, S., and FUJISHIRO, R., *J. chem. Thermodyn.*, **5**, 777 (1973)
169 BIRDI, G. S., and MAHL, B. S., *J. chem. Thermodyn.*, **6**, 918 (1974)
170 GRACIA, M., OTIN, S., and LOSA, C. G., *J. chem. Thermodyn.*, **6**, 701 (1974); **7**, 293 (1975)
171 DAS, S. K., DIAZ PEÑA, M., and McGLASHAN, M. L., *Pure appl. Chem.*, **2**, 141 (1961)
172 HILL, R. J., O'KANE, E., and SWINTON, F. L., *J. chem. Thermodyn.*, **10**, 1153, 1205 (1978)
173 SWINTON, F. L., in *Chemical Thermodynamics, Specialist Periodical Reports*, Ed. McGlashan, M. L., Vol. 2, 147, Chemical Society, London (1978)
174 THORP, N., and SCOTT, R. L., *J. phys. Chem., Ithaca*, **60**, 670 (1956); CROLL, I. M., and SCOTT, R. L., *J. phys. Chem., Ithaca*, **62**, 954 (1958); SIMON, M., and KNOBLER, C. M., *J. chem. Thermodyn.*, **3**, 657 (1971)
175 BEDFORD, R. G., and DUNLAP, R. D., *J. Am. chem. Soc.*, **80**, 282 (1958); DUNLAP, R. D., BEDFORD, R. G., WOODBREY, J. C., and FURROW, S. D., *J. Am. chem. Soc.*, **81**, 2927 (1959); WILLIAMSON, A. G., and SCOTT, R. L., *J. phys. Chem., Ithaca*, **65**, 275 (1961)
176 DYKE, D. E. L., ROWLINSON, J. S., and THACKER, R., *Trans. Faraday Soc.*, **55**, 903 (1959)
177 CROLL, I. M., and SCOTT, R. L., *J. phys. Chem., Ithaca*, **68**, 2853 (1964)
178 SCOTT, R. L., *J. phys. Chem., Ithaca*, **62**, 136 (1958)
179 DUNCAN, W. A., SHERIDAN, J. P., and SWINTON, F. L., *Trans. Faraday Soc.*, **62**, 1090 (1966)
180 GAW, W. J., and SWINTON, F. L., *Trans. Faraday Soc.*, **64**, 637, 2023 (1968)
181 POWELL, R. J., SWINTON, F. L., and YOUNG, C. L., *J. chem. Thermodyn.*, **2**, 105 (1970)

182 MATTINGLEY, B. I., HANDA, Y. P., and FENBY, D. V., *J. chem. Thermodyn.*, 7, 169 (1975)

183 ARMITAGE, D. A., and MORCOM, K. W., *Trans. Faraday Soc.*, 65, 688 (1969)

184 CHAND, A., and FENBY, D. V., *J. chem. Thermodyn.*, 7, 403 (1975)

185 DUNCAN, W. A., and SWINTON, F. L., *Trans. Faraday Soc.*, 62, 1082 (1966); OTT, J. B., GOATES, J. R., and CARDON, D. L., *J. chem. Thermodyn.*, 8, 505 (1976)

186 FENBY, D. V., *Rev. pure appl. Chem.*, 22, 55 (1972); SWINTON, F. L., in *Molecular Complexes*, Ed. Foster, R., Vol. 2, Paul Elek, London (1974)

187 KOGAN, I. V., and MORACHEVSKII, A. G., *Zh. prikl. Khim., Leningr.*, 45, 1885 (1972); CHINIKAMALA, A., HOUTH, G. N., and TAYLOR, Z. L., *J. chem. Engng Data*, 18, 322 (1973)

188 FENBY, D. V., and SCOTT, R. L., *J. phys. Chem.*, *Ithaca*, 71, 4103 (1967); FENBY, D. V., McLURE, I. A., and SCOTT, R. L., *J. phys. Chem.*, *Ithaca*, 70, 602 (1966)

189 HOWELL, P. J., SKILLERNE DE BRISTOWE, B. J., and STUBLEY, D., *J. chem. Thermodyn.*, 4, 225 (1972); SKILLERNE DE BRISTOWE, B. J., and STUBLEY, D., *J. chem. Thermodyn.*, 5, 121, 865 (1973)

190 FENBY, D. V., and RUENKRAIRERGSA, S., *J. chem. Thermodyn.*, 5, 227 (1973); RUENKRAIRERGSA, S., FENBY, D. V., and JONES, D. E., *ibid.*, 347; LEONG, C. Y., JONES, D. E., and FENBY, D. V., *J. chem. Thermodyn.*, 6, 609 (1974)

191 WILLIAMSON, A. G., in *Chemical Thermodynamics, Specialist Periodical Reports*, Ed. McGlashan, M. L., Vol. 2, 174, Chemical Society, London (1978)

192 ARMITAGE, D. A., BEAUMONT, T. G., DAVIS, K. M. C., HALL, D. J., and MORCOM, K. W., *Trans. Faraday Soc.*, 67, 2548 (1971)

193 HALL, D. J., MORCOM, K. W., and BRINDLEY, J. M. T., *J. chem. Thermodyn.*, 6, 1133 (1974)

194 MEYER, R., MEYER, M., and METZGER, J., *Thermochim. Acta*, 9, 323 (1974)

195 ANDREWS, A. W., HALL, D., and MORCOM, K. W., *J. chem. Thermodyn.*, 3, 527 (1971); MATTINGLEY, B. I., HANDA, Y. P., and FENBY, D. V., *J. chem. Thermodyn.*, 7, 169 (1975); ARMITAGE, D. A., BRINDLEY, J. T. M., and MORCOM, K. W., *J. chem. Thermodyn.*, 10, 581 (1978); MURRAY, R. S., and MARTIN, M. L., *ibid.*, 613, 701, 711; BRINDLEY, J. M. T., OSBORNE, C. G., and MORCOM, K. W., *J. chem. Thermodyn.*, 11, 687 (1979)

196 ALDERSLEY, S. C., CALADO, J. C. G., and STAVELEY, L. A. K., *J. inorg. nucl. Chem.*, 41, 1269 (1979)

197 RODRIGUEZ, A. T., *PhD Thesis*, Univ. Sheffield (1979)

198 BROWN, I., and SMITH, F., *Aust. J. Chem.*, 7, 269 (1954); 8, 62 (1955); BROWN, I., and FOCK, W., *Aust. J. Chem.*, 9, 180 (1956)

199 ORYE, R. V., and PRAUSNITZ, J. M., *Trans. Faraday Soc.*, 61, 1338 (1965)

200 PALMER, D. A., and SMITH, B. D., *J. chem. Engng Data*, 17, 71 (1972)

201 TANAKA, R., MURAKAMI, S., and FUJISHIRO, R., *J. chem. Thermodyn.*, 6, 209 (1974)

202 MAGIERA, B., and BROSTOW, J., *J. phys. Chem.*, *Ithaca*, 75, 4041 (1971)

203 MURAKAMI, S., AMAYA, K., and FUJISHIRO, R., *Bull. chem. Soc. Japan*, 37, 1776 (1964)

204 JOHARI, G. P., *J. chem. Engng Data*, 13, 541 (1968)

205 DIAZ PEÑA, M., CRESPO COLIN, A., and COMPOSTIZO, A., *J. chem. Thermodyn.*, 10, 337, 1101 (1978)

206 MATTINGLEY, B. I., and FENBY, D. V., *J. chem. Thermodyn.*, 7, 307 (1975)

207 RADOJKOVIĆ, N., TASIĆ, A., GROZDANIĆ, D., DJORDJEVIĆ, B., and MALIĆ, D., *J. chem. Thermodyn.*, 9, 349 (1977)

208 KIYOHARA, O., BENSON, G. C., and GROLIER, J.-P. E., *J. chem. Thermodyn.*, 9, 315 (1977)

209 GROLIER, J.-P. E., BENSON, G. C., and PICKER, P., *J. chem. Engng Data*, 20, 243 (1975)

210 GROLIER, J.-P., KIYOHARA, O., and BENSON, G. C., *J. chem. Thermodyn.*, 9, 697 (1977); KIYOHARA, O., HANDA, Y. P., and BENSON, G. C., *J. chem. Thermodyn.*, 11, 453 (1979); URDANETA, O., HAMAM, S., HANDA, Y. P., and BENSON, G. C., *ibid.*, 851; URDANETA, O., HANDA, Y. P., and BENSON, G. C., *ibid.*, 857

211 MARSH, K. N., FRENCH, H. T., and ROGERS, H. P., *J. chem. Thermodyn.*, 11, 892 (1979)

212 CLEVER, H. L., PIRKLE, Q. R., ALLEN, B. J., and DERRICK, M. E., *J. chem. Engng Data*, 17, 31 (1972)

213 DERRICK, M. E., and CLEVER, H. L., *J. phys. Chem.*, *Ithaca*, 75, 3728 (1971);

DERRICK, M. E., and CLEVER, H. L., *J. Colloid & Interface Sci.*, **39**, 593 (1972); CLEVER, H. L., and HSU, K.-Y., *J. chem. Thermodyn.*, **10**, 213, 225 (1978)
214 REEDER, K. S., BLOCK, T. E., and KNOBLER, C. M., *J. chem. Thermodyn.*, **8**, 133 (1976); HANDA, Y. P., KNOBLER, C. M., and SCOTT, R. L., *J. chem. Thermodyn.*, **9**, 451 (1977)
215 ANDREWS, A. W., and MORCOM, K. W., *J. chem. Thermodyn.*, **3**, 513, 519 (1971)
216 GROLIER, J.-P. E., BATTER, D., and VIALLARD, A., *J. chem. Thermodyn.*, **6**, 895 (1974)
217 GUILLEN, M. D., and GUTIERREZ LOSA, C., *J. chem. Thermodyn.*, **10**, 567 (1978)
218 DESHPANDE, D. D., and OSWAL, S. L., *J. chem. Thermodyn.*, **7**, 155 (1975)
219 BRUCE, G. R., and MALCOLM, G. N., *J. chem. Thermodyn.*, **1**, 183 (1969)
220 FINDLAY, T. J. V., and KAVANAGH, P. J., *J. chem. Thermodyn.*, **6**, 367 (1974)
221 BEATH, L. A., O'NEILL, S. P., and WILLIAMSON, A. G., *J. chem. Thermodyn.*, **1**, 51, 293 (1969)
222 MARKGRAF, H. G., and NIKURADSE, A., *Z. Naturforsch.*, **A9**, 27 (1954)
223 KATO, M., and SUZUKI, N., *J. chem. Thermodyn.*, **10**, 435 (1978)
224 MEYER, R. J., and GIUSTI, G. L., *J. chem. Thermodyn.*, **9**, 1101 (1977)
225 BROWN, I., and SMITH, F., *Aust. J. Chem.*, **8**, 62 (1955); **13**, 30 (1960); **15**, 9 (1962)
226 LETCHER, T. M., and BAYLES, J. W., *J. chem. Engng Data*, **16**, 266 (1971); *J. chem. Thermodyn.*, **4**, 159, 551 (1972)
227 CHAND, A., HANDA, Y. P., and FENBY, D. V., *J. chem. Thermodyn.*, **7**, 401 (1975); HANDA, Y. P., FENBY, D. V., and JONES, D. E., *ibid.*, 337
228 GARRETT, P. R., POLLOCK, J. M., and MORCOM, K. W., *J. chem. Thermodyn.*, **3**, 135 (1971); **5**, 569 (1973)
229 GARRETT, P. R., and POLLOCK, J. M., *J. chem. Thermodyn.*, **9**, 561, 1045 (1977)
230 FINDLAY, T. J. V., and KENYON, R. S., *Aust. J. Chem.*, **22**, 865 (1969); SHARMA, B. R., and SINGH, P. P., *J. chem. Thermodyn.*, **5**, 361 (1973); PHUTELA, R. C., ARORA, P. S., and SINGH, P. P., *J. chem. Thermodyn.*, **6**, 801 (1974)
231 DESHPANDE, D. D., and PANDYA, M. V., *Trans. Faraday Soc.*, **63**, 2149 (1967); DESHPANDE, D. D., and BHATGADDE, L. G., *J. phys. Chem.*, *Ithaca*, **72**, 261 (1968)
232 VELASCO, I., OTIN, S., and GUTIERREZ-LOSA, C., *J. Chim. phys. phys.-chim. biol.*, **76**, 381 (1979)
233 HEPLER, L. G., and FENBY, D. V., *J. chem. Thermodyn.*, **5**, 47 (1973); MATSUI, T., HEPLER, L. G., and FENBY, D. V., *J. phys. Chem.*, *Ithaca*, **77**, 2397 (1973)
234 FRANCIS, A. W., *Critical Solution Temperatures*, American Chemical Society, Washington DC (1961); *Liquid–Liquid Equilibriums*, Interscience, New York (1963); *J. chem. Engng Data*, **10**, 45, 145, 260, 327 (1965); **11**, 96, 234, 557 (1966); **12**, 269, 380 (1967); *Chem. Engng Sci.*, **22**, 627, 707, 737 (1967)
235 KORTÜM, G., and FREIER, H.-J., *Monatsh. Chem.*, **85**, 693 (1954); *Chemie-Ingr-Tech.*, **26**, 670 (1954)
236 RÖCK, H., and SCHNEIDER, G., *Z. phys. Chem. Frankf. Ausg.*, **8**, 154 (1956); SCHNEIDER, G., *Z. phys. Chem. Frankf. Ausg.*, **24**, 165 (1960); HOSSEINI, S. M., and SCHNEIDER, G., *Z. phys. Chem. Frankf. Ausg.*, **36**, 137 (1963)
237 KEYES, D. B., and HILDEBRAND, J. H., *J. Am. chem. Soc.*, **39**, 2126 (1917)
238 KOHLER, F., LIEBERMANN, E., SCHANO, R., AFFSPRUNG, H. E., MORROW, J. K., SOSNKOWSKA-KEHIAIAN, K., and KEHIAIAN, H., *J. chem. Thermodyn.*, **7**, 241 (1975)
239 KISER, R. W., JOHNSON, G. D., and SHETLAR, M. D., *J. chem. Engng Data*, **6**, 338 (1961); SAVINI, C. G., WINTERHALTER, D. R., and VAN NESS, H. C., *J. chem. Engng Data*, **10**, 168 (1965); SINOR, J. E., and WEBER, J. H., *J. chem. Engng Data*, **5**, 243 (1960); RENON, H., and PRAUSNITZ, J. M., *Chem. Engng Sci.*, **22**, 299 (1967)
240 STAVELEY, L. A. K., and SPICE, B., *J. chem. Soc.*, 406 (1952)
241 SCATCHARD, G., and TICKNOR, L. B., *J. Am. chem. Soc.*, **74**, 3724 (1952); SCATCHARD, G., GOATES, J. R., and McCARTNEY, E. R., *ibid.*, 3721; SCATCHARD, G., and SATKIEWICZ, F. G., *J. Am. chem. Soc.*, **86**, 130 (1964)
242 GOATES, J. R., SNOW, R. L., and JAMES, M. R., *J. phys. Chem.*, *Ithaca*, **65**, 335 (1961); GOATES, J. R., SNOW, R. L., and OTT, J. B., *J. phys. Chem.*, *Ithaca*, **66**, 1301 (1962); MRAZEK, R. V., and VAN NESS, H. C., *A.I. Ch. E.Jl.*, **7**, 190 (1961); WILLIAMSON, A. G., and SCOTT, R. L., *J. phys. Chem.*, *Ithaca*, **64**, 440 (1960); BROWN, I., and FOCK, W., *Aust. J. Chem.*, **14**, 387 (1961); BROWN, I., and SMITH, F., *Aust. J. Chem.*, **15**, 1 (1962)
243 MOELWYN-HUGHES, E. A., and MISSEN, R. W., *J. phys. Chem.*, *Ithaca*, **61**, 518 (1957); PARASKEVOPOULOS, G. C., and MISSEN, R. W., *Trans. Faraday Soc.*, **58**, 869 (1962);

OTTERSTEDT, J.-E. A., and MISSEN, R. W., *ibid.*, 879; WOLFF, H., and HOPPEL, J. E., *Ber. Bunsenges. phys. Chem.*, **72**, 1173 (1968)

244 RAMALHO, R. S., and RUEL, M., *Can. J. chem. Engng*, **46**, 456 (1968); NGUYEN, T. H., and RATCLIFF, G. A., *J. chem. Engng Data*, **20**, 252, 256 (1975); SAYEGH, S. G., and RATCLIFF, G. A., *J. chem. Engng Data*, **21**, 71 (1976)

245 FRENCH, H. T., RICHARDS, A., and STOKES, R. H., *J. chem. Thermodyn.*, **11**, 671 (1979)

246 STOKES, R. H., and BURFITT, C., *J. chem. Thermodyn.*, **5**, 623 (1973); *J. chem. Thermodyn.*, **7**, 803 (1975); MARSH, K. N., and BURFITT, C., *ibid.*, 955; MARSH, K. N., and ALLAN, W. A., *J. chem. Thermodyn.*, **9**, 1109 (1977); STOKES, R. H., ADAMSON, M., and RICHARDS, A., *J. chem. Thermodyn.*, **11**, 303 (1979)

247 HSU, K. Y., and CLEVER, H. L., *J. chem. Engng Data*, **20**, 268 (1975); HWANG, S. C., and ROBINSON, R. L., *J. chem. Engng Data*, **22**, 319 (1977); CHRISTENSEN, J. J., IZATT, R. M., EATOUGH, D. J., and HANSEN, L. D., *J. chem. Thermodyn.*, **10**, 829 (1978)

248 BARKER, J. A., *J. chem. Phys.*, **20**, 794, 1526 (1952); *J. chem. Phys.*, **21**, 1391 (1953); BARKER, J. A., and SMITH, F., *J. chem. Phys.*, **22**, 375 (1954)

249 BROWN, I., FOCK, W., and SMITH, F., *J. chem. Thermodyn.*, **1**, 273 (1969)

250 ANAND, S. C., GROLIER, J.-P. E., KIYOHARA, O., HALPIN, C. J., and BENSON, G. C., *J. chem. Engng Data*, **20**, 184 (1975); BENSON, G. C., and KIYOHARA, O., *J. chem. Engng Data*, **21**, 362 (1976)

251 WILLIAMSON, A. G., in *Chemical Thermodynamics, Specialist Periodical Reports*, Ed. McGlashan, M. L., Vol. 2, 95, Chemical Society, London, (1978)

252 FREYMANN, R., *C.r. hebd. Séanc. Acad. Sci., Paris*, **193**, 928 (1931); *C.r. hebd. Séanc. Acad. Sci., Paris*, **195**, 39 (1932)

253 VAN THIEL, M., BECKER, E. D., and PIMENTEL, G. C., *J. chem. Phys.*, **27**, 486 (1959)

254 LUCK, W. A. P., in *Water: A Comprehensive Treatise*, Ed. Franks, F., Vol. 2, Chap. 4, Plenum, New York (1973)

255 PRIGOGINE, I., *The Molecular Theory of Solutions*, Chap. 15, North-Holland, Amsterdam (1957)

256 LANGMUIR, I., *The Distribution and Orientation of Molecules, 3rd Colloid Symp. Monogr.*, Chemical Catalog Co, New York (1925)

257 ABRAMS, D. S., and PRAUSNITZ, J. M., *A.I.Ch.E.Jl.*, **21**, 116 (1975); FREDENSLUND, A., JONES, R. L., and PRAUSNITZ, J. M., *ibid.*, 1086

258 FREDENSLUND, A., GMEHLING, J., and RASMUSSEN, P., *Vapour Liquid Equilibria using UNIFAC*, Elsevier, Amsterdam (1977)

259 WILSON, G. M., and DEAL, C. H., *Ind. & Eng. Chem., Fundam.*, **1**, 20 (1962); DERR, E. L., and DEAL, C. H., *Adv. Chem. Ser.*, No. 124, 11 (1973); TOCHIGI, K., and KOPINA, K., *J. chem. Engng Japan*, **9**, 267 (1976); *Ind. Engng Chem.*, **60**, 28 (1968)

260 KOJIMA, K., and TOCHIGI, K., *Prediction of Vapour–Liquid Equilibria by the ASOG Method*, Elsevier, Amsterdam (1979)

261 KEHIAIAN, H. V., *Chemical Thermodynamic Data on Fluids and Fluid Mixtures*, 121, IPC Scientific and Technical Press, London (1979)

262 CAMPBELL, A. N., KARTZMARK, E. M., and CHATTERJEE, R. M., *Can. J. Chem.*, **44**, 1183 (1966); MORCOM, K. W., and TRAVERS, D. N., *Trans. Faraday Soc.*, **61**, 230 (1965); BECKER, F., and KIEFER, M., *Z. Naturforsch.*, **A25**, 7 (1969); STAVELEY, L. A. K., TUPMAN, W. I., and HART, K. R., *Trans. Faraday Soc.*, **51**, 323 (1955); SANNI, S. A., FELL, C. J. D., and HUTCHISON, H. P., *J. chem. Engng Data*, **16**, 424 (1971)

263 BECKER, F., and KIEFER, M., *Z. Naturforsch*, **A26**, 1040 (1971); BECKER, F., KIEFER, M., RHENSIUS, P., and SCHAFER, H. D., *Z. phys. Chem. Frankf. Ausg.*, **92**, 169 (1974); SANNI, S. A., and HUTCHISON, P., *J. chem. Engng Data*, **18**, 317 (1973); McGLASHAN, M. L., and RASTOGI, R. P., *Trans. Faraday Soc.*, **54**, 496 (1958); BEATH, L. A., and WILLIAMSON, A. G., *J. chem. Thermodyn.*, **1**, 51 (1969); BRUCE, G. R., and MALCOLM, G. N., *ibid.*, 183

264 NAGATA, I., and HAYASHIDA, H., *J. chem. Eng. Japan*, **3**, 161 (1970); BENNETT, J. E., and BENSON, G. C., *Can. J. Chem.*, **43**, 1912 (1965); RASTOGI, R. P., NATH, J., and MISRA, R. R., *J. chem. Thermodyn.*, **3**, 307 (1971)

265 CRESWELL, C. J., and ALLRED, A. L., *J. phys. Chem., Ithaca*, **66**, 1469 (1962)

266 NAKANISHI, K., WADA, H., and TOUHARA, H., *J. chem. Thermodyn.*, **7**, 1125 (1975)

267 GORDY, W., *J. chem. Phys.*, **7**, 93 (1939); GORDY, W., and STANFORD, S. C., *J. chem. Phys.*, **8**, 170 (1940)

268 CHUN, K. W., DRUMMOND, J. C., SMITH, W. H., and DAVISON, R. R., *J. chem.*

Engng Data, **20**, 58 (1975); CHAND, A., and FENBY, D. V., *J. chem. Engng Data*, **22**, 289 (1977); NAKANISHI, K., ASHITANI, K., and TOUHARA, H., *J. chem. Thermodyn.*, **8**, 121 (1976); KOWALSKI, B., ORSZÁGH, A., and CZERNIK, S., *ibid.*, 425; OBA, M., MURAKAMI, S., and FUJISHIRO, R., *J. chem. Thermodyn.*, **9**, 407 (1977)

269 THACKER, R., and ROWLINSON, J. S., *Trans. Faraday Soc.*, **50**, 1036 (1954)
270 HUSSEIN, M. A., and MILLEN, D. J., *J. chem. Soc., Faraday Trans. 2*, **70**, 685, 693 (1974)
271 FRANK, H. S., and EVANS, M. W., *J. chem. Phys.*, **13**, 507 (1945); KAUZMANN, W., *Adv. protein Chem.*, **14**, 1 (1959)
272 McAULIFFE, C., *J. phys. Chem., Ithaca*, **70**, 1267 (1966)
273 ARNETT, E. M., KOVER, W. B., and CARTER, J. V., *J. Am. chem. Soc.*, **91**, 4028 (1969)
274 TANFORD, C., *The Hydrophobic Effect: Formation of Micelles and Biological Membranes*, Wiley, New York (1973)
275 VIERK, A. L., *Z. anorg. Chem.*, **261**, 283 (1950); MORCOM, K. W., and SMITH, J. M., *J. chem. Thermodyn.*, **1**, 503 (1969)
276 RAMALO, R. S., and DROLET, J. F., *J. chem. Engng Data*, **16**, 12 (1971); NICHOLSON, D. E., *J. chem. Engng Data*, **5**, 309 (1960); HANSON, D. O., and VAN WINKLE, M., *ibid.*, 30
277 MALCOLM, G. N., and ROWLINSON, J. S., *Trans. Faraday Soc.*, **53**, 921 (1957)
278 BLANDAMER, M. J., HIDDEN, N. J., MORCOM, K. W., SMITH, R. W., TRELOAR, N. C., and WOOTTEN, M. J., *Trans. Faraday Soc.*, **65**, 2633 (1969); MORCOM, K. W., and SMITH, R. W., *Trans. Faraday Soc.*, **66**, 1073 (1970); NAKAYAMA, H., and SHINODA, K., *J. chem. Thermodyn.*, **3**, 401 (1971); MATSUMOTO, Y., TOUHARA, H., NAKANISHI, K., and WATANABE, N., *J. chem. Thermodyn.*, **9**, 801 (1977)
279 MURTI, P. S., and VAN WINKLE, H., *J. chem. Engng Data*, **3**, 72 (1958); DAWE, R. A., and NEWSHAM, D. M. T., *J. chem. Engng Data*, **18**, 44 (1973); GOODWIN, S. R., and NEWSHAM, D. M. T., *J. chem. Thermodyn.*, **3**, 325 (1971); WILSON, A., and SIMONS, E. L., *Ind. Engng Chem.*, **44**, 2214 (1952); KATAYAMA, T., *Chem. Engng, Tokyo*, **26**, 361 (1962)
280 HILL, A. E., *J. Am. chem. Soc.*, **45**, 1143 (1923); HILL, A. E., and MALISOFF, W. M., *J. Am. chem. Soc.*, **48**, 918 (1926)
281 JÄNECKE, E., *Z. phys. Chem.*, **A164**, 401 (1933)
282 MATOUS, J., HRNCIRIK, J., NOVÁK, J. P., and SOBR, J., *Coll. Czech. chem. Comm.*, **35**, 1904 (1970); **37**, 2653 (1972); GLEW, D. N., and WATTS, H., *Can. J. Chem.*, **51**, 1933 (1973)
283 LARKIN, J. A., *J. chem. Thermodyn.*, **7**, 137 (1975); PEMBERTON, R. C., and MASH, C. J., *J. chem. Thermodyn.*, **10**, 867 (1978)
284 MARSH, K. N., and RICHARDS, A. E., *Aust. J. Chem.*, **33**, 2121 (1980)
285 BOSC, E., *Z. phys. Chem.*, **65**, 458 (1909); BUTLER, J. A. V., THOMSON, D. W., and MacLENNAN, W. H., *J. chem. Soc.*, 674 (1933); GROLIER, J.-P. E., and VIALLARD, A., *J. Chim. Phys.*, **67**, 1582 (1970); ABELLO, L., *J. Chim. Phys.*, **70**, 1355 (1973)
286 COX, H. L., and CRETCHER, L. H., *J. Am. chem. Soc.*, **48**, 451 (1926); COX, H. L., and NELSON, W. L., *J. Am. chem. Soc.*, **49**, 1080 (1927)
287 OGORODNIKOV, S. K., KOGAN, V. B., and MOROZOV, A. I., *Zh. prikl. Khim. Leningr.*, **35**, 685 (1962); REMM, K., and BITTRICH, H.-J., *Z. phys. Chem.*, **251**, 109 (1972); AMAYA, K., and FUJISHIRO, R., *Bull. chem. Soc. Japan*, **30**, 940 (1957); NAKAYAMA, H., and SHINODA, K., *J. chem. Thermodyn.*, **3**, 401 (1971)
288 STEDMAN, D. F., *Trans. Faraday Soc.*, **24**, 289 (1928)
289 KOHLER, F., *Monatsh. Chem.*, **82**, 913 (1951); COPP, J. L., *Trans. Faraday Soc.*, **51**, 1056 (1955)
290 ANDON, R. J. L., and COX, J. D., *J. chem. Soc.*, 4601 (1952); COX, J. D., *ibid.*, 4606; ANDON, R. J. L., COX, J. D., and HERINGTON, E. F. G., *J. chem. Soc.*, 3188 (1954); *Discuss. Faraday Soc.*, **15**, 168 (1953); *Trans. Faraday Soc.*, **53**, 410 (1957)
291 SCATCHARD, G., KAVANAGH, G. M., and TICKNOR, L. B., *J. Am. chem. Soc.*, **74**, 3715 (1952); KUBASCHEWSKI, O., and WEBER, W., *Z. Elektrochem.*, **54**, 200 (1950)
292 WILSON, R. Q., MUNGER, H. P., and CLEGG, J. W., *Chem. Engng Prog. Symp. Ser.*, **48**, No. 3, 115 (1952)
293 COPP, J. L., and EVERETT, D. H., *Discuss. Faraday Soc.*, **15**, 174 (1953); BITTRICH, H.-J., and KRAFT, G., *Z. phys. Chem.*, **227**, 359 (1964)
294 CORNISH, R. E., ARCHIBALD, R. C., MURPHY, E. A., and EVANS, H. M., *Ind. Engng Chem.*, **26**, 397 (1934); RIVENO, F., *Bull. Soc. chim. Fr.*, 1606 (1963); SCHMELZER, J., and QUITZSCH, K. *Z. phys. Chem.*, **52**, 28 (1973)
295 TAYLOR, J. B., and ROWLINSON, J. S., *Trans. Faraday Soc.*, **51**, 1183 (1955)

296 FRICKE, R., and LÜKE, J., *Z. Elektrochem.*, **36**, 309 (1930); GEE, G., *Q. Rev. chem. Soc.*, **1**, 265 (1947)
297 LAL, M., and SWINTON, F. L., *Physica*, **40**, 446 (1968)
298 PRIGOGINE, I., BINGEN, R., and BELLEMANS, A., *Physica*, **20**, 633 (1954)
299 WALTERS, G. K., and FAIRBANK, W. N., *Phys. Rev.*, **103**, 262 (1956)
300 ZINOVEVA, K. N., and PESHKOV, V. P., *Soviet Phys. JETP*, **5**, 1024 (1957); **10**, 22 (1960); **17**, 1235 (1963)
301 TEDROW, P. M., and LEE, D. M., in *Low Temperature Physics*, Eds. Daunt, J. G., Edward, D. O., Milford, F. J., and Yaqub, M., Part A, 248, Plenum, New York (1965)
302 LE PAIR, C., TACONIS, K. W., OUBOTER, R. DeB., and DAS, P., *Physica*, **28**, 305 (1962); **31**, 764 (1965); *Cryogenics*, **3**, 112 (1963)
303 BETTS, D. S., *Contemp. Phys.*, **9**, 97 (1968)
304 BARDEEN, J., BAYM. G., and PINES, D., *Phys. Rev.*, **156**, 207 (1967); EBNER, C., *ibid.*, 222
305 OUBOTER, R. DeB., TACONIS, K. W., LE PAIR, C., and BEENAKKER, J. J. M., *Physica*, **26**, 853 (1960)
306 BROUWER, J. P., VAN DEN MEIJDENBERG, C. J. N., and BEENAKKER, J. J. M., *Physica*, **50**, 93 (1970)
307 BROUWER, J. P., VOSSEPOEL, A. M., VAN DEN MEIJDENBERG, C. J. N., and BEENAKKER, J. J. M., *Physica*, **50**, 125 (1970)
308 EWING, M. B., McGLASHAN, M. L., and TZIAS, P., *J. chem. Thermodyn.*, **13**, 527 (1981)

Chapter 6
Fluid mixtures at high pressures

6.1 Introduction

The early theoretical work of van der Waals[1] on critical phenomena in both pure and mixed fluids was instrumental in encouraging many talented research workers to undertake experimental work in this field in the latter part of last century. Between 1890 and 1910 Kamerlingh Onnes, Kuenen, Keesom and others at Leiden[2] carried out extensive experimental investigations of fluid phase behaviour at moderate pressures especially in the region of critical loci, and, together, discovered phase relationships of surprising complexity and variety. Interest in such phenomena waned during the next several decades and was only revived after World War 2 when Russian workers, notably Krichevskii and Tsiklis[3], published the results of measurements where the experimental pressures were extended to 10^3 bar and beyond. Even in the 1950s and 1960s textbooks concerned with the thermodynamics of multicomponent systems were written that paid scant attention to the properties of supercritical mixtures. The situation has changed considerably over the past decade and there is now great interest in the phase behaviour of fluid mixtures at high pressures.

This upsurge of interest has been occasioned by the importance that high pressure phase equilibria has now assumed in various areas of physical science. Among the more important of these can be listed:

(a) The need to develop, often at cryoscopic temperatures, new and highly-efficient separation techniques in the natural gas, oil and petrochemical industries.

(b) The increasing use of supercritical fluids as solvents in many novel and economically-important chemical processes. Examples can be quoted that are as diverse as the use of the pharmacologically-acceptable supercritical carbon dioxide in the decaffeination of raw coffee[4], to the use of toluene in the conversion of coal to liquid hydrocarbon by supercritical dissolution followed by subsequent hydrogenation[5].

(c) The need of geochemists and geophysicists to understand the various high-temperature, high-pressure processes that occur in the formation and migration of minerals and other geological deposits within the earth's crust.

It has already been explained in Chapters 4 and 5 why excess functions cannot be used to represent the thermodynamic properties of fluid mixtures at pressures exceeding a few bars. At higher pressures, the phase relationships for mixtures can best be expressed by (p, T, x) surfaces presented in either graphical or tabular form or often by (p, T), (T, x) or (p, x) projections of these surfaces on the appropriate plane. Experimental high-pressure (p, V, x) surfaces are rarer because of the difficulty of measuring fluid densities at elevated pressures.

Experimental techniques currently used for the determination of temperature, pressure and density in fluid systems have been exhaustively reviewed in a recent IUPAC publication[6] and this volume also contains excellent reviews of the critical state[7a] and of the experimental investigation of high-pressure phase equilibria[7b]. Another, more detailed, survey of this latter topic has been given by Young[8].

The investigation of phase equilibria in binary mixtures in which the critical points of the two pure components are connected by a continuous critical line can usually be undertaken using apparatus operating up to 100 or 200 bar. In many cases, the complete phase diagram can be determined by employing simple equipment in which the mixture is confined over mercury. Isothermal pressure measurements at different degrees of compression enable dew, bubble and critical points to be determined. These transition points can be determined either visually or by an analysis of the resultant (p, V) isotherms. Although efficient stirring of the fluid phases usually presents a formidable problem, such apparatus is capable of high accuracy, although the solubility of the mercury in the other phases is a possible source of error[9]. Such error can be eliminated by using gallium as a confining fluid of low vapour pressure, but its ease of oxidation poses handling problems[10]. Recently-designed equipment of this type has been described by Kay[11] and by McGlashan[12].

Fluid mixtures which exhibit liquid–liquid immiscibility or in which there are no continuous loci of critical points usually require measurements at pressures well above 100 bar before their phase diagrams can be fully elucidated. A different type of apparatus is required because the confining liquid must be eliminated. Such apparatus can be constructed in one of three configurations.

The first of these is the single-pass flow design in which the more volatile component in the gaseous state is passed through a liquid mixture at a measured temperature and pressure. The effluent gaseous phase is sampled and analysed, as is the residual liquid phase (or phases). Such an apparatus is particularly useful for studying mixtures containing high concentrations of the volatile component and the major uncertainty is the possible lack of attainment of equilibrium within the liquid sample[13].

This uncertainty is greatly reduced in the vapour recirculation apparatus, the second method in common use[14]. In this design the vapour is removed from the sample chamber by a built-in recirculating pump, passed through a vapour-sampling loop, and then injected back into the liquid. After a suitable interval, the vapour and liquid are sampled and analysed and the temperature and pressure are monitored continuously. The single-pass and recirculation methods are often used by the same research group to study the same mixture over different concentration regions and dual-purpose, hybrid apparatuses are common[15].

At extremes of pressure, above 10^4 bar, the only practical technique is the static one in which the fluid is confined in a sample chamber that is often equipped with windows so that direct visual identification of the presence of the several fluid phases is possible. Stirring, and thus the achievement of equilibrium, is often the major problem, and frequently rocking the sample chamber is the only feasible method of agitation. Nevertheless such apparatus has been used successfully to obtain phase diagrams of great complexity under extreme conditions[16].

Because of the great variety of phase behaviour revealed by experimentalists, a comprehensive discussion of their results is inhibited unless the different types of diagram are classified in some way. The classification used in this chapter is that due to Scott and van Konynenburg[17], who showed that the van der Waals equation, when applied to mixtures, predicted qualitatively almost all of the types of phase equilibrium that have been discovered experimentally and, indeed, they postulated some thermodynamically-allowable diagrams that have not yet been verified in the laboratory. Scott and van Konynenburg's work is discussed more fully in Chapter 8, while in this chapter we make use of their classification without direct reference to liquid mixture theory.

The six principal classes of phase diagram are best differentiated by consideration of their (p, T) projections as shown in *Figure 6.1*. It is seen from this figure that type I mixtures have a continuous gas–liquid critical line and complete miscibility of the liquid phases at all temperatures. The additional feature of type II is the presence of liquid–liquid immiscibility at relatively low temperatures, the locus of UCSTs remaining distinct from the gas–liquid critical line. Types III, IV and V are distinguished by the absence of a continuous gas–liquid critical curve, type IV exhibiting both a UCST and a LCST and type V a LCST only. Mixtures conforming to type VI are invariably composed of complex molecules, in the sense that hydrogen bonding or other strong intermolecular forces are present, and display both a UCST and a LCST at temperatures well removed from the gas–liquid critical temperature of the more volatile component. Type I mixtures are described in Section 6.3, types II and VI in Section 6.4, type III in Section 6.5 and types IV and V in the final section of this chapter, Section 6.6. First, we derive some important thermodynamic relationships concerning the gas–liquid critical state of binary mixtures.

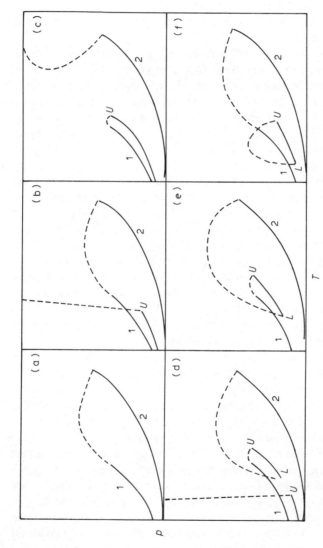

Figure 6.1 The six principal types of phase diagram for binary mixtures shown as (p, T) projections. Solid curves 1 and 2 are the vapour pressure curves of the two pure components, the other solid curves are three-phase (LLG) lines. U and L are upper critical and lower critical end points respectively, and the dashed lines are critical loci. (a) Type I, argon + krypton; (b) type II, xenon + hydrogen chloride; (c) type III, neon + krypton; (d) type IV, methane + 1-hexene; (e) type V, methane + n-hexane; (f) type VI, water + 2-butanone

6.2 Gas–liquid critical states of binary mixtures

If a homogeneous liquid mixture in equilibrium with an equal volume of vapour is heated, and if the pressure is adjusted to maintain the equality of volumes, then the system comes eventually to a critical point at which all the intensive properties of the coexisting phases have become the same. Such a critical point is physically but not thermodynamically analogous to the gas–liquid critical point of a pure substance.

It will become clear later in this chapter that there is often no absolute distinction between gas–liquid and liquid–liquid critical points in mixtures, for both are points of incipient material but not of mechanical instability. Thus the critical points described in this section, although having many of the qualities of the gas–liquid critical points of pure substances, are described by the equations of Section 4.8.

The proof of the statement that these critical points are not ones of mechanical instability can be obtained only by discussing the behaviour of the system on a (a, v, x) surface, at constant temperature, for the molar Helmholtz free energy is (classically) a continuous differentiable function of v and x in regions of both material and mechanical instability. The (g, p, x) surface, at constant temperature, cannot properly be used until it has been shown that the system is mechanically stable.

The conditions of equilibrium at constant temperature of two phases, prime and double prime, may be written

$$p' = p'' \tag{6.1}$$

$$\mu_1' - \mu_2' = \mu_1'' - \mu_2'' \tag{6.2}$$

$$\mu_2' = \mu_2'' \tag{6.3}$$

or, in terms of a and its derivatives,

$$a_v' = a_v'' \quad \text{(cf. 3.23)} \tag{6.4}$$

$$a_x' = a_x'' \quad \text{(cf. 3.24)} \tag{6.5}$$

$$a' - v'a_v' - a'a_x' = a'' - v''a_v'' - x''a_x'' \quad \text{(cf. 3.25)} \tag{6.6}$$

If the mixture is to be stable under fluctuations of density and composition, then the determinant

$$D = \begin{vmatrix} a_{2v} & a_{xv} \\ a_{xv} & a_{2x} \end{vmatrix} \quad \text{(cf. 3.26)} \tag{6.7}$$

must be positive. This condition requires that a_{2x} is positive (a condition that is always satisfied in simple systems such as those discussed in this section and is formally analogous to U_{2v} being positive in a system of one component) and that a_{2v} is positive (the condition of mechanical stability and the analogue of thermal stability in a system of one component). It requires also that

$$D = -(\partial p/\partial v)_x(\partial^2 g/\partial x^2)_p > 0 \tag{6.8}$$

This equation alone suggests that D could be zero by the system becoming either mechanically or materially unstable.

However, it is easy to show that only one of these limits is reached at a critical point (with the exception of a special case discussed below). Consider the Taylor expansion of a about its value at the critical point in powers of $\delta x = (x - x^c)$ and $\delta v = (v - v^c)$. Substitution of the first terms of this expansion and of its differentiated forms into equations (6.4) and (6.5) gives

$$a_{xv}\Delta x + a_{2v}\Delta v = 0 \tag{6.9}$$

$$a_{2x}\Delta x + a_{xv}\Delta v = 0 \tag{6.10}$$

where Δx is $x' - x''$. Either of these equations gives the slope of the tie-line joining (v', x') to (v'', x'') on the (v, x) projection of the (a, v, x) surface. At the critical point, this tie-line becomes the tangent to the curve bounding the unstable region, that is, to the curve along which D is zero. The equation for the tangent is, therefore, any one of the three

$$a_{xv}\, dx + a_{2v}\, dv = 0 \tag{6.11}$$

$$a_{2x}\, dx + a_{xv}\, dv = 0 \tag{6.12}$$

$$D_x\, dx + D_v\, dv = 0 \tag{6.13}$$

The first two may be written

$$\left[-\left(\frac{\partial p}{\partial x}\right)_v\right]dx + \left[-\left(\frac{\partial p}{\partial v}\right)_x\right]dv = 0 \tag{6.14}$$

$$\left[\left(\frac{\partial^2 g}{\partial x^2}\right)_p - \left(\frac{\partial p}{\partial x}\right)_v^2 \bigg/ \left(\frac{\partial p}{\partial v}\right)_x\right]dx + \left[-\left(\frac{\partial p}{\partial x}\right)_v\right]dv = 0 \tag{6.15}$$

If these two equations are to represent the same tangent, then it is clear that $(\partial^2 g/\partial x^2)_p$ must be zero, whatever the value of $(\partial p/\partial v)_x$. The third equation, (6.13), gives

$$a_{3x} - 3a_{2xv}\left(\frac{a_{xv}}{a_{2v}}\right) + 3a_{x2v}\left(\frac{a_{xv}}{a_{2v}}\right)^2 - a_{3v}\left(\frac{a_{vx}}{a_{2v}}\right)^3 = 0 \tag{6.16}$$

or

$$(\partial^3 g/\partial x^3)_p = 0 \tag{6.17}$$

Figure 6.2 shows the (v, x) projection of the (p, v, x) surface, which may be derived from the (a, v, x) surface by differentiation. Three boundary curves are shown. The outermost is that of the coexisting or equilibrium volumes and was called the *connodal* or *binodal curve* by van der Waals and his school. The next is the boundary curve for material instability which is tangential to the equilibrium curve at the critical point. It is the *spinodal curve* of van der Waals. The innermost is the boundary curve for mechanical stability, that is, the locus of the maxima and minima of the lines

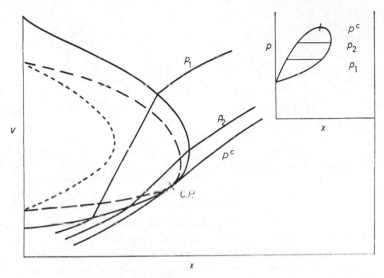

Figure 6.2 *The* (v, x) *projection of a* (p, v, x) *surface at a temperature above the critical point of the components*

Boundary curves:

——————— *coexistence curve*

– – – – *limit of material stability*

- - - - - *limit of mechanical stability*

Three isobars are shown, the highest of which passes through the critical point (CP). The inset shows the (p, x) *projection of the surface*

representing the hypothetical continuous variation of p with v, at constant x, through the two-phase region. This curve does not go near the critical point but meets that of material instability at $x = 0$. This meeting follows directly from equation (6.7) which shows that D can be zero for a pure substance only if a_{2x} is zero, since a_{2v} becomes infinite as x approaches zero (or unity).

The separation between the curves of mechanical and material stability justifies the treatment of the critical point of binary mixtures by means of a (g, p, x) surface at constant temperature or a (g, T, x) surface at constant pressure, and equations (4.132)–(4.139) are applicable equally to a gas–liquid and to a liquid–liquid critical point in a binary mixture. The following conclusions follow from these classical equations:

(1) The derivatives $(\partial p/\partial x)'_{T,\sigma}$ and $(\partial T/\partial x)'_{p,\sigma}$ vanish at the critical point and have opposite signs to $(\partial p/\partial x)''_{T,\sigma}$ and $(\partial T/\partial x)''_{p,\sigma}$ respectively. The equation for the temperature derivatives is (4.139); the analogous one for the pressure derivatives is

$$\left(\frac{\partial x}{\partial p}\right)'_{T,\sigma} = -\left(\frac{\partial x}{\partial p}\right)''_{T,\sigma} = -\frac{6v^c_{2x}}{\Delta x \, g^c_{4x}} \tag{6.18}$$

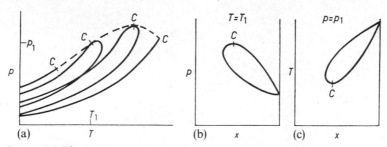

Figure 6.3 The (p, T) projection and (p, x) and (T, x) sections of the (p, T, x) surface. The critical line is shown dashed, and critical points are marked C

The (p, x) and (T, x) loops for a typical system in which $p_2^c > p_1^c$ and $T_2^c > T_1^c$ are shown in *Figure 6.3*(b) and (c). The critical point is always at an extreme value of p (at constant T) and of T (at constant p).

(2) The derivative $(\partial p/\partial T)_{x,\sigma}$ is generally non-zero at the critical point, which is neither at the maximum temperature nor at the maximum pressure of the (p, T) loop at constant x. This is a necessary consequence of the rounded ends of the (p, x) and (T, x) loops. All critical points lie on the envelope of the (p, T) loops, as is readily seen in *Figure 6.3*(a) by considering neighbouring loops of infinitesimally different compositions. *Figure 6.4* combines *Figure 6.3*(a), (b) and (c) into a three-dimensional sketch.

(3) The critical point is not generally at the end of the (v, x) curve of *Figure 6.2*, as the tie-lines do not connect phases of equal composition. An azeotrope is an exception to this and is discussed separately below. The derivative $(\partial p/\partial x)_{T,\sigma}$ is infinite at the end of this loop and so, by equation (4.113)

$$(\partial v/\partial x)_{p,T} = \Delta v/\Delta x \tag{6.19}$$

That is, the isobar that passes through this point (p_2 in *Figure 6.1*) is collinear with the tie-line.

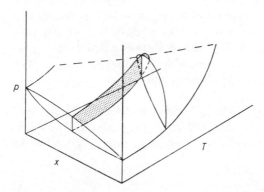

Figure 6.4 The (p, T, x) surface. The two sections drawn have a common critical point

It is seen that there is nothing thermodynamically unusual about the shape of the (p, x), (T, x) or (p, T) loops. Their rounded ends are a natural and necessary consequence of the assumption that the Helmholtz free energy is a continuous differentiable function of v and x. Nevertheless, the shape of these loops gives rise to experimental behaviour which is, at first sight, odd.

Consider the head of a (p, T) loop of constant composition such as that shown in *Figure 6.5(a)*. The critical point, C, is neither the point of maximum pressure, A, nor that of maximum temperature, B. If a mixture of this composition is compressed isothermally at a temperature below C, then normal condensation occurs. The first drop of liquid appears at the dew point (dp) and the last bubble of gas vanishes at the bubble point (bp). If, however, the temperature is between C and B, then complete condensation never occurs. The critical point, C, is the meeting point of the dew and bubble curves, and so a line of increasing pressure at a temperature between C and B cuts the (p, T) loop at two dew points.

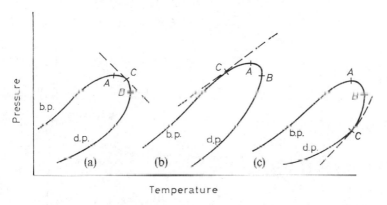

Figure 6.5 The three types of (p, T) loop

On compression, the amount of liquid increases to a maximum and then falls again to a second dew point. This evaporation of a liquid by an isothermal increase of pressure is the opposite of the normal behaviour of a mixture and of the invariable behaviour of a pure substance. The phenomenon was therefore called *retrograde condensation* by Kuenen[18] who was the first to observe it under equilibrium conditions and to interpret it correctly.

Figure 6.6 shows some of Kuenen's original observations on a mixture of carbon dioxide and chloromethane with a mole fraction of the former of 0.41, and at a temperature of 378 K. These conditions correspond to those of *Figure 6.5(a)* where this temperature lies between the critical point, C, of 375 K and the maximum of the (p, T) loop, B, of 379.7 K. The sudden onset of retrograde behaviour at 82.7 bar is very marked. However, the critical point need not necessarily lie between A and B but can lie outside them on either side, as is shown in *Figure 6.5(b)* and (c). If the order of the points is

C, A, B, then retrograde behaviour between two dew points occurs again, as is clear from *Figure 6.5*(b). Isotherms that pass through two dew points were said by Kuenen to show *retrograde behaviour of the first kind.* An isotherm that passes through two bubble points exhibits *retrograde behaviour of the second kind.* This is less common but occurs when the order of the three points is *A, B, C* as in *Figure 6.5*(c). Similarly, unorthodox behaviour is found on changing the temperature at constant pressure. Retrograde isobars can join a pair of dew points, as in *Figure 6.5*(b), or one of bubble points, as in *Figure 6.5*(a) and (c).

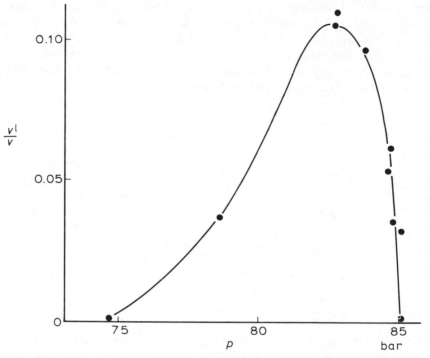

Figure 6.6 Retrograde condensation. The proportion of the volume occupied by the liquid phase in carbon dioxide + methyl chloride on isothermal compression at 378.2 K. The dew points are at 74.5 and 85.0 bar (calculated from the results of Kuenen[18])

There are no agreed names for the points *A, B* and *C* of *Figure 6.5*. Point *C* is the only one properly called a critical point, since it is a state of the system in which the two phases have become identical. It was called a *plait point* by van der Waals and his school, since it is the point at which a plait or fold develops in the (*a, v, x*) surface. Kuenen calls point *B* the *critical point of contact,* which he unfortunately sometimes shortens to *critical point.* This is not a happy choice, as *B* is not a critical point in the usual sense of that term. The names cricondenbar or maxcondenbar have been suggested for point *A* and cricondentherm or maxcondentherm for point *B*[19].

An interesting situation arises if the critical point is at the end of the (v, x) loop, as in *Figure 6.7*. The coexistence curve on this figure is the (v, x) projection of a line in a three-dimensional (p, v, x) space. The line has the following properties. First, the value of the pressure at the critical point is an extremum—usually a maximum. This follows directly from the fact that the tie-lines are perpendicular to the p-axis and converge to zero length at the critical point. Secondly, the bubble-point pressure always exceeds the dew-point pressure. That is, for any given composition, x^*, the equilibrium pressure $p''(x^*)$ is greater than $p'(x^*)$ in *Figure 6.7*. These two properties of the (p, v, x) line require that $(\partial p/\partial x)'_{T,\sigma}$ and $(\partial p/\partial x)''_{T,\sigma}$ both approach zero at the critical point. It is clear from the diagram that $(\partial v/\partial x)'_{T,\sigma}$ and $(\partial v/\partial x)''_{T,\sigma}$ are both infinite at this point and therefore $(\partial p/\partial x)^c_{v,T}$ is zero and $(\partial v/\partial x)^c_{p,T}$ is infinite. Hence,

$$\left(\frac{\partial p}{\partial v}\right)^c_{x,T} = -\frac{(\partial p/\partial x)^c_{v,T}}{(\partial v/\partial x)^c_{p,T}} = 0 \tag{6.20}$$

or

$$a^c_{xv} = 0 \qquad a^c_{2x} = 0 \qquad D^c = 0 \tag{6.21}$$

Thus the critical point is now both materially and mechanically unstable and the fluid near this point behaves as a pure substance. For example, mechanical stability near the critical point requires that a^c_{3v} is also zero. Such a system cannot be discussed on a (g, p, x) surface, since the infinity in $(\partial v/\partial x)_{p,T}$ produces a discontinuity.

This behaviour is found whenever an azeotrope persists up to the critical point. The (p, x) loop now ends in a cusp and not with a rounded end, for

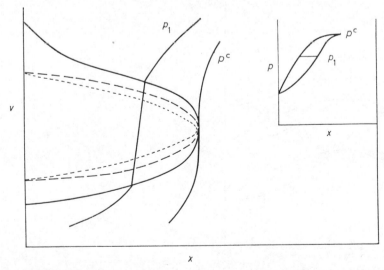

Figure 6.7 The gas–liquid critical point of an azeotrope. The (v, x) projection and, inset, the (p, x) projection. The conventions are those of Figure 6.2

the only one of the three functions (6.4)–(6.6) that retains any first-order terms is a_x for which the leading terms are

$$a_x = a_x^c + a_{2x}^c(\delta x) + \tfrac{1}{2}a_{3x}^c(\delta x)^2 + a_{2xv}^c(\delta x)(\delta v) + \tfrac{1}{2}a_{x2v}^c(\delta v)^2 + \cdots$$
(6.22)

or

$$0 = a_{2x}^c[\delta x' - \delta x''] + \tfrac{1}{2}a_{x2v}^c[(\delta v')^2 - (\delta v'')^2] + \cdots$$
(6.23)

Hence $\delta x'$ is equal to $+\delta x''$ and is of the order of $(\delta v)^2$. This differs from the normal critical mixture for which $\delta x'$ is equal to $-\delta x''$ and is of the order of δv. The shape of the cusp is shown in *Figure 6.7*.

  If the critical point moves farther round the (v, x) loop to larger volumes, then the critical point again becomes normal, that is, non-azeotropic, but is now at a minimum of the (p, x) loop, not a maximum. The system now forms an azeotrope at a pressure above the critical for that temperature and at a different composition. The transition, with changing temperature, from normal behaviour, through the critical azeotrope, to this inverted behaviour is shown in *Figure 6.8*, which is based on a typical simple type I system carbon dioxide + ethene[20]. The critical line in such a diagram is the locus of the extrema of the (p, x) loops (or cusp) and not, as was often misleadingly shown in early published diagrams[21], the envelope of these loops. Such an envelope has little significance, for only on the (p, T) projection is it the critical line. Thus the coincidence of an azeotropic and a critical point leads to simultaneous mechanical and material instability, whereas one of these points without the other produces one type of instability only.

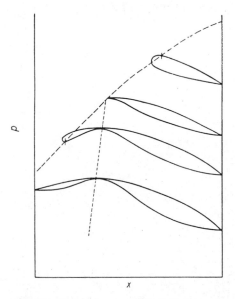

Figure 6.8 Intersection of an azeotropic line (dotted) and a critical line (dashed). Four isotherms are shown

The conclusions of this section are qualitatively correct, but if we tried to extract more detailed results from the classical equations we should meet the same quantitative errors as found in Chapter 3 and in Section 4.8; that is, the detailed shapes of the thermodynamic lines are not those predicted by the classical assumption that the Helmholtz free energy is an analytic function of v and x at a critical point. We return to this point in Section 8.9.

6.3 High-pressure phase equilibria: type I mixtures

The usual conditions under which a binary mixture will conform to type I behaviour, with a continuous gas–liquid critical line and the absence of liquid–liquid immiscibility, are that the two substances should be of similar chemical types and/or their critical properties should be comparable in magnitude. Typical examples are thus argon + krypton[22], methane + ethane[23], methane + nitrogen[24], carbon dioxide + oxygen[25], benzene + cyclohexane[26] and n-hexane + *cis*-decalin[27]. These mixtures are composed solely of non-dipolar molecules and whereas the majority of type I systems are in this category a few are known that contain polar substances, e.g., sulphur dioxide + chloromethane[28] and hydrogen chloride + dimethylether[29].

Mixtures of substances belonging to a particular homologous series deviate from simple type I behaviour only when their size difference, and thus the values of their critical properties, exceeds a certain ratio. For example, in the n-alkane series, with methane as one component a change from type I to type V occurs first with n-hexane[30], whereas, with ethane as the lighter component, n-nonodecane[31,32] is the first to exhibit partial miscibility in the liquid state. With propane, this behaviour would probably occur when the second component is a long-chain hydrocarbon in the range $n-C_{40}$ to $n-C_{50}$ but the phase diagram is complicated by the freezing of the heavier component at temperatures approaching the gas–liquid critical temperature for propane (369.82 K).

Type I mixtures can be conveniently subdivided first by considering the shape of the continuous critical line that connects the critical points of the two pure components and further, by noting the presence or absence of azeotropy. *Figure 6.9* represents the several possible shapes of the (p, T) projection of the critical curve. Mixtures approximating to curve (c), where the critical locus is almost linear in both the (p, T) and (T, x) planes, are usually formed from substances with very similar critical properties and whose mixtures, at lower pressures, exhibit only small departures from Raoult's law. Examples are carbon dioxide ($T^c = 304.1$ K, $p^c = 73.8$ bar) + nitrous oxide[33] ($T^c = 309.6$ K, $p^c = 72.4$ bar), and benzene ($T^c = 562.2$ K, $p^c = 49.0$ bar) + toluene[34] ($T^c = 591.8$ K, $p^c = 41.0$ bar).

Mixtures corresponding to curve (b) where the critical curve is convex upwards and frequently exhibits a maximum in both the (p, T) and (p, x) planes are extremely common among type I systems and occur whenever there are moderately large differences between the critical temperatures or

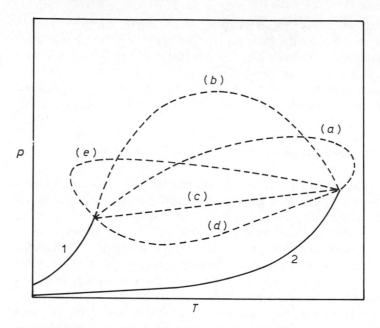

Figure 6.9 Five possible shapes of continuous critical loci for type I mixtures:
(a) *cycloheptane + tetraethylsilane;* (b) *argon + methane;* (c) *carbon dioxide + nitrous oxide;* (d) *propane + hydrogen sulphide;* (e) *acetone + n-pentane*

volumes of the pure components. There is, for example, a clear increase in curvature along the series n-hexane + n-heptane[34], + n-octane[34], + n-decane[27], + n-tridecane[27] and + n-tetradecane[27]. When the difference in critical temperature and critical volume exceeds a certain value, the critical locus becomes discontinuous and type V or type IV behaviour results. The phase diagrams of many mixtures of simple molecules frequently correspond to type I, curve (*b*) behaviour. Typical examples are argon + krypton[22], which is illustrated in *Figure 6.10*(a), argon + methane[35] and krypton + methane[36].

　　Mixtures whose critical curve conforms to the shape of curve (*a*) are extremely rare. Because the critical line goes initially to higher temperatures before bending back to the critical point of the more volatile component, there are maxima in both the (p, T) and (T, x) planes. This means that there is partial phase immiscibility over a certain range of concentration, such immiscibility occurring at temperatures higher than the critical temperatures of both pure components. By common definition the substances should thus be considered to be in the gaseous state and phase behaviour such as this is one example of so-called gas–gas immiscibility. This phenomenon is much commoner in type III mixtures and is discussed in more detail in Section 6.5. When, as in the present case, this behaviour is associated with a continuous critical line, for example, in the binary mixtures hydrogen chloride + diethylether[29] and cycloheptane + tetra-ethylsilane[37], it has been termed by Scott[38] 'gas–gas immiscibility of the third kind'. Type I mixtures whose critical loci are everywhere concave upwards and may exhibit minima as in curve (d) illustrated in *Figure 6.9* are indicative of large positive deviations from Raoult's law and hence of weak, unlike intermolecular interactions. Such behaviour is thus found for binary mixtures of a polar with a non-polar substance, and for some mixtures of aromatic hydrocarbons with aliphatic or alicyclic hydrocarbons. Typical examples are ethane + hydrogen chloride[39], propane + hydrogen sulphide[40], methanol + benzene[41], n-hexane + benzene[34], n-hexane + toluene[42] and cyclohexane + benzene[26].

The last type of critical locus, shown in *Figure 6.9*, curve (e), where the (p, T) projection extends through a temperature minimum, is observed for several mixtures and is usually associated with the occurrence of a positive azeotrope extending up to the critical line. Examples are acetone + n-pentane[43], + n-hexane[43] and + n-heptane[43], water + 1-propanol[44] and carbon dioxide + ethane[44,45]. The (p, T) projection for this latter mixture is illustrated in *Figure 6.10*(b). The azeotropic line is tangential to the projection of the critical locus but does not meet it at the temperature minimum.

An example of a mixture that forms a negative critical azeotrope is shown in *Figure 6.10*(c). Such behaviour is rarely found experimentally for it is associated with strong intercomponent interactions and negative deviations from Raoult's law. The required degree of interaction seldom persists up to temperatures of the gas–liquid critical region.

Continuous critical loci are not confined to the five simple types shown in *Figure 6.9*. The mixture, ammonia + 2,2,4-trimethylpentane[46], for example, is found to possess a highly convoluted critical locus where the (p, T)

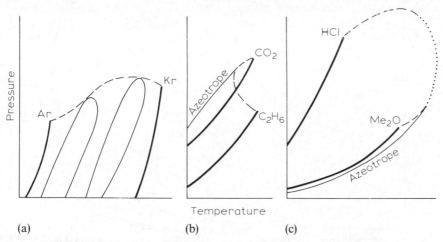

Figure 6.10 The (p, T) diagrams of three systems. The first is normal, the second has a positive, the third a negative azeotrope. The critical lines are dashed (or dotted, where not determined experimentally). Two (p, T) loops for mixtures of fixed composition are shown in the first diagram, and the azeotropic lines in the second and third

projection, commencing at the critical point of pure ammonia, first exhibits a temperature minimum, then a pressure minimum and finally a pressure maximum.

In principle, there is no restriction on even more complicated behaviour, as, for example, has recently been claimed for the type I system methane + n-pentane[47]. *Figure 6.11* illustrates the (p, x) diagram for this system at high methane concentrations and at temperatures near the methane critical temperature (190.6 K). Three isotherms are shown at 190.6, 192.2 and 198.2 K. At the highest temperature, the isotherm is unexceptional, exhibiting ordinary retrograde condensation. The dew-point curve develops a double S-shape at 192.2 K and, over a very restricted range of composition, quadruple dew points are found so that, on isothermal compression, the

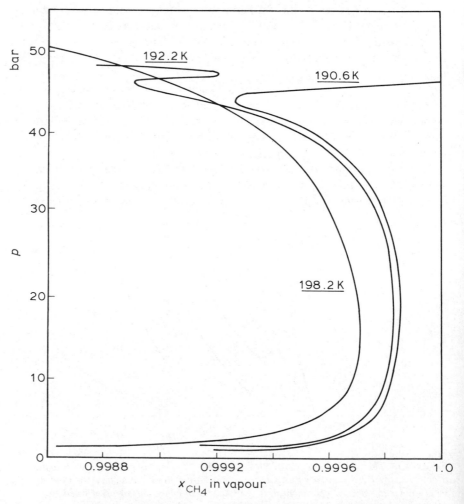

Figure 6.11 Isotherms near the critical temperature of methane for methane + n-pentane mixtures illustrating double-, triple- and quadruple-valued dew points

retrograde condensation phenomenon occurs twice. Triple dew points appear at the lowest temperature illustrated and these are likely to persist to temperatures just below the methane critical temperature. Triple dew points have been detected under those latter conditions in methane + n-butane[48] mixtures, also at very high methane concentrations and over a very narrow range of composition.

6.4 High-pressure phase equilibria: type II and type VI mixtures

Type II and type VI mixtures are characterized by the presence of liquid–liquid immiscibility at temperatures below the gas–liquid critical temperature of the more volatile component. There is thus one additional critical line. In the case of type II systems, the extra line is the locus of liquid–liquid UCSTs extending down, with either positive or negative slopes, to low pressures where it terminates at an upper critical end point (UCEP) and a third, gaseous, phase is formed. A three-phase line (LLG) extends from the UCEP to lower pressures and temperatures.

The additional critical line in type VI mixtures has a section that represents the locus of LCSTs and terminates on the three-phase line at a lower critical end point (LCEP) with $T_{UCEP} > T_{LCEP}$. The two sections of the liquid–liquid critical line may merge at higher temperatures and pressures as is shown in *Figure 6.1*(f) or may diverge as the pressure or temperature is increased. In both type II and type VI mixtures, the phase diagrams may be complicated by the presence of azeotropic lines but the essential features of both are the continuous gas–liquid critical line and the fact that the gas–liquid and liquid–liquid critical phenomena remain quite distinct. It is likely that many type I mixtures would conform to type II behaviour at sufficiently low temperatures if the presence of a solid phase did not first intervene.

The effect of pressure on UCSTs and LCSTs was studied systematically by Timmermans, first in association with Kohnstamm[49] at Amsterdam and later with his own group at Brussels[50]. More recently, Franck[51] and Schneider have made major contributions in this area and the latter, in particular, has written several comprehensive reviews[52].

Equation (4.147) relates the pressure dependence of a liquid–liquid critical temperature to the shape of the v^E and s^E curves. If neither v^E nor s^E has a point of inflection, then the sign of (dT^c/dp) is that of the ratio $(v^E/s^E)^c$. At a UCST the excess entropy is generally positive and the second derivative s_{2x} is always negative. Hence, if v^E is also positive, an increase of pressure leads to an increase in the critical temperature, whereas a negative v^E gives rise to a negative value of (dT^c/dp). The opposite behaviour is encountered at a LCST where s^E is necessarily negative. Examples of all four types of pressure dependence are now known and are referred to later in this section where it will be seen that (dT^c/dp) itself can change sign with applied pressure as a result of a reversal of the sign of v^E.

The (p, x) graphs at low and medium pressures in the region of two liquid phases may be of four kinds, as shown in *Figure 6.12*. The first type occurs when the pressure of the three-phase line, LLG, lies between the saturated vapour pressures of the pure components, so no azeotrope is formed. In the second, a homogeneous positive azeotrope is formed outside the composition range of immiscibility. In the third, a heterogeneous azeotrope (or hetero-azeotrope) is formed in which the pressure of the LLG line is the highest at which vapour can exist at that particular temperature. The composition of the vapour lies between those of the immiscible liquids and not, as in *Figure 6.12*(a) and (b), outside them. *Figure 6.12*(d) shows the formation of a homogeneous negative azeotrope at a composition well removed from those of the immiscible liquids. The (T, x) projections corresponding to the diagrams illustrated in *Figure 6.12* are shown in *Figure 6.13*.

Examples of all four types of behaviour are known but, although the liquid–liquid critical loci have been followed up to high pressures so that their shapes are well established, the gas–liquid critical region has rarely been examined with the same thoroughness so that it is sometimes impossible to place a particular mixture with absolute certainty within the proper class, II to VI.

Mixtures that conform to the (p, x) diagram of *Figure 6.12*(a) are composed of partially miscible liquids in which the pure components have very different volatilities. Examples are n-pentane + nitrobenzene[53] which has a negative value of (dT^c/dp) and carbon dioxide + n-octane[54], + n-decane[55],

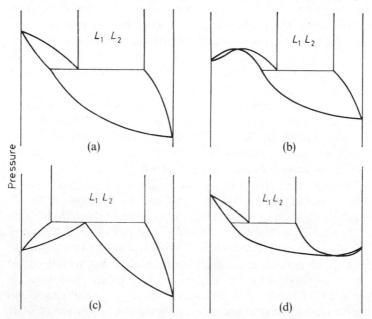

Figure 6.12 The four types of (p, x) curve found with partially miscible liquids. The area L_1L_2 is the region of two liquid phases

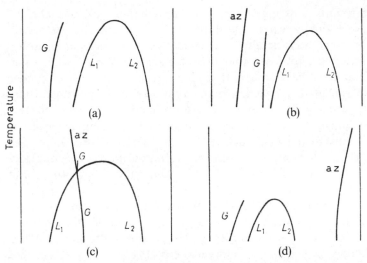

Figure 6.13 The (T, x) projections of the four types of system shown in Figure 6.12. It is assumed that all the liquid boundary curves, L_1 and L_2, meet at a UCST. The lines marked G are the compositions of the gaseous phases in equilibrium with the immiscible liquids, and those marked 'az' the compositions of the homogeneous azeotropes

+n-undecane[55] and +2-octanol[56], in which the UCST increases with applied pressure.

The second type of diagram, Figure 6.12(b), is less common but occurs for a few aqueous mixtures such as water + phenol[53,57] and water + 2-butanone[58]. The positive azeotrope in water + phenol mixtures disappears as the temperature is raised so there is no critical azeotropy. The simple mixture carbon dioxide + ethane[59] is almost certainly in this category for it forms a positive azeotrope and, by comparison with other carbon dioxide + n-alkane mixtures, would exhibit a UCST at low temperatures but for the formation of the solid phase of carbon dioxide which intervenes before the UCST is reached. Another simple mixture in this category is xenon + hydrogen chloride[60] where the existence of the positive azeotrope is well established above 159 K and, judging by the shape of the isotherms at that temperature, a UCST is expected to form at slightly lower temperatures.

The third type, shown in Figure 6.12(c), is probably the commonest and is found whenever two liquids of similar vapour pressure are only partially miscible. Mixtures of a fully- or near fully-fluorinated aliphatic or alicyclic fluorocarbon with the corresponding hydrocarbon are the classic examples of this type, exemplified by methane + tetrafluoromethane[61] and cyclo-hexane + perfluorocyclohexane[62]. The former mixture has been studied at pressures up to 1400 bar and the experimentally observed value of (dT_{UCST}/dp) of +0.013 K bar^{-1} is in accord with the relatively large positive value of v^E noted in Table 5.4. Trifluoromethane + methane[61] and + tetrafluoro-methane[63] and ethane + tetrafluoromethane[63] and + trifluormethane[63] all behave similarly. Other mixtures of this type are methanol + cyclohexane[57,64] and, perhaps surprisingly, water + naphthalene[65], +diphenyl[66], +tetralin[67], +*trans*-decalin[67] and +*cis*-decalin[67]. The fact that the critical temperatures of these hydrocarbons approach that of water leads to the probable existence of a continuous gas–liquid critical line although this has only been confirmed experimentally for the first-named mixture. All five hydrocarbons are thus wholly miscible with water in temperature and pressure ranges that are associated unequivocally with the liquid state. These systems all revert to type (b) in the immediate neighbourhood of the critical solution point. This must happen necessarily unless, by chance, the composition of the azeotrope is exactly equal to that of the liquid phases at their critical point. In general, this does not occur (*Figure 6.13*(c)).

Examples corresponding to *Figure 6.12*(d) are extremely rare since the occurrence of this behaviour at low vapour pressures implies that the excess Gibbs function is negative at the azeotrope but positive at the immiscibility. The best example is probably acetic acid + triethylamine[71]. A negative azeotrope is observed at high acid concentrations but two, partially-miscible liquid phases are formed at high amine concentrations with a UCST at 403 K and $x_{acid} = 0.25$. Although the gas–liquid critical line has not apparently been investigated experimentally, the mixture is likely to be type II for the critical temperatures, 592.7 K for the acid and 535.6 K for the amine, are relatively close together and both are well above the UCST so that a continuous critical locus is highly probable. This mixture has extremely large negative values of all the excess functions at 298 K; the largest values in the region of the negative azeotrope, $x_{acid} = 0.69$, are $g^E \simeq -6.2 \, kJ \, mol^{-1}$, $h^E \simeq -13 \, kJ \, mol^{-1}$ and $v^E \simeq -12 \, cm^3 \, mol^{-1}$. The large negative value of g^E for a mixture that undergoes phase separation implies a very strong degree of inter-component interaction so that it cannot be considered to be a simple binary mixture. There is good circumstantial evidence that strong attractive forces between highly polar, equimolar, acetic acid:triethylamine complexes and hydrogen-bonded acetic acid dimers lead to the formation of 3:1 aggregates. These highly-associated quaternary aggregates are only partially miscible with additional amine, and thus phase separation occurs below the UCST.

Aqueous solutions of hydrogen chloride[68], hydrogen bromide[68] and sulphur dioxide[68,69] are other examples of systems in which the properties of the mixtures are quite different at the two extremes of composition and whose low temperature phase behaviour corresponds to that depicted in *Figure 6.12*(d). Weak solutions are ionic and the mixtures form negative azeotropes because of the large enthalpy of solvation of the ions. However, mixtures rich in hydrogen chloride, hydrogen bromide or sulphur dioxide are little ionized and Roozeboom showed in 1888 that the three liquids were only partly miscible with water at high concentrations and therefore at high vapour pressures. Water + hydrogen chloride has been reinvestigated more recently by Kao[68] and by Marshall and Jones[70]. The latter workers examined the critical locus in the neighbourhood of the water critical point only. Because the critical temperature of hydrogen chloride (324.7 K) is low and lies beneath that of the UCST, this mixture, and the other two aqueous systems mentioned above, is type III, in character with a discontinuous critical locus.

It is convenient to consider type VI mixtures in this section for the only feature that such systems present in addition to those already displayed by type II mixtures is the presence of a LCST that occurs at lower temperatures than the UCST. Type VI systems were not included in the classification of Scott and van Konynenburg[17] for they are not predicted by the van der Waals equation or any other theory that assumes a spherically-symmetric intermolecular potential. It is shown later in Chapter 8 that mixture theories incorporating angle-dependent potentials predict the formation of type VI as well as all the other classes of phase diagram.

The (p, T, x) diagram of a typical type VI system is shown in *Figure 6.14* and the (p, T) projection in *Figure 6.1(f)*. The mixture is seen to form both a UCST and a LCST and the positive value of $(dT_{LCST}/dp)_x$ and the negative value of $(dT_{UCST}/dp)_x$ leads to the formation of a closed dome of immiscibility. The two critical curves meet at an upper critical pressure, above which the two liquid phases are miscible in all proportions. Such points in (p, T, x)-space, where $(dT^c/dp) = \pm\infty$, have been called hypercritical solution points. Type VI behaviour is only found in mixtures of chemically-complex substances where one or both of the pure components exhibit self-association due to H-bonding and, in the mixture, there is the added complication of strong inter-component H-bonding interactions. Water (or D_2O) is, almost always, one of the components. The region or regions of

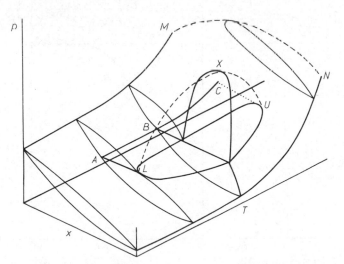

Figure 6.14 The (p, T, x) surface for a typical type VI mixture where LX and UX, the loci of the LCST and UCST (shown dashed) meet at X, an upper critical pressure, and a closed dome of immiscibility is formed. The gas–liquid critical locus is also shown dashed, and M and N are the gas–liquid critical points of the two pure substances. ABC is the (p, T) projection of the three-phase line

immiscibility usually occur at low temperatures, well removed from those of the gas–liquid critical lines, and so the latter are found to be continuous.

Four possible types of immiscibility phenomenon are shown in *Figure 6.15* as (T, p) projections. The first, for water + 2-butanone[72], forms a closed dome of immiscibility as in *Figure 6.14*, where the two components become wholly miscible at pressures above 1060 bar. Other mixtures of this type are water + 2-butanol[72] and + 2-butoxyethanol[73,74].

The diagram for heavy water + 3-methylpyridine[74] is shown in *Figure 6.15(b)*. This mixture, together with water + 4-methylpiperidine[74], are the only known examples of a binary mixture where the variation of the excess volume with pressure is such that the values (dT^c/dp) are of opposite sign to those depicted in *Figure 6.15(a)*. The loci of the UCST and the LCST are thus divergent with the result that the region of immiscibility increases with increasing pressure and the system forms a 'tube of immiscibility' in (p, T, x) space.

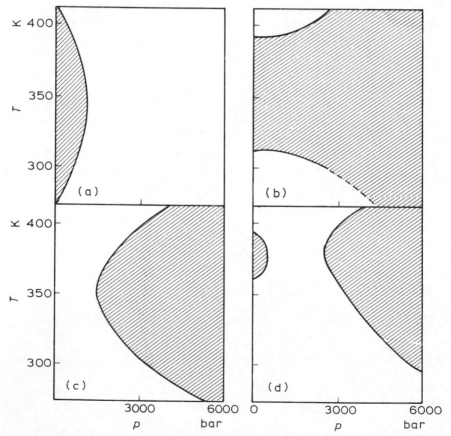

Figure 6.15 *Four different types of high-pressure liquid–liquid immiscibility observed in type II mixtures. Two-phase regions are shown shaded. (a) water + 2-butanone; (b) heavy water + 3-methylpyridine; (c) water + 3-methylpyridine; (d) heavy water + 2-methylpyridine*

Figure 6.15(c) illustrates the phase diagram for water + 3-methyl-pyridine[74] where the phenomenon of so-called 'high-pressure immiscibility' is seen. Here the system is wholly miscible at low pressures but two liquid phases appear at around 1400 bar and the zone of immiscibility increases rapidly with further rise in pressure. The dome of immiscibility shown in *Figure 6.14* is thus inverted for such mixtures. Other mixtures exhibiting this behaviour are water + 2-methylpyridine[75], + 4-methyl-pyridine[75] and heavy water + 4-methylpyridine[75]. Schneider has shown how a smooth transition from *Figure 6.15*(b) to *Figure 6.15*(c) behaviour occurs in the ternary mixture water + heavy water + 3-methylpyridine[75].

The fourth type of diagram, shown in *Figure 6.15*(d), is that of heavy water + 2-methylpyridine, in which the behaviour of *Figures 6.15*(a) and (c) are combined. This mixture is wholly miscible at intermediate pressures but only partially miscible at low and high pressures. The thermodynamic behaviour of such mixtures is interesting for, with the usual assumptions concerning the absence of inflection points in the excess function–composition relationships, the excess enthalpy must change sign from negative to positive with increasing temperature at constant pressure whereas the excess volume must change sign in the same sense with increasing pressure when measured at constant temperature. An unusual feature of such extremely non-ideal mixtures is the fact that both the excess enthalpy and the excess volume must pass through zero in the wholly-miscible region which exists at intermediate pressures. The excess Gibbs function must also be positive in the region of the surfaces of immiscibility and less positive or negative in the intervening regions of miscibility, and so may well also be zero at certain points in (p, T, x)-space. Such behaviour has been recorded for the excess volume of water + 3-methylpyridine[76] although, in this case, the low-pressure region of immiscibility is not detected and is presumably displaced to negative pressures.

6.5 High-pressure phase equilibria: type III mixtures

When the mutual immiscibility of the two components in a binary mixture becomes sufficiently great, the locus of the liquid–liquid critical-solution points moves to higher temperatures and, eventually, interacts with the gas–liquid critical curve. The latter becomes discontinuous and type II behaviour is transformed into type III. Four possible kinds of type III behaviour are depicted in *Figure 6.16* as (p, T) projections.

In all cases, the critical line extending from the gas–liquid critical point of the more volatile component ends at an upper critical end point where the gaseous phase and the liquid phase, richer with respect to the more volatile substance, have the same composition. At lower temperatures, the two liquids are immiscible as indicated by the three-phase L_1L_2G line culminating at the UCEP. The critical line connecting the UCEP to the gas–liquid critical point is frequently quite short and often extends monotonically from

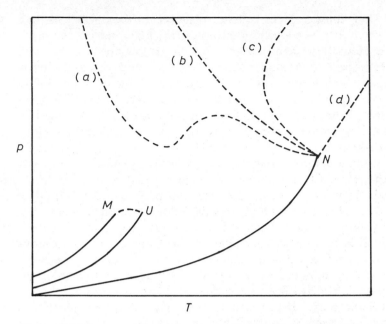

Figure 6.16 Four possible types of critical locus in type III mixtures. The trend in shape from (a) to (d) indicates an increasing degree of immiscibility between the two components. M and N are the critical points of the two pure components and U is the upper critical end point, the termination of the three-phase line L_1L_2G

one point to the other although it can exhibit various degrees of curvature and maxima or minima are observed in certain cases.

Ethane + methanol[77] is a typical type III mixture where the high-pressure critical locus exhibits first a pressure maximum as it leaves the critical point of pure methanol followed by a pressure minimum at lower temperatures, after which it continues with negative slope up to high pressures where it finally merges with solid phases. The phase diagram is illustrated schematically as a (p, T, x) sketch in *Figure 6.17*(a) and as the (p, T) projection in *Figure 6.17*(b). In the latter diagram, the short length of the critical line connecting the UCEP to the critical point of ethane is seen, the two points differing by only 3.2 K.

Many binary mixtures have been discovered whose phase diagrams correspond to that of *Figure 6.17*. Mixtures of carbon dioxide with the lowest n-alkanes are wholly miscible and correspond to simple type I behaviour. Liquid–liquid immiscibility at a UCST first becomes apparent at CO_2 + n-heptane when type II diagrams are formed and the UCST moves to higher temperatures as the chain length increases. Type III behaviour sets in at CO_2 + n-tridecane[54,57,74,75] where the minimum in the critical line is so low that its (p, T) projection appears to cut the vapour-pressure curve of pure methane. The minimum moves to higher temperatures and pressures as the size of the hydrocarbon is increased and, for example, occurs at 313 K and 180 bar for CO_2 + n-hexadecane[54] and at 370 K and

370 bar for CO_2 + squalane[78], a highly-branched C_{30} alkane. The critical locus is displaced to still higher temperatures and pressures for CO_2 + squalene[79] and the minimum is now absent so that a (p, T) diagram results that corresponds to *Figure 6.16*(b). CO_2 + 2,5-hexanediol[56] exhibits similar behaviour.

A comparable pattern has been noted for low molecular weight hydrocarbons with various polar molecules. Alwani and Schneider[56] have studied the phase diagrams for a series of such mixtures with ethane as the common component. The ethane + methanol system has been discussed earlier and diagrams corresponding to *Figure 6.16*(b) have been found for ethane + N,N-dimethylformamide, + nitromethane, + 2,5-hexanediol, and + N-methylacetamide, the critical line moving to higher temperatures and pressures along the series. A similar trend has been observed for mixtures of ethene with these same four polar substances. Several binary mixtures where both the substances are hydrocarbons are found to behave comparably. Methane + methylcyclopentane[80] and + 1-heptene conform to *Figure 6.16*(a) and methane + toluene[80] to *Figure 6.16*(b). Methane + squalane[81] closely resembles the last mixture.

In all of these mixtures, the projection of the three-phase line on the (p, T) plane lies entirely between the vapour pressure curves, as is shown in *Figures 6.16* and *6.17*, but, under certain circumstances, it can be displaced to lower temperatures and/or higher pressures and lie everywhere above the vapour-pressure curve of the more volatile component. Two alternatives resulting in different phase behaviour are possible. The three-phase line can terminate either at a UCEP that lies at a lower temperature than the critical temperature of the volatile component, or it can continue and end at a higher temperature. The consequences of these two situations are illustrated by the two (p, T, x) surfaces depicted in *Figure 6.18*. Note that, for clarity, the critical line commencing at the critical point of the less volatile substance has been omitted from this figure. The distinctive feature of both diagrams in *Figure 6.18* is the presence of hetero-azeotropy. Aqueous

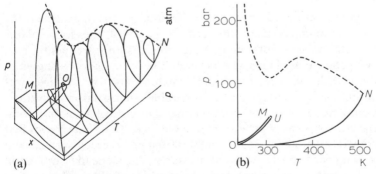

(a)   (b)

Figure 6.17 The (p, T, x) surface and (p, T) projection of the system ethane + methanol[77]. The critical lines are shown dashed. M and N are the critical points of pure ethane and methanol respectively, and U is the upper critical end point where the gas phase and the liquid phase rich in ethane have the same composition

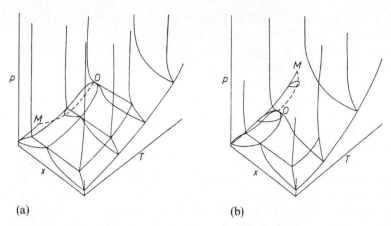

(a)   (b)

Figure 6.18 The (p, T, x) surfaces of heterogeneous azeotropes in the critical region. In (a) the critical end point, O, lies between the critical temperatures of the pure components (e.g., water + diethylether[88]) and in (b) it is at a temperature lower than those of both the components (e.g., water + n-hexane[89])

mixtures of diethylether[82], propane[83] and 1-butene[84] are found to conform to *Figure 6.18*(a) whereas *Figure 6.18*(b) behaviour has been observed for aqueous mixtures of many other hydrocarbons such as ethane[84,85,107], propane[85,107], n-butane[85,87,107], n-pentane[88], n-hexane[86,89,104], cyclohexane[66,85,89,107] and benzene[86,88,89,90,104].

A remarkable feature of these systems is that the pressure of the three-phase line near the UCEP is greater than the sum of the vapour pressures of the pure components at the same temperature. Thus, for n-hexane + water at 473 K : pure n-hexane 18.2 bar, pure water 15.5 bar: sum 33.7 bar; mixture on three-phase line 35.5 bar. Such behaviour is commonly thought to be thermodynamically impossible (that is, materially unstable), but this would be so only if the vapours in equilibrium with the pure and mixed liquids were perfect gases. Clearly, they are not, and so thermodynamic propriety is maintained.

One distinctive feature of many type III mixtures is that when the critical lines that commence at the critical point of the less volatile component are followed to sufficiently high pressures, they frequently pass through a temperature minimum and develop a positive slope, as shown in *Figure 6.16*(c). One of the first such mixtures to be studied was nitrogen + ammonia, a system of considerable industrial importance. Krichevskii and his colleagues[3,91] found that the critical line that commences at the critical point of pure ammonia (405.5 K, 113.5 bar) moves first to lower temperatures but at 360 K and 1100 bar it becomes vertical and moves back to higher temperatures and pressures. The slope (dT^c/dp) is still positive at the highest point reached, 453 K and 15 kbar. The (p, T, x) diagram of this system is shown in *Figure 6.19* and its appearance suggests that sufficiently high pressures will cause the fluids to separate at all temperatures, even at those above the critical point of pure ammonia. Here, the heterogeneous region is bounded at constant pressure by a UCST and at constant temperature by a lower critical solution pressure.

Such behaviour is another example of gas–gas immiscibility first mentioned in Section 6.3 and mixtures whose phase diagram corresponds to that depicted in *Figure 6.19*, where there is a temperature minimum in the critical line, are said to exhibit gas–gas immiscibility of the second kind. Mixtures of substances where the mutual miscibility is even lower, and whose critical curve extends directly from the critical point of the less volatile component with a positive slope, as in *Figure 6.16*(d), exhibit gas–gas immiscibility of the first kind.

This behaviour was first predicted by Kamerlingh Onnes and Keesom[92] on the basis of the van der Waals equation of state, and they were the first to use the phrase gas–gas immiscibility. This usage has been criticized subsequently by various authors but the argument is purely a semantic one for there is a continuous change from type II mixtures to type III, *Figure 6.16*(a), (b), (c) and (d), as the mutual weakness of the unlike intermolecular interactions increases. The densities of the coexisting fluid phases at these high pressures are usually comparable to the densities of two partially-miscible liquid phases coexisting in the vicinity of an 'ordinary' low-pressure UCST or LCST.

A consideration of the phase diagram illustrated in *Figure 6.19* indicates that, at high pressures, the (p, x) loops remain open at the top at all

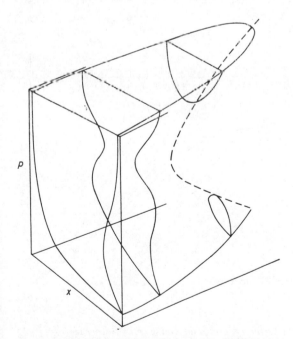

Figure 6.19 The (p, T, x) surface of type III mixtures such as nitrogen + ammonia and methane + ammonia, at temperatures near to the critical point of the less volatile component

temperatures and the suggestion[93,108] that the critical loci for such mixtures will pass through first a temperature maximum and then a pressure maximum and curve backwards to terminate on a three-phase (LLG) line is almost certainly wrong. No temperature maximum has yet been found on a gas–gas critical curve.

Almost 60 mixtures have now been investigated that exhibit gas–gas immiscibility of one kind or another. Schneider in his 1970 review[52] lists over 40. Examples that conform to *Figure 6.16*(c) include argon + ammonia[94], methane + ammonia[95], ethane + water[96], carbon dioxide + water[97] and benzene + water[90]. Many other water + hydrocarbon binaries are of this type[52]. Neon + krypton[98] is chemically one of the simplest of such mixtures and some (p, x) isotherms are shown in *Figure 6.20* in the region of the temperature minimum on the critical locus. This minimum occurs at 164.73 K and 1257 bar at a krypton mole fraction of 0.375. The typical 'waisted' profile of the isotherms at temperatures just below this minimum

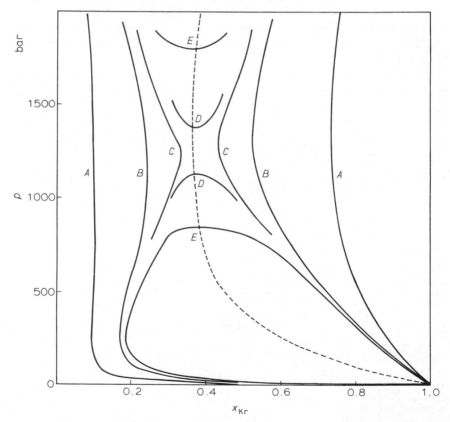

Figure 6.20 Gas–gas immiscibility of the second kind. Some (p, x) isotherms for neon + krypton[98] mixtures in the neighbourhood of the temperature minimum on the critical locus. The critical locus is shown dashed. A, 148.15 K isotherm; B, 163.15 K; C, 164.64 K; D, 164.92 K; E, 166.15 K

is clearly seen. The isotherms at still lower temperatures terminate at a critical end point $(F_1 F_2 S)$ where one of the phases is a pure solid.

There are many mixtures, such as neon + argon[99], neon + nitrogen[100], hydrogen + methane[101], hydrogen + carbon dioxide[102] and hydrogen + carbon monoxide[103], whose critical loci resemble that illustrated in *Figure 6.16*(b) and are everywhere concave upwards, but reach a critical end point with the appearance of a solid phase before a temperature minimum is attained so that gas–gas immiscibility is not detected. It is possible that, if such systems were to be examined over wider regions of (p, T, x)-space, the fluid–fluid critical locus would re-emerge at extremely high pressures and at temperatures above that of the critical temperature of the less volatile component. Such a possibility has been discussed by Streett and Hill[99], who tried to detect such a region of fluid–fluid immiscibility in the neon + argon system but were unsuccessful and remark that the onset of the crystallization of the phase richer with respect to the less volatile component may always rule out the experimental observation of such theoretically-possible equilibria.

Well over 20 aqueous binary mixtures are now known that exhibit gas–gas immiscibility of the second kind. In many of these, the second component is a low molecular weight aliphatic or aromatic hydrocarbon and Jockers and Schneider[67] list many of these together with some water + partially-fluorinated aromatic hydrocarbon mixtures such as water + fluorobenzene and + 1,4-difluorobenzene.

The series water + n-alkane has been discussed recently by De Loos *et al* [104] and gas–gas immiscibility of the second kind has now been confirmed for all mixtures from water + methane to water + n-heptane and a smooth trend noted in the variation of the pressure minimum with carbon number, the minimum decreasing from 990 bar for water + methane to 294 bar for water + n-heptane. These workers re-examined the water + propane system and showed that the earlier measurements of Sanchez and Coll[86] are seriously in error. They also confirmed that the measurements of Tsiklis and Maslennikova[87], who detected gas–gas immiscibility of the first kind for water + n-butane, are totally wrong and that the more recent measurements of Bröllos *et al.*[66] on this mixture are to be preferred. The enormous discrepancy between the absolute magnitudes and the shapes of the experimental critical loci for this mixture produced by these two latter groups, both of high reputation, is salutary in emphasizing the experimental difficulties in the investigation of mixtures under these high pressures.

The interesting phenomenon of barotropic inversion is observed in several mixtures that also exhibit gas–gas immiscibility, although this is not a prerequisite for the occurrence of the barotropic effect. This phenomenon was first detected by Kamerlingh Onnes[92] in 1906 when he compressed helium in the presence of liquid hydrogen at 20 K. The mixture behaves 'normally' at low pressures in that the helium-rich gaseous phase is less dense than the hydrogen-rich liquid phase. At a pressure of around 30 bar, a density inversion takes place and the gas phase becomes more dense than the liquid and sinks to the bottom of the containing vessel. One reason for the effect is that the more volatile component, helium in this case, has the greater molecular weight. The inversion pressure is a unique function of temperature in the two-phase region.

Other mixtures in which this effect has been observed include neon + methane[105], ammonia + argon[94] and + nitrogen[3,91] and water + argon[107] and + carbon dioxide[97]. It is likely that such density inversion occurs for solid–fluid equilibria as well as fluid–fluid and one interesting speculation by Streett and others[106] concerns the possibility that favourable conditions for this phenomenon exist within the outer layers of the atmosphere of the Jovian planets that are believed to consist primarily of hydrogen + helium mixtures. A floating mass of hydrogen-rich solid suspended in a hydrogen–helium fluid layer of equal density has been suggested as a possible explanation for the formation of Jupiter's Great Red Spot.

Most mixtures that exhibit gas–gas immiscibility of the first kind contain helium as one component, indeed the only known example that does not do so is water + argon[107]. Helium + xenon is a typical system and some of the (p, x) isotherms measured by de Swaan Arons and Diepen[108] are illustrated in *Figure 6.21*. The locus of the fluid–fluid LCST is seen to commence at the

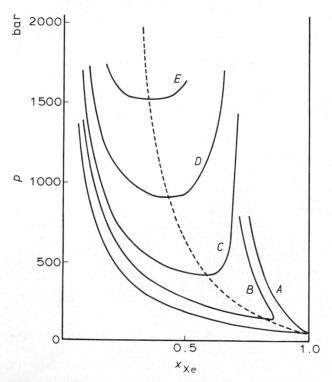

Figure 6.21 Gas–gas immiscibility of the first kind. Some (p, x) isotherms for helium + xenon mixtures[108]. The critical locus is shown dashed. A, 283.15 K isotherm; B, 291.15 K; C, 298.15 K; D, 310.65 K; E, 325.65 K

xenon gas–liquid critical point and to move directly to higher pressures and temperatures. The phase diagrams for such mixtures are thus particularly straightforward and require little comment.

De Swaan Arons and Diepen also made density measurements on the helium + xenon system and some of their results for the 303.65 K isotherms are given below where x', x'' and ρ', ρ'' are the mole fractions and densities of the coexisting fluid phases.

p/bar	x'_{Xe}	ρ'/g cm^{-3}	x''_{Xe}	ρ''/g cm^{-3}
800	0.26	0.80	0.63	1.85
700	0.33	0.93	0.60	1.62
650	0.41	1.10	0.57	1.43
627	0.49	1.291	0.49	1.291

This temperature is 15 K higher than the critical temperature of pure xenon and the pressures are some ten to 15 times greater than the xenon critical pressure, so it is interesting to note that the densities of the immiscible 'gaseous' phases are very comparable to the critical density of pure xenon (1.11 g cm^{-3}) and many times higher than the critical density of helium (0.07 g cm^{-3}). The reasons why such authorities as Streett[109] prefer to use the term 'fluid–fluid' immiscibility rather than 'gas–gas' are immediately obvious.

6.6 High-pressure phase equilibria: type IV and type V mixtures

The (p, T) projections for type IV and type V mixtures are illustrated in *Figure 6.1* where it is seen that they possess the common feature that the critical line beginning at the gas–liquid critical point of the less volatile component ends at a LCEP where it connects to the three-phase (LLG) line. The critical locus thus changes its character continuously from gas–liquid to liquid–liquid and provides a clear example of the confusion that can arise from a careless use of the words *gas* and *liquid* in the critical region of a mixture. The three-phase line is often quite short and ends at higher temperatures and pressures at a UCEP that connects, through a short critical line, with the gas–liquid critical point of the more volatile component. Type IV mixtures have the added complication of an 'ordinary' UCST at lower temperatures and it is extremely likely that most type V systems would show this additional region of partial miscibility if the onset of solid phases did not intervene.

Kuenen[110] was the first to discover type V behaviour in mixtures of ethane with ethanol, 1-propanol and 1-butanol. These three alcohols are completely miscible with ethane at room temperature but separation occurs near the critical temperature of pure ethane (305.42 K). *Figure 6.22* illustrates the (p, T, x) surface and the (p, T) projection for ethane + 1-propanol. The three-phase line is seen to extend only over some 3 K. Other similar systems include carbon dioxide + nitrobenzene[111] (range of immiscibility, 10 K), carbon dioxide + 2-nitrophenol[53,112] (14 K), ethane + 1,3,5-trichlorobenzene[112,113] (6.5 K) and ethane + 1,4-dichlorobenzene[112,113] (0.5 K).

More recently, it has been discovered that binary hydrocarbon mixtures can conform to type V (and to type IV) behaviour once the two components differ sufficiently in their critical properties. For n-alkane mixtures, with methane as one component, n-hexane is the first to exhibit partial miscibility. Davenport and Rowlinson[30] examined this mixture in the region of partial miscibility and also detected LCSTs in methane + 2-methylpentane, + 3-methylpentane, + 2,3-dimethylbutane, + 2,2-dimethylpentane, + 2,4-dimethylpentane, and + 2,2,4-trimethylpentane. The methane + n-hexane system has been examined in more detail by Chen, Chappelear and Kobayashi[114] who have established the LCEP at 182.46 K, 34.15 bar and $x_{CH_4} = 0.9286$ and the UCEP at 195.91 K, 52.06 bar and $x_{CH_4} = 0.99763$. Similar type V behaviour has been reported[17] for methane + 4-methyl-1-pentene and + *trans*-4-methyl-2-pentene. The volume of mixing for methane + 2-methylpentane[115] has been measured in the wholly-miscible region at temperatures below the LCST and extremely large negative values were found. The contraction is so great (~ 5 per cent at 155 K) that the apparent partial molar volume of methane at infinite dilution is negative. Negative methane partial molar volumes were also detected for the wholly-miscible, type I system, methane + 2-methylbutane and are likely to be present in many of the other mixtures noted above.

Davenport and Rowlinson[30] proposed the term *limited immiscibility* to describe the solution properties of such type V mixtures and the additional

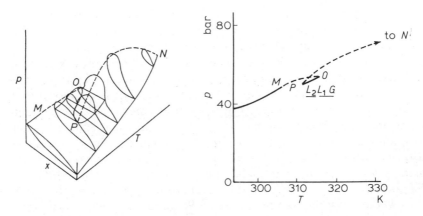

Figure 6.22 The (p, T, x) surface and (p, T) projection of ethane + 1-propanol[110]. Here, the three-phase region that lies between the critical end points O and P is wholly above the critical point of pure ethane, M. Often, with systems of this type, the end point P lies 10–20 K below M

term *absolute immiscibility* for another class of mixture, such as methane + n-heptane, + 2-methylhexane, + toluene, and + 2,3,4-trimethylpentane, where the critical locus extends down to regions of (p, T, x)-space where solid phases are formed as depicted in *Figure 6.23*. The metastable LCST shown in this figure can occasionally be detected experimentally if the liquid phases can be supercooled sufficiently before the onset of crystallization. There is thus no temperature between the solid–liquid boundary and the gas–liquid critical curve at which the pair are completely miscible. The words *limited* and *absolute* are used in an analogous way to their use in describing azeotropy.

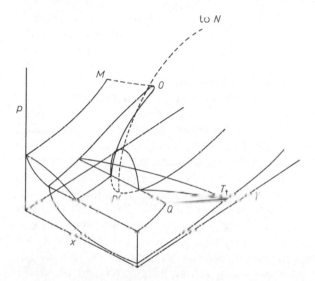

Figure 6.23 The (p, T, x) surface of a type V mixture that exhibits absolute immiscibility (e.g., methane + n-heptane[30]). The labelling is as in Figure 6.22 except that P' is now a metastable LCST at a temperature lower than the quadruple line, Q, where the four phases $(G + L + L + S)$ are in equilibrium. T_t is the triple point of the less volatile component

When ethane is substituted for methane the pattern of phase behaviour is very similar, and n-nondecane[31] is the first n-alkane to show limited immiscibility where, for this system, the two critical end points are separated by only 1.2 K. The separation of the LCEP and the UCEP is increased to 2.9 K for ethane + n-$C_{20}H_{42}$. The higher n-alkanes, such as n-$C_{28}H_{58}$, melt above the critical temperature of ethane and therefore show no immiscibility but, as branching tends to lower the melting temperature of hydrocarbons, LCSTs are found for such mixtures as ethane + 2,6,10,15,19,23-hexamethyltetracosane[81] ($C_{30}H_{62}$, squalane). The onset of liquid immiscibility has been studied by examining ternary mixtures[31] of the type ethane + hexadecane + n-eicosane. The principle of congruence suggests that such systems can be considered to be pseudo-binary because of the close similarity of chemical type of the two long-chain molecules. The physical properties of the 'second component' can thus be varied continuously by altering the relative amounts of the two heavier substances.

The ratio of chain length in the binary mixture of normal paraffins which first shows liquid immiscibility is 6 for methane solutions, 9.5 for ethane and over 13 for propane. If the solvent has six or more carbon atoms and is thus a liquid at room temperature, then phase separation of this type only occurs if the solute is a polymer. This has now been demonstrated for many polymer–solvent mixtures and it is quite possible for the LCST to be 100 K or more below the critical point of the pure solvent although well above its normal boiling point[116]. For example, the LCST in n-pentane + polyisobutene is found at 349 K, which is only 40 K above the normal boiling point of n-pentane and over 120 K below its critical temperature. LCSTs in polymer + solvent mixtures are, like their counterparts in low molecular weight systems, extremely sensitive to small variations in the chain length and branching of either component.

Liquid nitrogen is a poorer solvent than liquid methane for the higher hydrocarbons and immiscibility starts here with ethane[117]. Such mixtures are much less studied than the equivalent binary hydrocarbon systems and, although there are some measurements on nitrogen + propane[118], + n-butane[119] and + n-hexane[120], detailed phase diagrams are still lacking.

Certain polymer–solvent mixtures were the first systems to be found to conform to type IV behaviour. A UCST that occurs at temperatures well below that of the LCST is found experimentally for many such mixtures and, indeed, such a phase separation (at a so-called theta or Flory temperature[121]) is predicted on statistical mechanical, combinatorial, grounds for all such mixtures. Often, in real mixtures, crystallization of the solvent obscures the low-temperature phase separation. The UCST and LCST for cyclohexane + polystyrene[116] occur at 303 and 453 K while for benzene + polyisobutene[122] the temperatures are 296 and 533 K. Liddell and Swinton[122] have measured the enthalpy of mixing at infinite dilution of this last-named mixture as a function of temperature in the intermediate, miscible region and observe the predicted change of sign from $+1080\,\mathrm{J\,mol^{-1}}$ at 300 K to $-260\,\mathrm{J\,mol^{-1}}$ at 453 K.

Davenport et al.[115] detected type IV behaviour in mixtures composed of simple molecules when they located a UCST in methane + 1-hexene at a temperature just above the melting point of pure 1-hexene, having already detected a LCST in this mixture. Scott and van Konynenburg[17] subsequently discovered three other simple mixtures that exhibit similar phase diagrams and all are listed in the table below where x_2 is the mole fraction of the heavier component.

Scott and van Konynenburg have followed the transition of the type IV methane + 3,3-dimethylpentane system to a typical type III mixture, methane + 2-methylhexane, by studying the ternary mixture, methane + 3,3-dimethylpentane + 2-methylhexane. This is another example of the use of

	LCST		UCST	
System	T/K	x_2	T/K	x_2
CH_4 + 1-hexene	179.6	0.09	133.8	0.12
CH_4 + 2,3-dimethyl-1-butene	190.5	0.10	120.4	0.145
CH_4 + 2-methyl-1-pentene	176.3	0.132	141.7	0.150
CH_4 + 3,3-dimethylpentane	169.0	0.118	124.3	0.130

mixtures of two solutes with similar physical properties to produce a pseudo-binary mixture where the thermodynamic behaviour of the 'solute' can be varied continuously over a small range. The loci of the LCST and of the UCST are found to merge at a temperature of 150 K.

An even more interesting use of pseudo-binary mixtures to produce a continuous spectrum of thermodynamic properties was the discovery by Creek, Knobler and Scott[123] of tricritical points in methane + (n-pentane + 2,3-dimethylbutane) and in methane + (2,2-dimethylbutane + 2,3-dimethylbutane). Methane is wholly miscible with both n-pentane and 2,2-dimethylbutane but forms a type V mixture with 2,3-dimethylbutane with the LCEP at 195.15 K and the UCEP at 198.47 K. The addition of either n-pentane or 2,2-dimethylbutane increases both the LCEP and the UCEP but their loci are convergent and meet at a tricritical point where the physical properties of all three phases in equilibrium become identical.

The tricritical points are located at $T = 202.72$ K, $p = 60.33$ bar, $x_1 = 0.955$, $x_2 = 0.031$ and $x_3 = 0.014$ for the first-named mixture and at $T = 201.72$ K, $p = 59.45$ bar, $x_1 = 0.944$, $x_2 = 0.051$ and $x_3 = 0.005$ for the second mixture, where x_1 and x_3 are the mole fractions of methane and of 2,3-dimethylbutane, respectively.

In such mixtures Z, the relative volume fraction of the two larger hydrocarbons, is a pseudo-field variable and the phenomenological theory outlined in Section 4.10 suggests that the temperature difference, ΔT, between the two corresponding critical end points and also the associated pressure difference, Δp, should be proportional to $(Z - Z_{\text{tricrit}})^{\beta_3}$. Values of the critical exponent of $\beta_3 = 1.51 \pm 0.03$ and of 1.42 ± 0.05 result from an analysis of the $\Delta T : Z$ and $\Delta p : Z$ data, respectively, in reasonable agreement with the classically predicted value of $\beta_3 = \frac{3}{2}$.

Van Konynenburg and Scott[17] and Furman and Griffiths[124] have shown how tricritical points of a different kind can occur in binary mixtures when the two components possess a certain symmetry and phase diagrams of considerable complexity can arise. This behaviour is predicted to occur in mixtures of two optically-active d,1-isomers. An essential condition is, of course, that the two isomers be only partially miscible, i.e., the mixture should exhibit large positive deviations from ideality. To date, all mixtures of this type have been found to behave ideally within the limits of currently-available experimental techniques.

226 Fluid mixtures at high pressures

References

1 VAN DER WAALS, J. D., *Die Kontinuität des gasförmigen und flüssigen Zustandes*, 2nd Edn, Vol. 1: *Single Component Systems* (1899); Vol. 2: *Binary Mixtures* (1900), Barth, Leipzig

2 KUENEN, J. P., *Theorie der Verdampfung und Verflüssigung von Gemischen*, Barth, Leipzig (1906); ROOZEBOOM, H. W. B., *Die heterogenen Gleichgewichte 2. Systeme aus zwei Komponenten*, Part 1, Vieweg, Brunswick (1904); *Systeme mit zwei flüssigen Phasen*, Part 2, Vieweg, Brunswick (1918)

3 KRICHEVSKII, I. R., *Acta phys.-chim. URSS*, **12**, 480 (1940); *idem.*, *Technique of Physicochemical Investigations at High and Superhigh Pressures*, 3rd Edn, Izd. Khimiya, Moscow (1965); TSIKLIS, D. S., and ROTT, L. A., *Russ. chem. Rev*, **36**, 351 (1967); TSIKLIS, D. S., *Handbook of Techniques in High-pressure Research and Engineering*, Plenum, New York (1968)

4 ZOSEL, K., *Angew. Chem.*, **17**, 702, 716 (1978)

5 WHITEHEAD, J. C., and WILLIAMS, D. F., *J. Inst. Fuel*, **48**, 182, 397 (1975)

6 LE NEIDRE, B., and VODAR, B. (Eds.), *Experimental Thermodynamics*, Vol. 2, Butterworths, London (1975)

7 (*a*) LEVELT SENGERS, J. M. H., ref. 6, p. 657; (*b*) SCHNEIDER, G. M., ref. 6, p. 787

8 YOUNG, C. L., in *Chemical Thermodynamics, Specialist Periodical Reports*, Ed. McGlashan, M. L., Vol. 2, 71, Chemical Society, London (1978)

9 JEPSON, W. B., RICHARDSON, M. J., and ROWLINSON, J. S., *Trans. Faraday Soc.*, **53**, 1586 (1957)

10 PAK, S. C., and KAY, W. B., *Ind. & Eng. Chem., Fundam.*, **11**, 255 (1972)

11 KAY, W. B., and PAK, S. C., *J. chem. Thermodyn.*, **12**, 673 (1980)

12 HUGILL, J. A., and McGLASHAN, M. L., *J. chem. Thermodyn.*, **10**, 85 (1978)

13 HIZA, M. J., and HERRING, R. N., *Adv. cryogen. Engng*, **8**, 158 (1963); KIRK, B. S., and ZIEGLER, W. T., *Adv. cryogen. Engng*, **10**, 160 (1965); CHEN, R. J. J., RUSKA, W. E. A., CHAPPELEAR, P. S., and KOBAYASHI, R., *Adv. cryogen. Engng*, **18**, 202 (1973)

14 MICHELS, A., DUMOULIN, E., and VAN DIJK, T. J. J., *Physica*, **27**, 886 (1961); AKERS, W. W., and EUBANKS, L. A., *Adv. cryogen. Engng*, **3**, 275 (1960); STREETT, W. B., *Cryogenics*, **5**, 27 (1965); STREETT, W. B., and CALADO, J. C. G., *J. chem. Thermodyn.*, **10**, 1089 (1978); STEAD, K. and WILLIAMS, J. M., *J. chem. Thermodyn.*, **12**, 265 (1980)

15 DUNCAN, A. G., and HIZA, M. J., *Adv. cryogen. Engng*, **15**, 42 (1970)

16 KRICHEVSKII, I. R., and TSIKLIS, D. S., *Zh. fiz. Khim.*, **17**, 115 (1943); RIGAS, T. J., MASON, D. F., and THODOS, G., *Ind. Engng Chem.*, **50**, 1297 (1958); SCHNEIDER, G. M., *Z. phys. Chem. Frankf. Ausg.*, **37**, 333 (1963); ALWANI, Z., and SCHNEIDER, G. M., *Ber. Bunsenges. phys. Chem.*, **73**, 294 (1969); TRAPPENIERS, N. J., and SCHOUTEN, J. A., *Physica*, **73**, 527 (1974)

17 SCOTT, R. L., and VAN KONYNENBURG, P. H., *Discuss. Faraday Soc.*, **49**, 87 (1970); VAN KONYNENBURG, P. H., and SCOTT, R. L., *Phil. Trans.*, **A298**, 495 (1980)

18 KUENEN, J. P., *Communs phys. Lab. Univ. Leiden*, No. 4, (1892)

19 SAGE, B. H., LACEY, W. N., and SCHAAFSMA, J. G., *Ind. Engng Chem.*, **26**, 214 (1934)

20 HASELDEN, G. G., NEWITT, D. M., and SHAH, S. M., *Proc. R. Soc.*, **A209**, 1 (1951)

21 ROOZEBOOM, H. W. B., *Die heterogenen Gleichgewichte 2. Systeme aus zwei Komponenten*, Part 1, Figs. 46–54, pp. 90–101, Vieweg, Brunswick (1904); KUENEN, J. P., *Theorie der Verdampfung und Verflüssigung von Gemischen*, Figs. 27 and 28, p. 77, Barth, Leipzig (1906); ZERNIKE, J., *Chemical Phase Theory*, Fig. 24, p. 126, Kluwer, Antwerp (1956)

22 SCHOUTEN, J. A., DEERENBERG, A., and TRAPPENIERS, N. J., *Physica*, **81A**, 151 (1975)

23 RUHEMANN, M., *Proc. R. Soc.*, **A171**, 121 (1939); WICHTERLE, I., and KOBAYASHI, R., *J. chem. Engng Data*, **17**, 9 (1972)

24 BLOOMER, O. T., and PARENT, J. D., *Chem. Engng Prog. Symp. Ser.*, **49**, 11 (1953); STRYJEK, R., CHAPPELEAR, P. S., and KOBAYASHI, R., *J. chem. Engng Data*, **19**, 334 (1974)

25 ZENNER, G. H., and DANA, L. I., *Chem. Engng Prog. Symp. Ser.*, **59**, 36 (1963); MUIRBROOK, N. K., and PRAUSNITZ, J. M., *A.I.Ch.E.Jl.*, **11**, 1092 (1965)

26 HISSONG, D., and KAY, W. B., *Proc. Am. Petrol. Inst., Refin. Div.*, **48**, 397 (1968)

27 PAK, S. C., and KAY, W. B., *Ind. & Eng. Chem., Fundam.*, **11**, 255 (1972)

28 CAUBET, F., *C. r. hebd. Acad. Sci., Paris*, **131**, 108 (1900); *Z. phys. Chem.*, **43**, 115 (1903)
29 KUENEN, J. P., *Phil. Mag.*, **1**, 593 (1901); *Z. phys. Chem.*, **37**, 485 (1901)
30 DAVENPORT, A. J., and ROWLINSON, J. S., *Trans. Faraday Soc.*, **59**, 78 (1963)
31 KOHN, J. P., KIM, Y. J., and PAN, Y. C., *J. chem. Engng Data*, **11**, 33 (1966); RODRIGUES, A. B., and KOHN, J. P., *J. chem. Engng Data*, **12**, 191 (1967)
32 WAGNER, J. R., McCAFFREY, D. S., and KOHN, J. P., *J. chem. Engng Data*, **13**, 22 (1968)
33 COOK, D., *Proc. R. Soc.*, **A219**, 245 (1953)
34 KAY, W. B., and HISSONG, D., *Proc. Am. Petrol. Inst., Refin. Div.*, **47**, 653 (1967)
35 CHRISTIANSEN, L. J., FREDENSLUND, A., and MOLLERUP, J., *Cryogenics*, **13**, 405 (1973)
36 CALADO, J. C. G., DEITERS, U., and STREETT, W. B., *J. chem. Soc., Faraday Trans. 1*, **77**, 2503 (1981)
37 HICKS, C. P., and YOUNG, C. L., *Trans. Faraday Soc.*, **67**, 1598 (1971)
38 SCOTT, R. L., *Proc. 3rd Int. Conf. Chem. Thermodyn.*, Baden, Vol. 2, 220 (1973)
39 QUINT, N., *Proc. Sect. Sci. K. ned. Akad. Wet.*, **2**, 40 (1899); DEVYATYKH, G. G., DUDOROV, V. Y., AGLIULOV, N. K., STEPANOV, V. M., and SMIRNOV, M. I., *Izv. Akad. Nauk SSSR, Ser. Khim.*, 2653 (1972)
40 BREWER, J., RODEWALD, N., and KURATA, F., *A.I.Ch.E.Jl.*, **7**, 13 (1961)
41 STORVICK, T. S., and SMITH, J. M., *J. chem. Engng Data*, **5**, 130 (1960), SKAATES, J. M., and KAY, W. B., *Chem. Engng Sci.*, **19**, 431 (1964)
42 WATSON, L. M., and DODGE, B. F., *Chem. Engng Prog. Symp. Ser.*, **48**, 73 (1952)
43 KAY, W. B., *J. phys. Chem., Ithaca*, **68**, 827 (1964)
44 KUENEN, J. P., and ROBSON, W. G., *Phil. Mag.*, **4**, 116 (1902)
45 KHAZANOVA, N. E., and LESNEVSKAYA, L. S., *Zh. fiz. Khim.*, **41**, 2373 (1967)
46 KAY, W. B., and WARZEL, F. M., *A.I.Ch.E.Jl.*, **4**, 296 (1958)
47 CHEN, R. J. J., CHAPPELEAR, P. S., and KOBAYASHI, R., *J. chem. Engng Data*, **19**, 50 (1974); CHU, T.-C., CHEN, R. J. J., CHAPPELEAR, P. S., and KOBAYASHI, R., *J. chem. Engng Data*, **21**, 41 (1976)
48 CHEN, R. J. J., CHAPPELEAR, P. S., and KOBAYASHI, R., *J. chem. Engng Data*, **19**, 53 (1974); KAHRE, L. C., ibid., 67
49 TIMMERMANS, J., and KOHNSTAMM, P., *Proc. Sect. Sci. K. ned. Akad. Wet.*, **12**, 234 (1909–10); **13**, 507 (1910–11); **15**, 1021 (1912–13); *J. Chim. phys.*, **20**, 491 (1923)
50 TIMMERMANS, J., *Les solutions concentrées*, Masson, Paris (1936)
51 TÖDHEIDE, K., and FRANCK, E. U., *Z. phys. Chem. Frankf Ausg.*, **37**, 387 (1963); DANNEIL, A. TÖDHEIDE, K., and FRANCK, E. U., *Chemie-Ingr-Tech.*, **39**, 816 (1967); FRANCK, E. U., in *Physical Chemistry: An Advanced Treatise*, Eds. Eyring, H., Henderson, D., and Jost, W., Vol. 1, Academic Press, New York (1971); *see also Landolt-Börnstein*, New Series, Vol. IV/4, Ed. Hellwege, K.-H., Chap. 4, Springer, Berlin (1980)
52 SCHNEIDER, G. M., *Ber. Bunsenges. phys. Chem.*, **70**, 497 (1966); *Adv. chem. Phys.*, **17**, 1 (1970); in *Water: A Comprehensive Treatise*, Ed. Franks, F., Vol. 2, Plenum, New York (1973); *Pure appl. Chem.*, **47**, 277 (1976); in *Chemical Thermodynamics, Specialist Periodical Reports*, Ed. McGlashan, M. L., Vol. 2, 71, Chemical Society, London (1978)
53 BÜCHNER, E. H., *Z. phys. Chem.*, **54**, 665 (1906); **56**, 257 (1906)
54 SCHNEIDER, G. M., *Ber. Bunsenges. phys. Chem.*, **70**, 10 (1966); SCHNEIDER, G. M., ALWANI, Z., HEIM, W., HORVATH, E., and FRANCK, E. U., *Chemie-Ingr-Tech.*, **39**, 649 (1967)
55 REAMER, H. H., and SAGE, B. H., *J. chem. Engng Data*, **8**, 508 (1963); **10**, 49 (1965)
56 ALWANI, Z., and SCHNEIDER, G. M., *Ber. Bunsenges. phys. Chem.*, **80**, 1310 (1976)
57 ARING, H., and VON WEBER, U., *J. prakt. Chem.*, **30**, 295 (1965); ROTH, K., SCHNEIDER, G. M., and FRANCK, E. U., *Ber. Bunsenges. phys. Chem.*, **70**, 5 (1966)
58 MARSHALL, A., *J. chem. Soc.*, **89**, 1350 (1906)
59 KUENEN, J. P., *Z. phys. Chem.*, **24**, 667 (1897)
60 CALADO, J. C. G., KOZDON, A. F., MORRIS, P. J., DA PONTE, M. N., STAVELEY, L. A. K., and WOOLF, L. A., *J. chem. Soc., Faraday Trans. 1*, **71**, 1372 (1975)
61 SIMON, M., and KNOBLER, C. M., *J. chem. Thermodyn.*, **3**, 657 (1971); PAAS, R., and SCHNEIDER, G. M., *J. chem. Thermodyn.*, **11**, 267 (1979)
62 DYKE, D. E. L., ROWLINSON, J. S., and THACKER, R., *Trans. Faraday Soc.*, **55**, 903 (1959); HICKS, C. P., and YOUNG, C. L., *Trans. Faraday Soc.*, **67**, 1605 (1971)
63 PETER, K. H., PAAS, R., and SCHNEIDER, G. M., *J. chem. Thermodyn.*, **8**, 731, 741 (1976)

64 WOOD, S. E., *J. Am. chem. Soc.*, **68**, 1963 (1946)
65 ALWANI, Z., and SCHNEIDER, G. M., *Ber. Bunsenges. phys. Chem.*, **73**, 294 (1969)
66 BROLLOS, K., PETER, K., and SCHNEIDER, G. M., *Ber. Bunsenges. phys. Chem.*, **74**, 682 (1970)
67 JOCKERS, R., and SCHNEIDER, G. M., *Ber. Bunsenges. phys. Chem.*, **82**, 576 (1978)
68 ROOZEBOOM, H. W. B., *Z. phys. Chem.*, **2**, 449 (1888); HAASE, R., NAAS, H., and THUMM, H., *Z. phys. Chem. Frankf. Ausg.*, **37**, 210 (1963); KAO, J. T. F., *J. chem. Engng Data*, **15**, 362 (1970)
69 SPALL, B. C., *Can. J. chem. Engng*, **41**, 79 (1963); BUTCHER, K. L., HANSON, C., and PLEWES, J. A., *Chem. Inds., Lond.*, 355 (1962); *Chem. Inds., Lond.*, 249 (1963)
70 MARSHALL, W. L., and JONES, E. V., *J. inorg. nuc. Chem.*, **36**, 2313 (1974)
71 VAN KLOOSTER, H. S., and DOUGLAS, W. A., *J. phys. Chem., Ithaca*, **49**, 67 (1945); KOHLER, F., LIEBERMANN, E., MIKSH, G., and KAINZ, C., *J. phys. Chem., Ithaca*, **76**, 2764 (1972)
72 TIMMERMANS, J., *J. Chim. phys.*, **20**, 491 (1923)
73 POPPE, G., *Bull. Soc. chim. Belg.*, **44**, 640 (1935)
74 SCHNEIDER, G. M., *Z. phys. Chem. Frankf. Ausg.*, **37**, 333 (1963)
75 SCHNEIDER, G. M., *Z. phys. Chem. Frankf. Ausg.*, **39**, 187 (1963)
76 ENGELS, P., and SCHNEIDER, G. M., *Ber. Bunsenges. phys. Chem.*, **76**, 1239 (1972)
77 KUENEN, J. P., *Phil. Mag.*, **6**, 637 (1903); MA, Y. H., and KOHN, J. P., *J. chem. Engng Data,*, **9**, 3 (1964); HEMMAPLARDH, B., and KING, A. D., *J. phys. Chem., Ithaca*, **76**, 2170 (1972)
78 LIPHARD, K. G., and SCHNEIDER, G. M., *J. chem. Thermodyn.*, **7**, 805 (1975)
79 ALWANI, Z., unpublished measurements
80 OEDER, D., and SCHNEIDER, G. M., *Ber. Bunsenges. phys. Chem.*, **73**, 229 (1969); **74**, 580 (1970)
81 PAAS, R., ALWANI, Z., HORVATH, E., and SCHNEIDER, G. M., *J. chem. Thermodyn.*, **11**, 693 (1979)
82 KUENEN, J. P., and ROBSON, W. G., *Z. phys. Chem.*, **28**, 342 (1899); SCHEFFER, F. E. C., *Z. phys. Chem.*, **84**, 728 (1913)
83 AZARNOOSH, A., and McKETTA, J. J., *J. chem. Engng Data*, **4**, 211 (1959); LI, C. C., and McKETTA, J. J., *J. chem. Engng Data*, **8**, 271 (1963); ANTHONY, R. G., and McKETTA, J. J., *J. chem. Engng Data*, **12**, 17 (1967)
84 LELAND, T. W., McKETTA, J. J., and KOBE, K. A., *Ind. Engng Chem.*, **47**, 1265 (1955); WEHE, A. H., and McKETTA, J. J., *J. chem. Engng Data*, **6**, 167 (1961)
85 DANNEIL, A., TÖDHEIDE, K., and FRANCK, E. U., *Chemie-Ingr-Tech.*, **39**, 816 (1967)
86 KOBAYASHI, R., and KATZ, D. L., *Ind. Engng Chem.*, **45**, 440 (1953); ROOF, J. G., *J. chem. Engng Data*, **15**, 301 (1970); SANCHEZ, M., and COLL, R., *An. Quim., Ser. A*, **74**, 132 (1978)
87 REAMER, H. H., OLDS, R. H., SAGE, B. H., and LACEY, W. N., *Ind. Engng Chem.*, **36**, 381 (1944); TSIKLIS, D. S., and MASLENNIKOVA, V. Y., *Dokl. Akad. Nauk. SSSR*, **157**, 426 (1964)
88 SCHEFFER, F. E. C., *Proc. Sect. Sci. K. ned. Acad. Wet.*, **15**, 380 (1912–13); **17**, 834 (1914–15); CONNOLLY, J. F., *J. chem. Engng Data*, **11**, 13 (1966)
89 SCHEFFER, F. E. C., *Proc. Sect. Sci. K. ned. Akad. Wet.*, **16**, 404 (1913–14); REBERT, C. J., and HAYWORTH, K. E., *A.I.Ch.E.Jl.*, **13**, 118 (1967); SULTANOV, R. G., and SORINA, V. G., *Zh. fiz. Khim.*, **47**, 1035 (1973)
90 ALWANI, Z., and SCHNEIDER, G. M., *Ber. Bunsenges. phys. Chem.*, **71**, 633 (1967)
91 KRICHEVSKII, I. R., and BOLSHAKOV, P. E., *Acta phys.-chim. URSS*, **14**, 353 (1941); KRICHEVSKII, I. R., and TSIKLIS, D. S., *Acta phys.-chim. URSS*, **18**, 264 (1943); TSIKLIS, D. S., *Dokl. Akad. Nauk SSSR*, **86**, 993 (1952)
92 KAMERLINGH ONNES, H., and KEESOM, W. H., *Communs phys. Lab. Univ. Leiden*, No. 96a, 96b, 96c (1906); *Suppl.*, No. 15, 16 (1907)
93 ZERNIKE, J., *Chemical Phase Theory*, 143–144, Kluwer, Antwerp (1956)
94 TSIKLIS, D. S., and VASIL'EV, Y. N., *Zh. fiz. Khim.*, **29**, 1530 (1955)
95 KRICHEVSKII, I. R., and TSIKLIS, D. S., *Zh. fiz. Khim.*, **17**, 126 (1943)
96 TÖDHEIDE, K., *Ber. Bunsenges. phys. Chem.*, **70**, 1022 (1966); DANNEIL, A., TÖDHEIDE, K., and FRANCK, E. U., *Chemie-Ingr-Tech.*, **39**, 816 (1967)
97 TÖDHEIDE, K., and FRANCK, E. U., *Z. phys. Chem. Frankf. Ausg.*, **37**, 387 (1963)
98 TRAPPENIERS, N. J., and SCHOUTEN, J. A., *Physica*, **73**, 546 (1974)

99 STREETT, W. B., and HILL, J. L. E., *J. chem. Phys.*, **54**, 5088 (1971)
100 STREETT, W. B., *Cryogenics*, **8**, 88 (1968)
101 TSANG, C. Y., CLANCY, P., CALADO, J. C. G., and STREETT, W. B., *Chem. Eng. Commun.*, **6**, 365 (1980)
102 STREETT, W. B., TSANG, C., DEITERS, U., and CALADO, J. C. G., *EFCE Publ. Ser.*, **11**, 39 (1980)
103 TANG, C. Y., and STREETT, W. B., *Fluid Phase Equilib.*, **6**, 261 (1981)
104 DE LOOS, T. W., WIJEN, A. J. M., and DIEPEN, G. A. M., *J. chem. Thermodyn.*, **12**, 193 (1980)
105 STREETT, W. B., and HILL, J. L. E., *Proc. XIIth Int. Congr. Refrig. Inds., Washington, DC*, **1**, 309 (1971)
106 STREETT, W. B., RINGERMACHER, H. I., and VERONIS, G., *Icarus*, **14**, 319 (1971)
107 LENTZ, H., and FRANCK, E. U., *Ber. Bunsenges. phys. Chem.*, **73**, 28 (1969)
108 DE SWAAN ARONS, J., and DIEPEN, G. A. M., *J. chem. Phys.*, **44**, 2322 (1966)
109 STREETT, W. B., *Can. J. chem. Engng*, **52**, 92 (1974)
110 KUENEN, J. P., and ROBSON, W. G., *Phil. Mag.*, **48**, 180 (1899)
111 KOHNSTAMM, P., and REEDERS, J. C., *Proc. Sect. Sci. K. ned. Akad. Wet.*, **14**, 270 (1911–12)
112 SCHEFFER, F. E. C., and SMITTENBERG, J., *Recl Trav. chim. Pays-Bas Belg.*, **52**, 1, 982 (1933)
113 TODD, D. B., and ELGIN, J. C., *A.I.Ch.E.Jl.*, **1**, 20 (1955); CHAPPELEAR, D. C., and ELGIN, J. C., *J. chem. Engng Data*, **6**, 415 (1961)
114 CHEN, R. J. J., CHAPPELEAR, P. S., and KOBAYASHI, R., *J. chem. Engng Data*, **21**, 213 (1976); LIN, Y.-N., CHEN, R. J. J., CHAPPELEAR, P. S., and KOBAYASHI, R., *J. chem. Engng Data*, **22**, 402 (1977)
115 DAVENPORT, A. J., ROWLINSON, J. S., and SAVILLE, G., *Trans. Faraday Soc.*, **62**, 322 (1966)
116 FREEMAN, P. I., and ROWLINSON, J. S., *Polymer*, **1**, 20 (1960); BAKER, C. H., BYERS BROWN, W., GEE, G., ROWLINSON, J. S., STUBLEY, D., and YEADON, R. E., *Polymer*, **3**, 215 (1962); ALLEN, G., and BAKER, C. H., *Polymer*, **6**, 81 (1965), MYRAT, C. D., and ROWLINSON, J. S., *ibid.*, 645
117 CHANG, S.-D., and LU, B. C.-Y., *Chem. Engng Prog. Symp. Ser.*, **63**, No. 81, 18 (1967); YU, P., ELSHAYAL, I. M., and LU, B. C.-Y., *Can. J. chem. Engng*, **47**, 495 (1969); STRYJEK, R., CHAPPELEAR, P. S., and KOBAYASHI, R., *J. chem. Engng Data*, **19**, 340 (1974)
118 ROOF, J. G., and BARON, J. D., *J. chem. Engng Data*, **12**, 292 (1967); SCHINDLER, D. L., SWIFT, G. W., and KURATA, F., *Hydrocarb. Process. Petrol. Refin.*, **45**, No. 11, 205 (1966)
119 AKERS, W. W., ATTWELL, L. L., and ROBINSON, J. A., *Ind. Engng Chem.*, **46**, 2539 (1954); LEHIGH, W. R., and McKETTA, J. J., *J. chem. Engng Data*, **11**, 180 (1966)
120 POSTON, R. S., and McKETTA, J. J., *J. chem. Engng Data*, **11**, 364 (1966)
121 FLORY, P. J., *Principles of Polymer Chemistry*, Cornell University Press, Ithaca (1953)
122 KRIGBAUM, W. R., and FLORY, P. J., *J. Am. chem. Soc.*, **75**, 5254 (1953); PATTERSON, D., DELMAS, G., and SOMCYNSKY, T., *Polymer*, **8**, 503 (1967); BARDIN, J. M., and PATTERSON, D., *Polymer*, **10**, 247 (1969); LIDDELL, A. N., and SWINTON, F. L., *Discuss. Faraday Soc.*, **49**, 115 (1970)
123 CREEK, J. L., KNOBLER, C. M., and SCOTT, R. L., *J. chem. Phys.*, **67**, 366 (1977); **74**, 3489 (1981)
124 FURMAN, D., and GRIFFITHS, R. B., *Phys. Rev.*, **A17**, 1139 (1978)

Chapter 7
The statistical thermodynamics of fluids

7.1 Intermolecular forces

The aim of this and of the final chapter is the understanding of the material in the previous six; that is, the interpretation of the thermodynamic properties in terms of the forces between the molecules. Even for the simplest systems this interpretation is a one-sided affair; we aim to calculate the properties from a presumed knowledge of the forces, or, with a more limited objective, we aim to show qualitatively that the kind of behaviour under discussion follows from a few minimal assumptions about the nature of the intermolecular forces. We obtain little quantitative information on the forces from any one set of thermodynamic properties. This inverse route, from the macroscopic properties to the microscopic forces, is now well-developed for dilute gases composed of simple molecules[1], although even here it is usually necessary to call also on non-thermodynamic information, such as the transport properties and collisional cross-sections from the scattering of molecular beams, or to supplement such 'statistical' information by measurements on a truly molecular scale, such as the spectroscopy of molecular dimers. The inverse route is of little value for pure liquids; here we postulate the form and strength of the intermolecular forces and then use statistical mechanical theory to see if the macroscopic consequences of this postulation match the thermodynamic evidence. However, as is stated briefly in the introduction to Chapter 4, the thermodynamic properties of liquid mixtures do tell us something of value about the relative strengths of the intermolecular potentials between like and unlike molecules (*see* Chapter 8).

Intermolecular forces are described at length by Maitland, Rigby, Smith and Wakeham[1], who are concerned principally with their dependence on the separation of the molecules, and by Gray and Gubbins[2], who give a detailed discussion of their dependence on orientation. Here we assume that the reader has at least an elementary knowledge of these forces and proceed to the development of the statistical mechanics. The only assumption we make in much of this development is that the forces satisfy the few conditions necessary for the system to have a conventional thermodynamics, e.g., for energy and free energy to depend on the volume but not on the shape of the fluids. These conditions are described by Münster[3]. The most important is that the forces are of short range: that is, that the energy of a pair of molecules (or of two fixed finite groups of molecules) fall off with r, the separation of the pair (or group), more rapidly than r^{-3}. This restriction excludes a simple discussion of ionic species, but allows for the discussion of dipolar molecules for reasons that are set out in Section 7.6.

It will often be convenient, and sometimes necessary, to restrict the discussion to systems in which the intermolecular energy is pair-wise additive; that is, the configurational potential energy $\mathcal{U}$ of a group of N molecules at $r_1 \ldots r_N$ ($= r^N$) and with orientations denoted symbolically by $\omega_1 \ldots \omega_N$ ($= \omega^N$), can be written

$$\mathcal{U}(r^N, \omega^N) = \sum_{i<j}^{N-1} \sum^{N} u(r_{ij}, \omega_i, \omega_j) \tag{7.1}$$

where $r_{ij} = |r_j - r_i|$ is the separation of molecules i and j. It is also often convenient to neglect the dependence of $\mathcal{U}$ or u on the ω_i. This is strictly justified only for fluids composed of the noble gases and their mixtures, but it is a reasonable approximation for simple molecular fluids such as N_2, O_2, CO and CH_4, and becomes a crude but nevertheless often invoked approximation for non-polar organic molecules such as C_2H_6, c-C_6H_{12}, C_6H_6 and CCl_4.

The first assumption, that of pair-wise additivity, always fails in dense fluids; the sum over the true pair potentials (as determined from the properties of the dilute gas) gives a total potential energy that is too negative by about 5–10 per cent for a typical configuration of a liquid at equilibrium. Nevertheless, it is often necessary if a theory is to be applied, and is less serious in the calculation of the properties of liquid mixtures since there is a substantial cancellation of multibody terms when calculating an excess function. The first assumption does not hinder the qualitative search for an understanding of the essence of the links between the intermolecular forces, on the one hand, and the thermodynamic properties and phase behaviour, on the other. Indeed, much of the quantitative testing of statistical theories is now carried out not with real liquids but with computer simulations of systems with pair-wise additive and often spherically symmetrical potentials (Sections 7.5, 7.6 and 8.6). The neglect of the dependence of u on ω does limit the theoretical understanding of the phase behaviour of some liquid mixtures (Section 8.9).

In this chapter we develop the statistical mechanics of pure fluids and in the next extend the results to mixtures. These are vast fields of which we attempt to cover only those parts that are related directly to the matter in hand—the interpretation of the thermodynamic properties and phase behaviour of liquids and, more particularly, of their mixtures.

7.2 Molecular distribution functions

The structure and properties of a fluid at equilibrium may be described by the canonical molecular distribution functions[2,4-6]. These define the probability of the occurrence of a particular arrangement of molecules in an assembly of fixed number, N, and fixed volume, V. The first is the probability of finding a molecule in a small volume element, $d\mathbf{r}_1$, at a position whose coordinates from any convenient origin are defined by the vector $\mathbf{r}_1$. This probability is proportional to $d\mathbf{r}_1$ and may be denoted $n^{(1)}(\mathbf{r}_1)$. It is independent of $\mathbf{r}_1$ in an isotropic fluid and is given by

$$n^{(1)} \equiv n = N/V \tag{7.2}$$

A pair distribution function, $n^{(2)}(\mathbf{r}_1, \mathbf{r}_2)$, defines the probability of finding simultaneously the centres of (spherical) molecules at $\mathbf{r}_1$ and $\mathbf{r}_2$. This satisfies the equation

$$\int n^{(2)}(\mathbf{r}_1, \mathbf{r}_2)\, d\mathbf{r}_2 = n^{(1)}(\mathbf{r}_1) \cdot (N-1) = N(N-1)/V \tag{7.3}$$

This equation follows from the fact that the integral represents the chance of finding a molecule at $\mathbf{r}_1$ (that is, $n^{(1)}$) multiplied by the total number of molecules, $(N-1)$, encountered in performing the integration over the volume of the fluid. Distribution functions of higher order may be defined similarly. Thus

$$n^{(h)}(\mathbf{r}_1 \ldots \mathbf{r}_h)\, d\mathbf{r}_1 \ldots d\mathbf{r}_h$$

is the probability of the simultaneous presence of h molecules in the h volume elements $d\mathbf{r}_1 \ldots d\mathbf{r}_h$. This function is related to $n^{(h+1)}$ by a generalization of equation (7.3)

$$\int n^{(h+1)} d\mathbf{r}_{h+1} = (N-h)n^{(h)} \tag{7.4}$$

In a perfect gas in which the molecules are point particles without mutual interaction

$$n^{(h)} = N!/V^h(N-h)! \simeq (N/V)^h \tag{7.5}$$

The second part of this equation holds if h is small compared with N.

These functions may be more accurately named configurational distribution functions, since they define only the positions of a group of molecules and tell nothing of their linear momenta or of their rotational or vibrational states. It is sometimes useful to define total distribution functions, $f^{(h)}$, which include this further information. It is often legitimate to factorize $f^{(h)}$ into two functions, one of which is $n^{(h)}$ and the other purely a molecular property, i.e., it is independent of the geometry of the sites $\mathbf{r}_1 \ldots \mathbf{r}_h$. It is always possible to do so for an assembly of structureless particles between which there are conservative central forces.

The noble gases are the only physical examples of such an assembly, and this section is restricted to them. (Polyatomic molecules are discussed in Section 7.7.) The total distribution function, $f^{(h)}$, for an assembly of monatomic molecules defines both the positions $\mathbf{r}_1 \ldots \mathbf{r}_h$ and the momenta $\mathbf{p}_1 \ldots \mathbf{p}_h$ of the h molecules. Clearly

$$\int \ldots \int f^{(h)} \, d\mathbf{p}_1 \ldots d\mathbf{p}_h = n^{(h)} \tag{7.6}$$

$$\iint f^{(h+1)} \, d\mathbf{r}_{h+1} \, d\mathbf{p}_{h+1} = (N - h) f^{(h)} \tag{7.7}$$

A simple and convenient starting point for a discussion of the statistical mechanics of a fluid is the probability of a given configuration of all N molecules in the assembly. This is denoted $n^{(N)}$ and is proportional to $\exp(-\mathcal{U}/kT)$, where $\mathcal{U}$ is again the energy of the given configuration that arises from the intermolecular forces. This type of proportionality is more familiar when applied to the state of single molecules than to that of an assembly of N molecules. The probability that a single molecule is in an energy state ε_i in an assembly of independent molecules (the *micro-canonical ensemble*) is proportional to $\exp(-\varepsilon_i/kT)$. This may be expressed

$$\frac{n_i}{N} = \left(\frac{1}{\psi}\right) e^{-\varepsilon_i/kT} \tag{7.8}$$

where the left-hand side is the ratio of the number of molecules of energy ε_i to the total number in the assembly (or the probability that any molecule chosen at random has this energy), and ψ, the reciprocal of the constant of proportionality, is the *molecular partition function*. In an ensemble of assemblies, in which each assembly has N molecules (the *canonical ensemble*), it follows similarly that the probablity of a configuration of energy $\mathcal{U}$ is

$$n^{(N)} = \left(\frac{1}{Q}\right) e^{-\mathcal{U}/kT} \tag{7.9}$$

where Q is again the reciprocal of the constant of proportionality. It is a function of the number, volume and temperature of the assembly. In a perfect gas, $n^{(N)}$ is given by the first part of equation (7.5) and $\mathcal{U}$ is zero for all configurations. Hence

$$Q = V^N/N! = (eV/N)^N \tag{7.10}$$

Stirling's approximation for a factorial has been used in the second part of this equation.

The distribution function $n^{(N)}$ is related to $n^{(N-1)}$ by equation (7.3), and $n^{(N-1)}$ is similarly related to $n^{(N-2)}$, etc. Hence by repeated integration, the general distribution function is given by

$$n^{(h)} = \frac{1}{Q(N-h)!} \int \ldots \int e^{-\mathcal{U}/kT} \, d\mathbf{r}_{h+1} \ldots d\mathbf{r}_N \tag{7.11}$$

and, in particular,

$$\frac{N}{V} = n^{(1)} = \frac{1}{Q(N-1)!} \int \cdots \int e^{-\mathcal{U}/kT} \, d\mathbf{r}_2 \ldots d\mathbf{r}_N \qquad (7.12)$$

or

$$Q = \frac{1}{N!} \int \cdots \int e^{-\mathcal{U}/kT} \, d\mathbf{r}_1 \ldots d\mathbf{r}_N \qquad (7.13)$$

This last equation may be integrated over all $d\mathbf{r}_1$ without any difficulty by referring all the vectors $\mathbf{r}_2 \ldots \mathbf{r}_N$ to molecule 1 as origin. The integration over $d\mathbf{r}_1$ thus gives simply the total volume, V, and so leads to equation (7.12). The integrations in equations (7.11)–(7.13) are taken over all configurations within the volume V that are accessible to the molecules. If $\mathcal{U}$ is put equal to zero in equation (7.13), then (7.10) is obtained again.

An alternative starting point to equation (7.9) is the corresponding equation for the total distribution function

$$f^{(N)} = \left(\frac{1}{Zh^{3N}}\right) e^{-\mathcal{H}/kT} \qquad (7.14)$$

where $\mathcal{H}$ is the sum of $\mathcal{U}$ and the molecular kinetic energies (not an enthalpy) for a given distribution of molecular positions and momenta. It is convenient to define the constant of proportionality with the factor of $(1/h^{3N})$, where h is Planck's constant. A set of integrations similar to equations (7.11)–(7.13) leads to

$$Z = \frac{1}{h^{3N}N!} \int \cdots \int e^{-\mathcal{H}/kT} \, d\mathbf{r}_1 \ldots d\mathbf{r}_N \, d\mathbf{p}_1 \ldots d\mathbf{p}_N \qquad (7.15)$$

The integration over the momenta may be made at once and leads to

$$Z = \left(\frac{2\pi mkT}{h^2}\right)^{3N/2} Q \qquad (7.16)$$

$$= (\psi^t/V)^N Q \qquad (7.17)$$

where ψ^t is the molecular translational partition function. The functions Z and Q are called, respectively, the *phase integral* (or partition function) and the *configuration integral* of the assembly.

The introduction of Planck's constant into equation (7.14) makes the classical translational partition function ψ^t into the limiting form of the quantal partition function at high temperatures. It also makes Z and ψ^t dimensionless quantities that may be properly called partition functions (or sums-over-states), but Q and (ψ^t/V) are not dimensionless. The factorization of equation (7.17) is not, therefore, the expression of one partition function as the product of two simpler ones, if such a function is defined as a sum-over-states, but is nevertheless the most common of the several possible ways[7] of factorizing Z. The total and configurational thermodynamic functions defined empirically in Section 2.6 are, respectively, the thermodynamic properties associated with Z and Q.

The relation of Q to the configurational thermodynamic functions is obtained most simply through the configurational energy U'. This is the average value of the intermolecular energy in an assembly of fixed number, volume and temperature. If $\mathscr{U}$ is the intermolecular energy of any arbitrary configuration of the molecules and $n^{(N)}$ the probability of that configuration, then U', the average value of $\mathscr{U}$, is given by

$$U' = \bar{\mathscr{U}} = \frac{1}{N!} \int \cdots \int \mathscr{U} n^{(N)} \, d\mathbf{r}_1 \ldots d\mathbf{r}_N \tag{7.18}$$

$$= \frac{1}{QN!} \int \cdots \int \mathscr{U} \, e^{-\mathscr{U}/kT} \, d\mathbf{r}_1 \ldots d\mathbf{r}_N \tag{7.19}$$

The factor of $(1/N!)$ is a normalization factor to prevent the repeated counting of essentially the same configuration of the indistinguishable molecules, as $d\mathbf{r}_1$, etc., occupy all possible positions in the volume V. Equation (7.19) is the usual equation for obtaining the average value of any function such as $\mathscr{U}$ in a canonical ensemble. The configurational free energy, A', is related to U' by

$$(\partial/\partial T)_V(A'/T) = -U'/T^2 \tag{7.20}$$

and so, by integration,

$$A' = -T \int \left[\frac{1}{T^2 N!} \int \cdots \int \mathscr{U} n^{(N)} \, d\mathbf{r}_1 \ldots d\mathbf{r}_N \right] dT \tag{7.21}$$

$$= -kT \int \left[\frac{1}{QN!} \int \cdots \int \frac{\mathscr{U}}{kT^2} e^{-\mathscr{U}/kT} \, d\mathbf{r}_1 \ldots d\mathbf{r}_N \right] dT \tag{7.22}$$

$$= -kT \int \left(\frac{\partial \ln Q}{\partial T} \right)_V dT \tag{7.23}$$

$$= -kT \ln Q \tag{7.24}$$

The constant of integration must be zero to satisfy the third law of thermodynamics, and it is shown below that A' cannot contain an additional term which is a function only of the volume if $(\partial A'/\partial V)_T$ is to approach $(-p)$ for a perfect gas as $\mathscr{U}$ approaches zero.

Equations (7.13), (7.19) and (7.24) are the basis of statistical theories of fluids. They relate the bulk properties A' and U' to the intermolecular energy $\mathscr{U}(\mathbf{r}_1 \ldots \mathbf{r}_N)$ of each molecular configuration. Equation (7.19) may be simplified if the intermolecular energy is a sum of pair potentials (equation (7.1)). The average configurational energy becomes

$$U' = \tfrac{1}{2} \int \int u(r_{12}) n^{(2)}(\mathbf{r}_1, \mathbf{r}_2) \, d\mathbf{r}_1 \, d\mathbf{r}_2 \tag{7.25}$$

The factor of $\tfrac{1}{2}$ is again a normalizing factor to prevent counting twice the contribution from a given pair of volume elements—once when the vector

r_1 lies in the first and r_2 in the second, and again when the positions are reversed. If molecule 1 is chosen as the origin of the coordinate system for r_2, then the integrand is independent of the position of molecule 1. The integration over dr_1 can, therefore, be made at once to give the volume V and

$$U' = \frac{V}{2} \int u(r_2) n^{(2)}(0, r_2) \, dr_2 \tag{7.26}$$

$$= \frac{V}{2} \int_0^\infty u(r) n^{(2)}(r) 4\pi r^2 \, dr \tag{7.27}$$

The pair distribution function is, therefore, the most important of the distribution functions and it is convenient to introduce a normalized function, or a *radial distribution function*, $g(r)$, which approaches unity for large values of r. This is related to $n^{(2)}$ by

$$n^{(2)}(r_1, r_2) = n^{(1)}(r_1) n^{(1)}(r_2) g[r_1, (r_2 - r_1)] \tag{7.28}$$

which reduces in an isotropic fluid to

$$n^{(2)}(r) = n^2 g(r) \tag{7.29}$$

The radial distribution function is unity for all values of r in a perfect gas.

The total free energy, A, is related to Z by an equation analogous to (7.24):

$$A = A_{\text{mol}} + A' = -kT \ln Z \tag{7.30}$$

$$A_{\text{mol}} = -kT \ln(\psi'/V)^N \tag{7.31}$$

Henceforth, the prime will be dropped from A, U, etc., since only configurational thermodynamic properties are needed for the determination of the equation of state, etc., of liquids and mixtures. The pressure is itself a configurational property, since (ψ'/V) and so A_{mol} are independent of the volume. It may be related to the pair distribution function by an equation similar to (7.27) for an assembly of pair potentials. Consider[8] a fluid confined to a cube of side l, a corner of which is the origin of the coordinate system, and let

$$r_i = t_i l \tag{7.32}$$

where t_i is a vector whose length is between 0 and 1

$$p = -\left(\frac{\partial A}{\partial V}\right)_T = \frac{kT}{Q}\left(\frac{\partial Q}{\partial V}\right)_T = \frac{kT}{3l^2 Q}\left(\frac{\partial Q}{\partial l}\right)_T \tag{7.33}$$

where

$$Q = \frac{l^{3N}}{N!} \int \ldots \int \exp\left[-\sum_{i<j}\sum u(t_{ij})/kT\right] dt_1 \ldots dt_N \tag{7.34}$$

where t_{ij} is the scalar distance between $\mathbf{t}_i$ and $\mathbf{t}_j$ and

$$\left(\frac{\partial Q}{\partial l}\right) = \frac{3NQ}{l} - \frac{l^{3N}}{(N-2)!} \int \cdots \int \left[\frac{\mathrm{d}u(lt_{12})}{\mathrm{d}(lt_{12})} \cdot \frac{t_{12}}{2kT}\right] \mathrm{e}^{-\mathscr{U}/kT} \, \mathrm{d}\mathbf{t}_1 \ldots \mathrm{d}\mathbf{t}_N \quad (7.35)$$

since there are $N(N-1)$ terms in the product of exponential terms in equation (7.34) and, on differentiation, each gives a term of the kind shown in equation (7.35). The introduction of $n^{(2)}$ from equation (7.11) and a change of the variable leads to

$$\left(\frac{\partial Q}{\partial l}\right) = \frac{3NQ}{l} - \frac{Q}{2l} \int \int \left[\frac{\mathrm{d}u(r)}{\mathrm{d}r} \cdot \frac{r}{kT}\right] n^{(2)}(\mathbf{r}_1, \mathbf{r}_2) \, \mathrm{d}\mathbf{r}_1 \, \mathrm{d}\mathbf{r}_2 \quad (7.36)$$

$$= \frac{3NQ}{l} - \frac{Ql^2}{2kT} \int_0^\infty r \frac{\mathrm{d}u(r)}{\mathrm{d}r} n^{(2)}(r) 4\pi r^2 \, \mathrm{d}r \quad (7.37)$$

Hence

$$p = \frac{NkT}{V} - \frac{1}{6} \int_0^\infty v(r) n^{(2)}(r) 4\pi r^2 \, \mathrm{d}r \quad (7.38)$$

where

$$v(r) = r(\mathrm{d}u(r)/\mathrm{d}r) \quad (7.39)$$

is the *intermolecular virial function*. It is related to a virial function, $\mathscr{V}$, of any arbitrary configuration by

$$\mathscr{V} = -\tfrac{1}{3} \sum_{i \neq j} \sum v(r_{ij}) \quad (7.40)$$

Hence, by equation (7.38), the average value in a canonical ensemble is

$$\bar{\mathscr{V}} = pV - NkT \quad (7.41)$$

7.3 Molecular correlation functions

The functions introduced in the last section are *distribution functions*, that is, they describe the probability of certain molecular distributions, in particular that of pairs of molecules. However, it is often more convenient to work instead with the closely related *correlation functions* which measure the departure of the distributions from their random values. These can again be defined for pairs, triplets and higher groups of molecules but, for simplicity, only the most important of them—the pair functions—are described here.

The *total correlation function*, $h(r)$, is the difference between the distribution function $g(r)$ and its random value of unity. That is, $h(r)$ is defined by

$$h(r) = g(r) - 1 \quad (7.42)$$

This function is called the total function to distinguish it from the *direct correlation function*, $c(r)$, which was introduced by Ornstein and Zernike[9,10] in 1914. They deduced, correctly, that $h(r)$ is generally of longer range than

$u(r)$ and sought, therefore, a correlation function which they hoped would have a range comparable with that of $u(r)$. They defined $c(r)$ by the integral equation

$$h(r_{12}) = c(r_{12}) + n \int c(r_{13}) h(r_{23}) \, d\mathbf{r}_3 \tag{7.43}$$

total = direct + indirect correlation function

The physical meaning of this definition is seen most readily by repeatedly replacing $h(r)$ by $[c(r) + \text{integral}]$ within the integral of equation (7.43). This gives

$$h(r_{12}) = c(r_{12}) + n \int c(r_{13}) c(r_{32}) \, d\mathbf{r}_3$$

$$+ n^2 \iint c(r_{13}) c(r_{34}) c(r_{42}) \, d\mathbf{r}_3 \, d\mathbf{r}_4 + \cdots \tag{7.44}$$

Thus the total function, $h(r_{12})$, is decomposed into a correlation of 1 and 2, both directly, through $c(r_{12})$, and indirectly, through all possible chains of direct correlation within the fluid.

It is convenient also to define three other functions which are useful in the description of molecular correlations:

$$f(r) = \exp[-u(r)/kT] - 1 \tag{7.45}$$

$$y(r) = g(r) \exp[u(r)/kT] = [1 + h(r)][1 + f(r)]^{-1} \tag{7.46}$$

$$\psi(r) = -kT \ln g(r) = u(r) - kT \ln y(r) \tag{7.47}$$

In the dilute gas (that is, at the level of the second virial coefficient), it is readily shown that both $h(r)$ and $c(r)$ reduce to $f(r)$, that $y(r)$ is unity and that $\psi(r)$ is equal to $u(r)$. The first of these three functions is called *Mayer's f-function*, the last the *potential of average force*. The usefulness of the second function, $y(r)$, is that it remains a continuous function of r even when $u(r)$, and hence $g(r)$, are discontinuous.

The *pressure*, or *virial equation* (7.38), can be rewritten in terms of $y(r)$ and $f'(r)$, the derivative of $f(r)$:

$$\frac{p}{nkT} = 1 + \tfrac{1}{6} n \int r f'(r) y(r) \, dr \tag{7.48}$$

The total correlation function $h(r)$ is related to the compressibility by the simple equation[a]

$$kT \left(\frac{\partial n}{\partial p} \right)_T = 1 + n \int h(r) \, d\mathbf{r} \tag{7.49}$$

---

[a] This equation requires the grand-canonical ensemble for its derivation, and so, for brevity, the reader is referred to a textbook on statistical mechanics. It differs from equation (7.48) in that it is not restricted to systems with pairwise-additive intermolecular potentials.

A similar equation for $c(r)$ can be obtained by integrating equation (7.43) over $d\mathbf{r}_2$

$$\int h(r)\,d\mathbf{r} = \int c(r)\,d\mathbf{r} + n \int c(r)\,d\mathbf{r} \int h(r)\,d\mathbf{r} \tag{7.50}$$

Substitution into equation (7.49) now gives

$$\frac{1}{kT}\left(\frac{\partial p}{\partial n}\right)_T = 1 - n \int c(r)\,d\mathbf{r} \tag{7.51}$$

The two forms of the compressibility equation (7.49) and (7.51), are entirely equivalent and it is a matter of convenience which is used. Generally, the second is preferred, since it yields, on integration, p as a function of n. The constant of integration is fixed by using the perfect-gas limit as a boundary condition. The pressure so obtained agrees with that from equation (7.48) if $c(r)$ and $y(r)$ are both exact and if $\mathscr{U}$ is a sum of pair potentials. There is a difference of pressure if either condition is broken, and this is a useful test of the consistency of any approximations to $c(r)$ and $y(r)$ that may be obtained from the theories discussed below.

The distribution and correlation functions, and hence the thermodynamic properties, can be written as integrals over sums of products of the functions f_{ij}. This is seen from equations (7.11)–(7.13) by replacing $\mathscr{U}$ by the sum of u_{ij} (equation (7.1)) and substituting f_{ij} for u_{ij} from equation (7.45):

$$\exp(-\mathscr{U}/kT) = \prod_{i>j}\prod \exp(-u_{ij}/kT) = \prod_{i>j}\prod (1+f_{ij}) \tag{7.52}$$

The right-hand side can be multiplied out to give sums of products of the f_{ij} and it is now trivial, in expressions such as equations (7.11)–(7.13), to integrate over all molecules not represented in a particular product. A typical term in Q is $f_{12}f_{13}f_{23}$, and here the integration over $d\mathbf{r}_4 \ldots d\mathbf{r}_N$ and [by the argument below equation (7.13)] over $d\mathbf{r}_1$ can be made at once to give a factor of V^{N-2}. In this way Q, $g(r)$, $c(r)$, etc. can be written as expansions in powers of the number density, n.

The corresponding expansion of the pressure, that is the *virial equation of state* [not to be confused with equation (7.38)], and functions derived from it, are useful only in the dilute or moderately dense gas. The expansions do not converge at the densities of liquids. However, a knowledge of the structure of the expansions of both correlation and thermodynamic functions is useful in the theory of dense fluids, and so is summarized here. The combinatorial algebra that arises from the substitution of equation (7.52) in the numerator and denominator of equation (7.11) is surprisingly formidable but can be found in the appropriate reviews[2,6,11].

The expansion of $n^{(2)}$, that is of g or h, is quoted below without derivation, and this one result is then used to generate the expansions of other functions, such as y and c. However, first it is necessary to define some terms and symbols.

Graph   A graph is the pictorial representation of an integral whose integrand is a product of f_{ij} functions, each of which is represented by a bond. If i and j remain fixed during the integration, they are called *root points* and are denoted by open circles. If the positions of i and j are the variables of the integral, then they are called *field points* and are denoted by filled circles. Thus

$$\underset{1 \qquad 2}{\circ\!\!-\!\!\!-\!\!\!-\!\!\circ} = f_{12}$$

 $= f_{12} \int f_{13} f_{23} \mathrm{d}\mathbf{r}_3$

In the graphs that express h_{12}, y_{12} and c_{12} the root points are molecules 1 and 2.

Articulation point   If a graph that is 'cut' at a field point falls into two or more parts, at least one of which contains no root points, then that field point is called an articulation point. Thus 3 is an articulation point in

All graphs with articulation points are excluded (by definition) from the sets defined below.

Connected graph   In a connected graph all field points are linked, directly or through others, to all root points.

Bridge point   A bridge point (or node) is a field point through which all paths from root point 1 to root point 2 have to pass, e.g., 3 in

3
1   2

Chain   A graph with at least one bridge point, and hence with no direct f_{12} bond.

Bundle   A graph with independent (or parallel) routes from 1 to 2, e.g.,

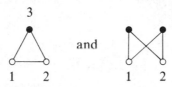

and

Clearly it has no bridge point.

Elementary graph   A connected graph which is neither a chain nor a bundle, e.g.,

An elementary graph cannot have a direct f_{12} bond; the addition of such a bond to an elementary graph turns it into a bundle.

Summation convention   Graphs are added to form sets, and in doing this each graph is first multiplied by n raised to a power equal to the number of field points, and divided by the symmetry number of the set of field points. Many authors now define the graphs themselves so as to include these two factors.

The starting point of the graphical expansions is the statement

$$h_{12} \equiv \text{all connected graphs} \tag{7.53}$$

where it is understood, here and later, that 1 and 2 are the root points. This expansion is

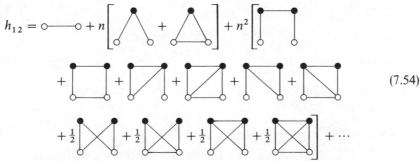

$$\tag{7.54}$$

From the statement (7.53) we have

$$y_{12} - 1 \equiv \text{all connected graphs without } f_{12} \text{ bonds} \tag{7.55}$$

This is most readily seen to be true by showing that (7.55) implies (7.53). The set of 'all connected graphs' can be obtained from (7.55) by adding the following three sets:

$$\text{all connected graphs} \equiv f_{12} + (y_{12} - 1) + f_{12}(y_{12} - 1)$$
$$= (1 + f_{12})y_{12} - 1 \tag{7.56}$$

which, by equation (7.46), is h_{12}. It is this absence of f_{12} bonds from y_{12} that leads to its important property of continuity, even for potentials for which f is discontinuous.

From (7.53) and the definitions above, it follows that h_{12} comprises f_{12} and all chains, bundles and elementary graphs. Hence, from equation (7.44), c_{12} contains f_{12}, the bundles and elementary graphs, since all higher terms in this equation have bridge points at molecule 3, molecules 3 and 4, etc. Furthermore, these higher terms contain *all* the chains, since each connected graph must appear once, and once only, and so cannot be present both in c_{12} and in the higher terms, which have bridge points at all possible field points $3 \ldots N$. Hence

$$c_{12} \equiv \text{all connected graphs without bridge points} \tag{7.57}$$

Therefore

$$h_{12} - c_{12} \equiv \text{all chains} \tag{7.58}$$

Let e_{12} be the set of all elementary graphs. Then

$$[\exp(h_{12} - c_{12} + e_{12}) - 1]]$$

is the set that includes all chains, all elementary graphs and all bundles of chains and elementary graphs. The factorial terms in the expansion of the exponential are the appropriate symmetry numbers of the graphs in these bundles. This set is, therefore, $(y_{12} - 1)$. That is

$$y_{12} = \exp(h_{12} - c_{12} + e_{12}) \tag{7.59}$$

This equation can be solved for c to give the useful result

$$c_{12} = h_{12} - \ln y_{12} + e_{12} \tag{7.60}$$

Hence the potential of average force is free from bundles since, from equation (7.47),

$$(u_{12} - \psi_{12})/kT = \ln y_{12} = (h_{12} - c_{12}) + e_{12} \tag{7.61}$$
$$\equiv \text{all chains and elementary graphs}$$

If a further set is defined by

$d_{12} \equiv$ all connected graphs free from f_{12} bonds
and without bridge points

$$= (y_{12} - 1) - (h_{12} - c_{12}) \tag{7.62}$$

then it follows from (7.57) that d is a sub-set of c. Furthermore, the whole of c can be obtained by adding to d all the graphs in fy, since this set and d are mutually exclusive and since all graphs in y that have bridge points lose them on inserting the f bond. Hence, from this argument, or directly from equations (7.46) and (7.62)

$$c_{12} = f_{12} y_{12} + d_{12} \tag{7.63}$$

The direct correlation function was defined by equation (7.43)—an integral equation which relates c to h and so to g and y. Equations (7.60) and (7.63) are two further relations between c and y, but their usefulness is diminished by the presence on the right-hand sides of the sets of graphs e_{12} and d_{12}. Nevertheless, they are exact and potentially useful consequences of the Ornstein–Zernike equation to which we turn in the next section.

These and the earlier equations obtained so far in this chapter are the formal preliminaries to any discussion of the theory of fluids. They show that a complete knowledge of the thermodynamic properties of a fluid can be obtained from a knowledge of Q or, if the system has only pair potentials, of $h(r)$ as functions of N, V and T. The equations are not themselves a useful theory of fluids, since they do not enable us to determine these functions. It is true that equations (7.11) and (7.13) provide, in principle, a direct method of determining Q and all $n^{(h)}$, but this direct attack on the problem is prohibitively difficult. The explicit calculation of $h(r)$ and $c(r)$ requires some approximations which are described in the next section.

7.4 The measurement and calculation of correlation functions

The pair distribution and correlation functions of the last two sections play a key role in the analysis of the structure and equilibrium properties of simple fluids. In this section we show briefly how they can be measured for real fluids, how they can be simulated for model fluids, and how they can be calculated theoretically by the use of suitable approximations in the equations of the last two sections.

The equilibrium pair correlation functions of a monatomic fluid are determined from the diffraction (or coherent quasi-elastic scattering) of a monochromatic beam of x-rays[12] or neutrons[13] with a wavelength of about 0.1 nm (1 Å). Molecular fluids pose greater problems since the intermolecular scattering has to be disentangled from the intramolecular, and if the molecule is heteroatomic there is more than one intermolecular distribution function[2,14].

For a monatomic fluid the scattered intensity at an angle θ to the incident beam, of radiation of wavelength λ, is a function of a reciprocal length, s, defined by

$$s = (4\pi/\lambda)\sin(\theta/2) \tag{7.64}$$

If $(1/s)$ is comparable with the range of $u(r)$, then the pattern of scattered radiation is determined by the total correlation function, $h(r)$. The intensity at 'angle' s, after subtraction of the single-atom scattering, is proportional to $H(s)$, a Fourier transform of $h(r)$ defined by the pair of equations

$$H(s) = n\int h(r)\exp(-i\mathbf{r}\cdot\mathbf{s})\,d\mathbf{r} \tag{7.65}$$

$$h(r) = \frac{1}{(2\pi)^3 n}\int H(s)\exp(i\mathbf{r}\cdot\mathbf{s})\,d\mathbf{s} \tag{7.66}$$

(There are several conventions about the insertion or omission of the factors n, and about the placing of the factor (2π).) These integrals can be changed to the 'real' form by expanding the exponentials, writing $\mathbf{r}\cdot\mathbf{s}$ as $rs\cos\theta$, where θ is the angle between $\mathbf{r}$ and $\mathbf{s}$, and integrating over angle variables after writing $d\mathbf{r}$ and $d\mathbf{s}$ in spherical polar coordinates

$$H(s) = 4\pi n\int_0^\infty h(r)\left(\frac{\sin(rs)}{rs}\right)r^2\,dr \tag{7.67}$$

$$h(r) = \frac{1}{2\pi^2 n}\int_0^\infty H(s)\left(\frac{\sin(rs)}{rs}\right)s^2\,ds \tag{7.68}$$

Thus an experimental determination of $H(s)$ leads directly to a knowledge of $h(r)$. The scattering at zero angle, which is almost inaccessible to measurement, is related to the compressibility of the fluid, as is seen by taking the limit $s = 0$ in equation (7.67). From equation (7.49)

$$H(0) = n\int h(r)\,d\mathbf{r} = kT\left(\frac{\partial n}{\partial p}\right)_T - 1 \tag{7.69}$$

The determination of $c(r)$ from $H(s)$ is achieved from the Fourier transforms of the Ornstein–Zernike equation (7.43). Multiply both sides by $\exp(-i\mathbf{r}_{12}\cdot\mathbf{s})$ and integrate over $d\mathbf{r}_{12}$:

$$\int h(r_{12})\exp(-i\mathbf{r}_{12}\cdot\mathbf{s})\,d\mathbf{r}_{12} = \int c(r_{12})\exp(-i\mathbf{r}_{12}\cdot\mathbf{s})\,d\mathbf{r}_{12}$$

$$+ n\int c(r_{13})h(r_{23})\exp(-i\mathbf{r}_{12}\cdot\mathbf{s})\,d\mathbf{r}_{12}\,d\mathbf{r}_3 \tag{7.70}$$

The integration over $d\mathbf{r}_{12}$ and $d\mathbf{r}_3$ in the last term is over all relative positions of the three molecules and so can be written also as an integral over $d\mathbf{r}_{13}$ and $d\mathbf{r}_{23}$. In the exponential we write $\mathbf{r}_{12} = \mathbf{r}_{13} + \mathbf{r}_{23}$, and so reduce the last integral to the product of two more simple integrals. From equation (7.65) we have

$$H(s) = C(s) + C(s)H(s) \tag{7.71}$$

or

$$C(s) = H(s)[1 + H(s)]^{-1}$$

where

$$C(s) = n \int c(r) \exp(-i\mathbf{r} \cdot \mathbf{s}) \, d\mathbf{r} \tag{7.72}$$

Hence

$$c(r) = \frac{1}{(2\pi)^3 n} \int C(s) \exp(i\mathbf{r} \cdot \mathbf{s}) \, d\mathbf{s} \tag{7.73}$$

$$= \frac{1}{2\pi^2 n} \int_0^\infty H(s)[1 + H(s)]^{-1} \left(\frac{\sin(rs)}{rs}\right) s^2 \, ds \tag{7.74}$$

This equation has been used by Pings[12], and some of his results for argon are shown in *Figure 7.1*.

A test of a theory of simple liquids is the degree to which it can reproduce structural information of the kind shown in this figure, and also thermodynamic properties of the kind given in the tables of Chapter 2. Such a test is a double one—it tests both our knowledge of intermolecular forces and our ability to calculate $g(r)$, $c(r)$, etc., from statistical mechanics. When there is the inevitable partial failure, it can be hard to know how to apportion the blame between these two heads. This difficulty can be overcome by using computer simulation to generate $g(r)$ and the thermodynamic properties for model systems with specified intermolecular potentials. A failure to reproduce such results must lie in the approximations introduced into the statistical mechanics (or be an 'experimental' failure in the computer simulation). We therefore supplement the results for argon, etc., with those for two model potentials.

Hard spheres

$$u(r) = \infty \quad (r < d) \qquad u(r) = 0 \quad (r > d) \tag{7.75}$$

Lennard-Jones (12,6) potential

$$u(r) = 4\varepsilon[(d/r)^{12} - (d/r)^6] \tag{7.76}$$

The second potential, for which the abbreviation LJ or LJ(12, 6) is used, is zero at $r = d$, the collision diameter, and has its minimum $u = -\varepsilon$ at $r^6 = 2d^6$. It is the most widely used model potential with any claim to realism, but results obtained with it cannot be equated with those for (say) argon, without incurring non-negligible errors[1]. The reduced scale of density, $\rho = Nd^3/V$, is generally used with both potentials, and the reduced scale of temperature, $\tau = kT/\varepsilon$, with the second.

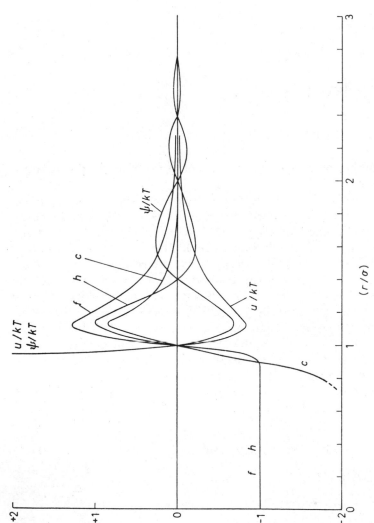

Figure 7.1 Potential functions and correlation functions. The potential u/kT and Mayer's function f are for a Lennard-Jones fluid at $kT/\varepsilon = 1.23$. The total correlation function, h, the direct correlation function, c, and the potential of average force, ψ/kT, are from the x-ray diffraction measurements of Mikolaj and Pings[12,15] on argon at 148 K and 40.7 cm^3 mol^{-1}

The two methods of computer simulation are known by the labels of Molecular Dynamics (MD) and Monte Carlo (MC) simulation. In the first, the evolution of an assembly of N molecules is followed by numerical solution of Newton's equations of motion. The system is one of fixed N, V and U and so is the simulation of a micro-canonical ensemble, but since the sequence of states is that of 'real time' both equilibrium and dynamic information can be obtained. In the second method, a sequence of states is generated such that each state occurs with a probability proportional to its Boltzmann factor, $\exp(-\mathscr{U}(\mathbf{r}^N)/kT)$. The sequence is (usually) specified by fixed values of N, V and T, and so the ensemble represented is canonical. The ordering of the steps of the sequence is arbitrary (that is, it contains no information) and so only thermodynamic properties can be calculated. The principles and practice of these techniques are described elsewhere[16].

For hard spheres, equation (7.75), the pair distribution function $g(r)$ has been calculated by MC simulation and tabulated in detail by Barker and Henderson[17]; a typical curve is shown in *Figure 7.2*. The pressure can be obtained at once from $g(d^+)$ via the virial equation (7.38), since $v(r)$ is zero unless $r = d$. Hence

$$pV/NkT = 1 + \tfrac{2}{3}\pi\rho g(d^+) \tag{7.77}$$

This function and several theoretical approximations to it are shown in *Figure 7.3*. At reduced densities above $\rho \sim 0.90$, the structure of a hard-sphere system changes dramatically as clusters of molecules start to solidify. Beyond $\rho \sim 0.96$ the process is complete and the system has the structure of

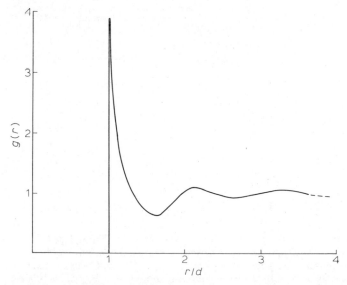

Figure 7.2 The radial distribution function for a system of hard spheres at a reduced density of $\rho = Nd^3/V = 0.8$, which is close to the density at which solidification starts, see Figure 7.3

a 12-coordinated close-packed crystal lattice[18], whose maximum density is $\rho = \sqrt{2}$. The ratio pV/NkT is about 10.3 at the melting point.

The most extensive calculations of $g(r)$ for a system of LJ molecules are those of Nicolas *et al.*[19] whose results amplify the earlier calculations of Hansen and Verlet[19]. They calculate the thermodynamic properties via equations (7.25) and (7.38) from these and earlier simulations of $g(r)$. The critical constants of this fluid are $\tau^c = kT^c/\varepsilon = 1.35$, $\rho^c = Nd^3/V^c = 0.35$ and a reduced critical pressure, $p^c d^3/\varepsilon = 0.142$. Thus the critical ratio $p^c V^c/NkT^c$ is 0.30, which is close to the value of 0.292 found for the noble gases (Section 7.5). The triple point is at $\tau^t = 0.68$ and a liquid density of $\rho^1 = 0.85$. A typical distribution function is shown in *Figure 7.4*. The oscillations at large separations resemble those for hard spheres (*Figure 7.2*) but there is a different shape near $r = d$. However, a comparison shows that the existence of a large first peak and the subsequent oscillations both have their principal origin in the repulsive forces; it is these that determine the structure of a simple liquid. On this simple deduction is built the whole set of perturbation theories discussed in Section 7.6.

The routes through $g(r)$ to the 'mechanical' properties, such as p and U, can usefully be supplemented by the direct calculation of the chemical

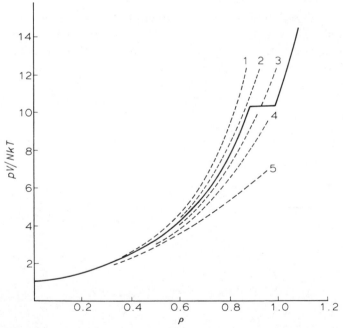

Figure 7.3 The equation of state of an assembly of hard spheres. The reduced density, ρ, is $(6/\pi)$ when the spheres occupy all space. The maximum attainable density is that of a regular close-packed structure, $\rho = \sqrt{2}$. The full lines (= machine calculations) show the transition to the solid state near $\rho = 0.95$. The five calculated curves are: 1 $(HNC)_p$, 2 $(PY)_c$, 3 $(PY)_p$, 4 $(HNC)_c$, and 5 the compressibility solution of Kirkwood's integral equation. The last resembles the Yvon–Born–Green equation at the densities at which both have solutions

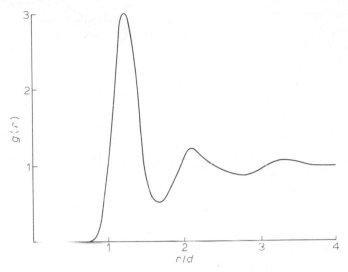

Figure 7.4 *The radial distribution function for a LJ(12,6) liquid near its triple point,* $\tau = 0.72$, $\rho = 0.85$

potential, and so other 'thermodynamic' properties such as entropy, by using Widom's potential distribution theorem which relates $\exp(\mu/kT)$ to the mean Boltzmann factor of the change of energy on introducing a fictitious test particle into a canonical ensemble. This method has been used recently for diatomic liquids[20].

We turn now to the attempts that have been made to calculate the structure and thermodynamic properties of simple liquids, and consider first the hard sphere and LJ fluids. Such theoretical links can be divided into two main groups. It is convenient to associate the first with the calculation of the total function, $h(r)$, $g(r)$ or $n^{(2)}(r)$, which are interrelated by equations (7.42) and (7.29), and the second with that of the direct function, $c(r)$.

The first is based on a family of integral equations[21] that relate $n^{(2)}$ to $n^{(3)}$, $n^{(3)}$ to $n^{(4)}$, etc. These may be obtained by differentiating the function (7.11) with respect to the vector $\mathbf{r}_1$. For the pair function

$$\frac{\partial n^{(2)}(\mathbf{r}_1, \mathbf{r}_2)}{\partial \mathbf{r}_1} = \frac{-1}{kTQ(N-2)!} \int \cdots \int \left(\frac{\partial \mathcal{U}}{\partial \mathbf{r}_1}\right) e^{-\mathcal{U}/kT} \, d\mathbf{r}_3 \ldots d\mathbf{r}_N \qquad (7.78)$$

Now

$$\left(\frac{\partial \mathcal{U}}{\partial \mathbf{r}_1}\right) = \frac{\partial u_{12}}{\partial \mathbf{r}_1} + \sum_{i=3}^{N} \frac{\partial u_{1i}}{\partial \mathbf{r}_1} \qquad (7.79)$$

and so the integrand of equation (7.78) is formed of two types of term. The first is the product of the exponential term and a function only of the positions of the first two molecules. This may be integrated at once over $d\mathbf{r}_3 \ldots d\mathbf{r}_N$ to give the pair function (7.11). The second type are $(N-2)$

identical terms, the first of which is a product of the exponential term and a function of the positions of the first three molecules. The integration over $d\mathbf{r}_4 \ldots d\mathbf{r}_N$ leads, therefore, to the triplet function. Thus

$$kT\left(\frac{\partial n^{(2)}(\mathbf{r}_1, \mathbf{r}_2)}{\partial \mathbf{r}_1}\right) + n^{(2)}(\mathbf{r}_1, \mathbf{r}_2)\left(\frac{\partial u_{12}}{\partial \mathbf{r}_1}\right)$$

$$+ \int n^{(3)}(\mathbf{r}_1, \mathbf{r}_2, \mathbf{r}_3)\left(\frac{\partial u_{13}}{\partial \mathbf{r}_1}\right) d\mathbf{r}_3 = 0 \tag{7.80}$$

This equation may be generalized for any function $n^{(h)}$ by starting with the appropriate equation (7.11). Thus there is a family of integro-differential equations, the first[a] of which, equation (7.80), gives $n^{(2)}$ in terms of $n^{(3)}$, the next gives $n^{(3)}$ in terms of $n^{(4)}$, and so on up to $n^{(N)}$. There is no hope of solving these equations save by the most drastic termination of the series. This is done by putting into the first equation an approximate value of $n^{(3)}$ and solving for $n^{(2)}$. The approximation is

$$n^{(3)}(\mathbf{r}_1, \mathbf{r}_2, \mathbf{r}_3) n^3 / n^{(2)}(\mathbf{r}_1, \mathbf{r}_2) n^{(2)}(\mathbf{r}_1, \mathbf{r}_3) n^{(2)}(\mathbf{r}_2, \mathbf{r}_3) = 1 \tag{7.81}$$

This is the *superposition approximation* of Kirkwood[23] which states that the probability of the simultaneous presence of molecules at $\mathbf{r}_1$, $\mathbf{r}_2$ and $\mathbf{r}_3$ is the product of the independent occurrence of pairs at $\mathbf{r}_1$ and $\mathbf{r}_2$, at $\mathbf{r}_1$ and $\mathbf{r}_3$, and at $\mathbf{r}_2$ and $\mathbf{r}_3$. The equation is not exact but its substitution into equation (7.80) gives one which can be solved. An integral equation for $n^{(2)}$ is obtained after integration by parts and can be written

$$kT \ln\left[\frac{n^{(2)}(r)}{n^2}\right] + u(r) = kT \ln y(r)$$

$$= \frac{\pi}{n^3} \int_0^\infty \int_{-s}^s (s^2 - t^2)\left(\frac{t + r}{rs}\right)[n^{(2)}(t + r) - n^2] \, dt \, n^{(2)}(s) v(s) \, ds \tag{7.82}$$

where s and t are dummy variables and $n^{(2)}(-r)$, $u(-r)$ and $v(-r)$ are defined to be equal to $n^{(2)}(r)$, $u(r)$ and $v(r)$. This is the form of the integral equation[6,21] of Yvon and of Born and Green. Kirkwood started from a slightly different but entirely equivalent set of integro-differential equations. These were also exact but the first of the set becomes an integral equation for $n^{(2)}$ that is not quite the same as equation (7.82) on using the super-position approximation.

Early attempts to solve equation (7.82), or the similar equation of Kirkwood, used an expansion of $n^{(2)}$ in powers of the density. This leads to a virial expansion for the pressure on substitution in equation (7.38). It is found that the superposition approximation gives correctly the known functions for the second and third virial coefficients but not the fourth and

---

[a] Equation (7.80) is the first useful equation of this family for a homogeneous fluid. The true first member is, however, the equation that relates the gradient of the singlet density n to an integral over $n^{(2)}$, which is a useful equation in the statistical mechanics of inhomogeneous fluids[22].

higher. This is clearly a consequence of the inadequacy of equation (7.81) when dealing with the simultaneous interaction of all molecules in a cluster of four or more. The approximation is exact for independent pairs and triplets, and even for a pair of triplets that have one or two molecules in common. It is not correct for larger clusters and is poor in the liquid[24].

Numerical solution of the integral equation is needed to obtain the pressure even for so simple a model as the hard-sphere fluid. The result is quantitatively poor although it is of historical interest that Kirkwood's solution[23] of this equation was the first indication that such a fluid has a phase transition, and the inspiration of the later experimental work on computers by Alder and by Wood.

It is possible to improve the original superposition approximation (7.81) but only at the cost of considerable complication. Thus Lee *et al.*[25] applied the closure to $n^{(4)}$, not to $n^{(3)}$, and so obtained an exact fourth virial coefficient and a good value for the fifth. Their results at 'liquid' densities were also good. However, this closure is difficult to use, and most recent work has been with the structurally more simple equations that arise in the study of the direct correlation function, $c(r)$.

The latter is defined by the Ornstein–Zernike equation (7.43). This is an integral equation for c but it has an explicit solution in terms of h, namely, equations (7.74) and (7.65). Clearly, if there were a second equation between c and h, then a simultaneous solution could be made to give both functions. The graphical expansions of Section 7.3 provide such equations but only with the penalty of the introduction of other unknown functions, such as the sets of graphs e_{12} and d_{12} of equations (7.60) and (7.63), respectively. The former, e_{12}, is the set of elementary graphs and the smallest which cannot be expressed explicitly as a closed function of f_{12} and y_{12}. A worthwhile approximation is, therefore, to assume that this set of graphs can be neglected and to write in place of equation (7.60)

$$c_{12} = h_{12} - \ln y_{12} \tag{7.83}$$

This approximation is named the *hyper-netted chain* (HNC) approximation[26], after the nature of the graphs retained on the right-hand side of (7.83). This and the Ornstein–Zernike equation (7.43) form a pair of simultaneous equations for c and y, since h and y are related by

$$h_{12} = (1 + f_{12})y_{12} - 1 \tag{7.84}$$

Substitution of (7.83) into (7.43) yields the HNC integral equation for y

$$\ln y_{12} = n \int [(1 + f_{13})y_{13} - 1 - \ln y_{13}][(1 + f_{23})y_{23} - 1] \, d\mathbf{r}_3 \tag{7.85}$$

which is simpler in form than the Yvon–Born–Green equation (7.82) but also has no known explicit solution, even for hard spheres.

The set of elementary graphs e_{12} is a sub-set of the set d_{12} defined by (7.62). Hence, to put $d_{12} = 0$ appears to be a grosser approximation than that of the HNC theory. It has, nevertheless, a simple physical interpretation, since if $d_{12} = 0$, then, from equation (7.63),

$$c_{12} = f_{12}y_{12} \tag{7.86}$$

or the range of $c(r)$ is no greater than that of $f(r)$, and hence of $u(r)$. This approximation, therefore, takes literally the original idea of Ornstein and Zernike that the direct correlation function is essentially one of short range. It can be shown[27] that (7.86) is, in fact, the only self-consistent way of requiring that the range of $c(r)$ shall not exceed that of $f(r)$. The results (*Figure 7.1*) show that $c(r)$ does indeed bear a strong resemblance to the general shape of $f(r)$, at least at large separations.

This approximation was first proposed, on quite different grounds, by Percus and Yevick[28] and is now known by their initials, PY. The specification of $c(r)$ in terms of the omitted graphs, $d(r)$, was first made by Stell[29].

The integral equation obtained by the substitution of (7.86) in the Ornstein–Zernike equation is simpler even than equation (7.85), namely,

$$y_{12} - 1 = n \int [f_{13}y_{13}][(1 + f_{23})y_{23} - 1] \, d\mathbf{r}_3 \tag{7.87}$$

It can be solved explicitly for an assembly of hard spheres[30], and the resulting expression for y_{12} between 0 and d suffices for the calculation of p both from the *pressure equation* (7.48) and the *compressibility equation* (7.51). The results are

$$\left(\frac{p}{nkT}\right)_p = \frac{1 + 2\eta + 3\eta^2}{(1 - \eta)^2} \tag{7.88}$$

$$\left(\frac{p}{nkT}\right)_c = \frac{1 + \eta + \eta^2}{(1 - \eta)^3} \tag{7.89}$$

where η is a reduced density which is the ratio of the true volume of the N molecules to the volume V, that is, $\eta = \pi N d^3/6V = (\pi/6)\rho$.

Figures 7.2 and *7.3* show that these equations are close to the experimental (i.e., computer) results for the fluid state and, although mutually inconsistent, are not wildly so. They are superior in accuracy and consistency to those obtained by solving numerically[31] the HNC equation (7.85). However, neither approximation shows the transition to the solid phase, and equations (7.88) and (7.89) suggest that p becomes infinite only as η approaches unity. This is a physically impossible high density, since η cannot exceed the reduced density of a regular close-packed array of spheres, namely, $\pi\sqrt{2}/6 = 0.7405$. An empirical but accurate equation is obtained[32] by adding one-third of equation (7.88) to two-thirds of equation (7.89).

However, these results lack physical meaning on quite different grounds before the limit $\eta = 1$ is reached. It is found that, if $\eta \gtrsim 0.8$, then the solution of equation (7.87) leads to negative values of y_{12} at certain separations r_{12}. Since g_{12} is a probability, it must be positive, and so negative solutions for y_{12} are inadmissible. There is no solution of the PY equation that is physically acceptable[33] beyond about $\eta = 0.8$; it does not describe the transition to the solid. Perhaps this is not surprising for an equation which implies a spherically symmetrical distribution of molecules around any chosen molecule.

Many attempts have been made to improve the HNC and PY approximations[21]. Consistency between the pressure and compressibility equations can be achieved by a suitable approximation for d_{12} and leads to an improvement in the higher coefficients of the virial expansion[27,34]. More powerful methods have been devised by developing sets of coupled integral equations which reduce to HNC and PY on truncation at the first equation, as was done in the Yvon–Born–Green hierarchy. By taking a pair of equations, we obtain the approximations known as HNC2 and PY2. There is no unique way of developing such equations, and so there is more than one approximation that can be called PY2. Wertheim[35] used what might be described as a superposition approximation for the elimination of a direct triplet function, c_3, whilst Percus[36] and Verlet[37] used a functional Taylor expansion, which also required explicit consideration of triplet correlation functions. All have their disadvantages. Wertheim's method is restricted to states of the fluid which can be approached from zero density along an isotherm which does not cut the vapour-pressure line, and so is inapplicable to the orthobaric liquid. Verlet's method is free from this disadvantage but is committed to an inherent lack of symmetry between molecules 1 and 2 in c_{12}, which is displeasing, and is troublesome in a mixture[38].

Figure 7.3 shows that the PY approximation is superior to HNC for hard spheres. PY2 (Verlet) is superior to both. However, although these results are encouraging, it is dangerous to assume that what is accurate for this assembly is necessarily equally so for more realistic models.

The PY and HNC equations can only be solved iteratively for the LJ potential. At typical liquid densities, this has been done for the PY approximation by Levesque[39], Watts[40], Mandel et al.[41], and Barker et al.[42,43]. The HNC approximation has been used by Madden and Fitts[44], and self-consistent combinations of PY and HNC by Mandel and Bearman[45]. On the whole, the PY approximation is better than the HNC, except at the lowest temperatures[44], but the pressures calculated from the virial and compressibility equations are inconsistent and generally inaccurate. A better route[43,46] to the thermodynamic properties is through the energy equation; that is, a calculation of U from $g(r)$ by means of equation (7.27), the calculation of A by means of the Gibbs–Helmholtz equation (7.20), and then to p and other properties by differentiation. Other integral equations and other model potentials have also been studied[43], but the consensus of opinion is now that the best way of using integral equations for dense liquids composed of spherical molecules is to restrict them to hard spheres or similar models, and to go from these systems to those with more realistic potentials by means of perturbation theory (Section 7.6), rather than attempt to find sufficiently accurate integral equations for the realistic systems.

The critical density is only about one-third of that of the liquid at the triple point, and greater accuracy might be expected here. We consider the LJ fluid since the hard-sphere fluid has no critical point. The PY equation has been solved several times in this region to obtain critical constants and to examine the singularities on the $A(V, T)$ or $p(V, T)$ surface. The best values[21,47–49] of the reduced critical density and temperature, obtained via the compressibility equation (the pressure equation gives $\beta_T^{-1} > 0$ at the critical point[47]) are $\rho^c = 0.280 \pm 0.002$ and $\tau^c = 1.315 \pm 0.002$. These are reasonably close to the values quoted above from computer simulation, namely, $\rho^c = 0.35$ and $\tau^c = 1.35$. The critical exponents β and γ (*see* Chapter 3) have their classical values of 0.5 and 1.0. For the HNC equation the best values[48,49] of the reduced critical constants, again via the compressibility equation, are $\rho^c = 0.27 \pm 0.01$ and $\tau^c = 1.402 \pm 0.001$. Over 20 years ago, M. S. Green[50] pointed out that the HNC equation would not yield a classical critical point, and this conclusion is confirmed by the numerical results[49], but unfortunately the singularity is unrealistic, with $\gamma < 1$. A more surprising result is the recent finding of K. A. Green *et al.*[51] that the Yvon–Born–(H. S.)Green equation, with the superposition approximation, behaves more realistically than the PY or HNC equations in the critical region, although its estimate of thermodynamic functions is generally worse. For convenience, they chose a square-well potential but there is no reason to suppose that their results would not be the same for the LJ or other more realistic potential. Their values of the critical exponents, $\beta = 0.330 \pm 0.008$, $\gamma = 1.24 \pm 0.04$ and $\delta = 4.4 \pm 0.2$, are almost the same as the best experimental values (Section 3.4). Why this should be is still not clear.

So far, the discussion of integral equations has been restricted to spherical potentials, but molecular fluids, with non-spherical potentials, have accounted for much of the effort in this direction in the last decade[52].

The simplest class of non-spherical potentials are those with a hard spherical core, equation (7.75), and an angular-dependent attractive potential, u_a, at $r > d$, e.g., hard spheres with point dipoles or point quadrupoles at their centres. The correlation functions now depend on the separation r_{12}, and on the angles θ_1, θ_2 and ϕ_{12} needed to describe the mutual orientation of, say, two point dipoles. Here, θ_1 is the angle of the first dipole μ_i with the vector $(\mathbf{r}_2 - \mathbf{r}_1)$, and ϕ_{12} is the azimuthal angle between the planes containing μ_1 and $(\mathbf{r}_2 - \mathbf{r}_1)$, and μ_2 and $(\mathbf{r}_2 - \mathbf{r}_1)$. The potential energy of the pair of dipoles is

$$u_a(r_{12}, \omega_1, \omega_2) = -(\mu_1 \mu_2 / 4\pi\varepsilon_0 r_{12}^3)(2\cos\theta_1 \cos\theta_2 - \sin\theta_1 \sin\theta_2 \cos\phi_{12})$$
$$(r_{12} > d) \qquad (7.90)$$

where ω_1 and ω_2 are abbreviations for the angular variables. The Ornstein–Zernike equation is now

$$h(r_{12}, \omega_1, \omega_2) = c(r_{12}, \omega_1, \omega_2)$$
$$+ (n/4\pi) \int c(r_{13}, \omega_1, \omega_3) h(r_{23}, \omega_1, \omega_3) \, d\mathbf{r}_3 \, d\omega_3 \qquad (7.91)$$

where $(1/4\pi)$ is the normalization factor for the integration over ω_3. (If the molecules lack the cylindrical symmetry of purely dipolar, or of linear molecules, then a third Eulerian angle is needed to describe this mutual orientation, and the factor of $(1/4\pi)$ becomes one of $(1/8\pi^2)$.)

A simple closure is now the *mean-spherical approximation* (MSA)[53], which is a generalization, and linearization, of the PY approximation. It is defined by

$$h(r_{12}, \omega_1, \omega_2) = -1 \qquad (r_{12} < d)$$

$$c(r_{12}, \omega_1, \omega_2) = -u_a(r_{12}, \omega_1, \omega_2)/kT \qquad (r_{12} > d) \qquad (7.92)$$

(The name arises from the resemblance of this approximation to the mean-spherical model of ferromagnetism.) The first part of equation (7.92) is exact since $y(r_{12})$ is zero within the hard-core, but the second is an approximation. Wertheim[54] solved these equations analytically to give $h(r_{12}, \omega_1, \omega_2)$ and from this result it is relatively straightforward[55] to obtain ΔA, the amount by which the free energy of the dipolar system differs from that of the parent hard-sphere system without dipoles. (Wertheim[56] has also described the implications of this and similar approximations for the dielectric constant of polar fluids, with which this book is not concerned.)

The correlation functions of the mean-spherical approximation can be tested[43,52] by comparison with computer simulations[57] of dipolar systems; the agreement is not good at short distances $r \gtrsim d$, but some of the error may be in the simulations. The present value of the theory is two-fold; first, it is one of the few approximations for which the integral equation can be solved analytically, and secondly, it has spawned a whole set of improved approximations, such as the *generalized mean-spherical approximations* (GMSA)[58] and *self-consistent Ornstein–Zernike approximation* (SCOZA)[59] in which, for spherical potentials, the direct correlation function is approximated by a Yukawa function, $\alpha r^{-1} \exp(-\gamma r)$, with the coefficients α and γ chosen to maintain consistency between the pressure and compressibility equations. Another approximation of the same kind is the *optimized random-phase approximation* (ORPA)[60] in which the outer part of the direct correlation function is written

$$c(r_{12}, \omega_1, \omega_2) = c_0(r_{12}) - u_a(r_{12}, \omega_1, \omega_2)/kT \qquad (r > d) \qquad (7.93)$$

where $c_0(r)$ is the correlation function for the hard-sphere system. In the PY approximation this is zero and equation (7.93) reduces to equation (7.92), but the true value of $c_0(r)$ is small and positive in this region. Similarly, a *linearization* of the *hyper-netted chain* approximation (LHNC)[61] gives

$$c_a(r_{12}, \omega_1, \omega_2) = h_a(r_{12}, \omega_1, \omega_2)h_0(r_{12})/g_0(r_{12})$$
$$- u_a(r_{12}, \omega_1, \omega_2)/kT \qquad (7.94)$$

a result which was earlier called the *single super-chain* approximation (SSC), from its graph-theoretical formulation[62], and later called the *generalized mean-field* approximation (GMF)[52,63]. The original mean-spherical approximation underestimates the angular dependence of $h(r_{12}, \omega_1, \omega_2)$:

this is improved[64] in the later treatments of which there may still be more to come.

Real polar molecules are polarizable and the statistical treatment of polarization bristles with difficulties, some of which arise from the fact that the forces to which it gives rise are never pair-wise additive[1,2]. A treatment that has sufficient rigour to yield not only the thermodynamic properties but also the dielectric constant requires a more sophisticated graph theory[56,62] than that set out in the last section. Perturbation theory is, on the whole, a more direct route to the thermodynamic properties.

The term 'molecular fluids' is now generally used to describe systems in which both the repulsive and attractive parts of the pair potentials lack spherical symmetry. In the real world, these include everything except the noble gases, but most of the experimental, computer simulation and theoretical work has been on linear molecules, particularly N_2, the halogens, and CS_2, and simple models for them. Some of the set of approximations that stem from the mean-spherical approximation can be applied to these systems, but the commonest way of discussing molecular fluids is by means of site–site distribution functions. For simplicity, this discussion is at first restricted to linear centro-symmetric molecules, e.g., N_2, Br_2, CS_2.

The total correlation function $h(r_{12}, \omega_1, \omega_2)$ is a function of too many variables for it to be easy to measure or tabulate. There are two common ways of overcoming this difficulty. First h, or any other function, X, of the same variables, can be written as an expansion[2,65–67] in normalized spherical harmonics, $Y_{lm}(\omega)$:

$$X(r_{12}, \omega_1, \omega_2) = 4\pi \sum_{l_1} \sum_{l_2} \sum_m \chi_{l_1 l_2 m}(r_{12}) Y_{l_1 m}(\omega_1) Y_{l_2 m}(\omega_2) \tag{7.95}$$

where l_1 and l_2 run over all positive integers (including zero) and m runs from $-l$ to $+l$, where l is the smaller of l_1 and l_2. The coefficients χ are functions of a single variable, r_{12}, the separation of the centres of mass. If the sum is dominated by only a few terms, then this is a convenient representation of the function X, since the coefficients χ can be shown readily as a function of r_{12} at each density and temperature. Unfortunately, the convergence is often poor. The spherical average of X is the first term, $\chi_{000}(r_{12})$. The multipole expansion of classical electrostatics[2] is conveniently expressed this way; thus the dipole–dipole energy equation (7.93) comprises the 110 and 111 terms, the quadrupole–quadrupole energy, the 220, 221 and 222 terms, etc. Some of these terms are related in fairly direct ways to particular experiments[68]; thus dielectric constant and infrared absorption are related to terms with $l = 1$, and light-scattering to those with $l = 2$, etc.

The principal way of obtaining the correlation functions of a molecular liquid is from the x-ray and neutron diffraction patterns. X-rays are scattered by the orbital electrons and neutrons by the nuclei, and the diffraction patterns are the interference of the scattering from a pair of centres. Hence, x-ray diffraction, and more directly neutron diffraction, yield not $h(r_{12}, \omega_1, \omega_2)$ but a set of functions $h_{\alpha\beta}(r_{12})$, where r_{12} is now the separation of a pair of centres, one at or around nucleus α and one at or around nucleus β. For a homonuclear diatomic molecule, e.g., N_2, there is one such function, $h_{NN}(r)$, which contains both the intra- and inter-molecular correlations. For a heteronuclear diatomic molecule, e.g., CO, there are three functions, $h_{CC}(r)$, $h_{CO}(r)$ and $h_{OO}(r)$, of which only the second has an intramolecular component. The observed scattering is the Fourier transform of a weighted sum of these functions and can be decomposed into its components only by making more than one kind of experiment. The x-ray scattering cross-sections of atoms increase regularly with the atomic numbers, but the coherent neutron cross-sections change dramatically from nucleus to nucleus even within a set of fixed atomic number, or chemical identity. Hence, by doing both x-ray and neutron experiments, and by doing the latter with isotopically substituted samples, it is, in principle, possible to decompose the transform of the scattering patterns into the *site–site correlation functions* $h_{\alpha\beta}(r)$. Even if all these are known, the result is not as complete a knowledge of the structure of the liquid as is provided by $h(r_{12},\omega_1,\omega_2)$ or, in principle, by a complete set of the coefficients of its harmonic expansion. Nevertheless, if the intermolecular potential is itself a function only of the separations of the same sites, then most, but not all, of the thermodynamic properties can be calculated from $h_{\alpha\beta}(r)$. Thus[67,69]

$$U = \tfrac{1}{4}n \sum_{\alpha\beta} \int u_{\alpha\beta}(r_{\alpha\beta}) g_{\alpha\beta}(r_{\alpha\beta}) \, d\mathbf{r}_{\alpha\beta}$$

$$p = nkT - \tfrac{1}{2}n^2 V \sum_{\alpha\beta} \overline{(\mathbf{r}_{\alpha\beta} + \mathbf{r}'_\alpha - \mathbf{r}'_\beta) \cdot \mathbf{r}_{\alpha\beta} \frac{\partial u_{\alpha\beta}(r_{\alpha\beta})}{\partial r_{\alpha\beta}}} \tag{7.96}$$

where the bar denotes a canonical average, as in equation (7.18), and $\mathbf{r}'_\alpha$ and $\mathbf{r}'_\beta$ are the vectors joining sites α and β with the centres of their respective molecules. It follows that the energy can be found directly from $g_{\alpha\beta}$, but not the pressure, although the latter can always be found indirectly, from, say, U at a series of densities and temperatures, by thermodynamic arguments. The alternative to equation (7.96) is the pair of equations

$$U = \tfrac{1}{2}n \sum_{l_1 l_2 m} \int u_{l_1 l_2 m}(r_{12}) g_{l_1 l_2 m}(r_{12}) \, d\mathbf{r}_{12} \tag{7.97}$$

$$p = nkT - \tfrac{1}{2}n^2 \sum_{l_1 l_2 m} \int \left[\mathbf{r}_{12} \cdot \frac{\partial}{\partial \mathbf{r}_{12}} u_{l_1 l_2 m}(r_{12}) \right] g_{l_1 l_2 m}(r_{12}) \, d\mathbf{r}_{12} \tag{7.98}$$

and

$$kT(\partial n/\partial p)_T = 1 - n \int h_{000}(r_{12}) \, d\mathbf{r}_{12} \tag{7.99}$$

where $u_{l_1 l_2 m}(r_{12})$ are the coefficients of the expansion of $u(r_{12},\omega_1,\omega_2)$, and $\mathbf{r}_{12}$ is the vector joining the centres of the molecules. The mean-square torque exerted on molecule 1 by its neighbours, $\overline{L^2}$, is a statistical average that has no counterpart in a system of spherical molecules[70]

$$\overline{L^2} = \tfrac{1}{2}nkT \sum_{l_1 l_2 m} l_1(l_1+1) \int u_{l_1 l_2 m}(r_{12}) g_{l_1 l_2 m}(r_{12}) \, d\mathbf{r}_{12} \tag{7.100}$$

Clearly, there is no contribution from terms with $l_1 = 0$. The mean-square torque can be observed by studying, for example, the isotope-separation factor between liquid and vapour[70].

Atom–atom distribution functions have been determined for a few simple molecules by the combination of x-ray and neutron studies described above. The best results are for N_2 (x-ray[71], neutron[72], computer simulation and theory[73,74]), Br_2 (x-ray[75], neutron[76], computer simulation and theory[74,77]), CS_2 (x-ray[78], neutron[79], computer simulation and theory[80]), and H_2O (x-ray[81], neutron[82], computer simulation and theory[83]). *Figures 7.5 and 7.6* show two typical sets of results, the atom–atom distribution function for BR_2 and the three atom–atom functions for H_2O. In each case, the intramolecular component has been removed. The former figure is fairly similar to that for a monatomic liquid or for a LJ fluid (*Figure 7.4*); that is, the liquid has a structure which is determined primarily by the packing of the cores of the molecules. It is hard to match the experimental curve exactly by computer simulation or by theoretical calculation since this would require a precise knowledge of the attractive and repulsive forces and of their dependence on orientation, of the quadrupole moment (and possibly of higher multipoles), of the polarizability and its anisotropy, and of the strength of the multi-body forces. This information is not yet available for any diatomic molecule[1], although the 'best estimates' are continually improving. A reasonable attempt to reproduce the distribution by computer simulation of a model with a quadrupole is shown in *Figure 7.5*.

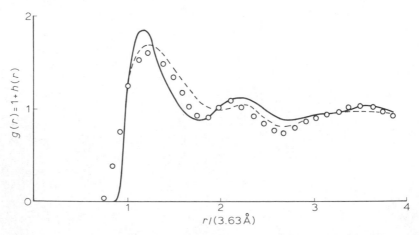

Figure 7.5 The intermolecular atom–atom distribution function for Br_2. The points are experimental[76], and the curves are attempts by Streett and Tildesley[74] to fit the results with two-centre LJ(12,6) potential, with (full line) and without (dashed line) a quadrupole moment

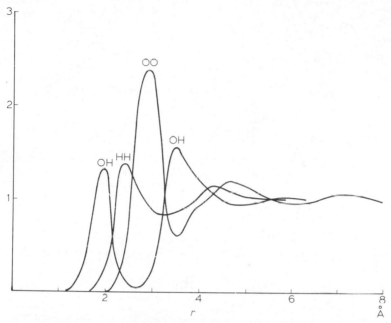

Figure 7.6 The intermolecular atom–atom distribution functions for water. The curves are sketches based on several sources[81–82]

The distribution functions for water are quite different, for the structure of this liquid is dominated by the attractive forces, that is, the hydrogen bonds[84]. Only the distribution function for pairs of oxygen atoms has any resemblance to that of the monatomic liquid. The function for the separation of O and H atoms has two large peaks, the inner of which at about 0.2 nm is the intermolecular hydrogen-bond length.

Theoretical calculations of the distribution functions $g_{\alpha\beta}(r)$ fall into two classes, perturbation calculations (Section 7.6) and calculations based on integral equations that follow from a closure of the Ornstein–Zernike equation, or from similar approximations. The PY and HNC approximations can themselves be used[85] but present great computational problems. More manageable and more widely used is the *reference interaction site model* (RISM) of Chandler and his colleagues[52,86], which resembles the PY approximation. It starts with an Ornstein–Zernike-like relation between the set $h_{\alpha\beta}$ and the set $c_{\alpha\beta}$, which is most readily written as a matrix equation between the set $H_{\alpha\beta}$ and the set $C_{\alpha\beta}$ in s-space. It closes this matrix equation with a PY-type closure which for rigid hard-core molecules (e.g., fused hard-sphere diatomics) is

$$h_{\alpha\beta}(r) = -1 \quad (r < d_{\alpha\beta}) \qquad c_{\alpha\beta}(r) = 0 \quad (r > d_{\alpha\beta}) \tag{7.101}$$

It differs from the full PY approximation in that the matrix equation between the $H_{\alpha\beta}$ and the $C_{\alpha\beta}$ is not precisely of the Ornstein–Zernike form. The sites, α, β, etc., are usually but not necessarily at the atomic sites. The

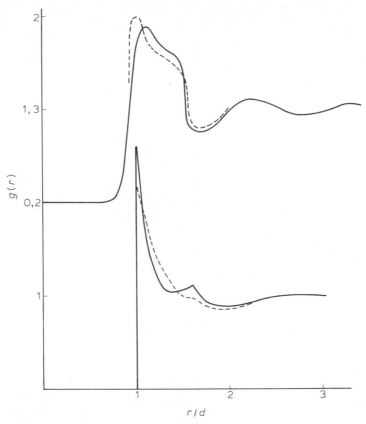

Figure 7.7 The intermolecular atom–atom distribution functions for two fused hard spheres (lower diagram) and for a diatomic LJ molecule with a bond length chosen to simulate N_2. The full lines are Monte Carlo simulations and the dashed lines RISM calculations based on two sites in each molecule[87, 88]

distribution functions for hard-core molecules have the usual discontinuities at $r = d_{\alpha\beta}$, and also cusps at larger separations which are related to the geometry of packing of the molecules in different orientations. *Figure 7.7* shows $g_{\alpha\alpha}(r)$ for a fluid of high density composed of homonuclear diatomic molecules formed from fused hard-spheres whose separation is 0.6 of the diameter, d, of each sphere[87], and the same function for homonuclear diatomic molecules formed from LJ sites[88] whose separation is chosen to represent N_2. The cusp is not present in the latter. The thermodynamic properties calculated from RISM are poor for hard-core molecules; it is primarily a theory of molecular *structure*.

In this section we have considered the measurement and computer simulation of distribution functions, and those theoretical methods for their calculation that follow directly from the fundamental statistical equations by means of well-defined approximations. These theories have had a limited success but are far from giving a quantitatively acceptable account of structure or thermodynamical properties. They are at their best for the simplest model fluids—hard spheres, and fused hard-spheres. Most recent work has built on these results by using them as the foundation of perturbation theories. The central idea of these theories is always that there is a reference system for which everything is known—structure and thermodynamics—and which is, in some sense, sufficiently close to the real system for the difference to be handled by means of an expansion in some small parameter. Before considering such theories, it is useful to interpolate a section which deals with one of the most important types of reference system, namely, that underlying the principle of corresponding states. This principle can be regarded as a device for obtaining the properties of one fluid from the known properties of another (the reference fluid) by means of a transcription that, at least formally, could be expressed as a perturbation expansion summed to all orders.

7.5 The principle of corresponding states

Consider N molecules each of two substances, 1 and 2, whose intermolecular pair potentials are related by the equations

$$u_{11}(r) = \varepsilon_{11}F(d_{11}/r) \tag{7.102}$$

$$u_{22}(r) = \varepsilon_{22}F(d_{22}/r) \tag{7.103}$$

where ε and d are an energy and a distance and F denotes some function common to both species. (Lennard-Jones potentials with the same indices, e.g., 12 and 6, for both substances are examples of potentials that are related by equations of this type.) Let the N molecules of the first substance be confined to a volume V at a temperature T, and those of the second to a volume (Vd_{22}^3/d_{11}^3) at a temperature $(T\varepsilon_{22}/\varepsilon_{11})$. If the two containing vessels have the same shape, then their linear dimensions are in the ratio (d_{11}/d_{22}). Hence, for each configuration of the first assembly, there is a corresponding configuration for the second, such that

$$\frac{\mathscr{U}_1[\ldots \mathbf{r}_i \ldots]}{kT} = \frac{\mathscr{U}_2[\ldots \mathbf{r}_i(d_{22}/d_{11})\ldots]}{k(T\varepsilon_{22}/\varepsilon_{11})} \tag{7.104}$$

and so, from equation (7.13)

$$Q_1(V, T) = (d_{22}/d_{11})^{-3N}Q_2(Vd_{22}^3/d_{11}^3, T\varepsilon_{22}/\varepsilon_{11}) \tag{7.105}$$

Thus the configuration integral Q_1 is calculable from a knowledge of Q_2 as a function of V and T. This equation may be written more neatly by introducing dimensionless ratios of the characteristic energies, f, and of the characteristic distances, g. The potential of either substance may then be written in the common form

$$u_{\alpha\alpha}(r) = f_{\alpha\alpha}u_{00}(r/g_{\alpha\alpha}) \tag{7.106}$$

where

$$f_{\alpha\alpha} = \varepsilon_{\alpha\alpha}/\varepsilon_{00} \qquad g_{\alpha\alpha} = d_{\alpha\alpha}/d_{00} \qquad (7.107)$$

and u_{00} is a common reference potential from which u_{11}, u_{22}, etc., may be obtained by choosing the appropriate scale factors f_{11} and g_{11}, etc. The useful name of *conformal substances* has been proposed for a group whose potentials are all related to each other and to a *reference substance* (subscript zero) by equation (7.106). The neater form of equation (7.105) is now

$$Q_1(V, T) = g_{11}^{3N} Q_0(V/g_{11}^3, T/f_{11}) \qquad (7.108)$$

and so for the configurational free energy

$$A_1(V, T) = f_{11} A_0(V/g_{11}^3, T/f_{11}) - 3NkT \ln g_{11} \qquad (7.109)$$

Differentiation with respect to the volume gives for a single phase

$$p_1(V, T) = (f_{11}/g_{11}^3) p_0(V/g_{11}^3, T/f_{11}) \qquad (7.110)$$

If the equation of state of the reference substance is

$$\phi_0(p, V, T) = 0 \qquad (7.111)$$

then that for substance 1 is

$$\phi_0(pg_{11}^3/f_{11}, V/g_{11}^3, T/f_{11}) = \phi_1(p, V, T) = 0 \qquad (7.112)$$

This is an expression of the *principle of corresponding states*.

If the state of the reference substance is represented by a (p, V, T) surface, then the states of all other conformal substances are represented by geometrically similar surfaces in which the axes of pressure, volume and temperature are multipled by (f/g^3), g^3 and f, respectively. It follows that all singular points on the surface, such as the solid, liquid and gas at the triple point and the fluid at the critical point, have values of p, V and T in the ratio of these scale factors. In particular

$$p_1^c = (f_{11}/g_{11}^3) p_0^c \qquad v_1^c = g_{11}^3 v_0^c \qquad T_1^c = f_{11} T_0^c \qquad (7.113)$$

so that equation (7.112) may be expressed in terms of the critical constants

$$\phi_1\left[\left(\frac{p}{p_1^c}\right), \left(\frac{v}{v_1^c}\right), \left(\frac{T}{T_1^c}\right)\right] = \phi_0\left[\left(\frac{p}{p_0^c}\right), \left(\frac{v}{v_0^c}\right), \left(\frac{T}{T_0^c}\right)\right] = 0 \qquad (7.114)$$

and

$$(pv/RT)_1^c = (pv/RT)_0^c \qquad (7.115)$$

These equations put the principle of corresponding states into the usual empirical form that has been used since the time of van der Waals. The simple dimensional analysis on which its statistical derivation is based needed only three assumptions: first, the phase integral can be factorized into a molecular and a configurational part; secondly, the latter may be treated by the methods of classical statistical mechanics; and thirdly, that the substances are conformal.

The choice of the critical constants as the characteristic parameters with which to reduce p, V and T to dimensionless ratios is an arbitrary one. There is no theoretical reason for preferring them to those of any other singular point, but there are two practical reasons. First, the empirical principle of corresponding states has been widely used for many years, particularly by engineers, and has invariably been based upon these scale factors. Secondly, small departures from equation (7.106) produce disproportionately large effects on the physical properties of the solid state. There are many sets of substances whose potentials are almost conformal and whose fluids obey equation (7.114) moderately well, but for which the properties of the solid states show little or no regularity. The solid state is particularly sensitive to small departures of the potential from spherical symmetry. The effect of these on the fluid states is discussed in the next section. The only practical disadvantage in the use of the critical constants is the difficulty of measuring V^c accurately. However, there are only two independent molecular parameters, f and g, and so it is possible to determine these from T^c and $(T^c/p^c)^{1/3}$, respectively.

The noble gases are the only substances whose molecules are spherical monatomic particles, and it is to these that one looks first for an example of a conformal set of substances. There are six noble gases, but only three of these, argon, krypton and xenon, can be used for a proper test of equation (7.114). Liquid helium and, to a lesser extent, liquid neon cannot be treated adequately by classical mechanics owing to the lightness of their molecules. Radon probably resembles the three gases below it, but too little is known of its properties for a searching test of the principle.

The reduced properties of argon, krypton and xenon are summarized in *Table 7.1*, in which ratios that should be the same for all three substances are shown in heavy type. The table is similar to one first published by Guggenheim[89] in 1945, but some of the more recent results on which it is based have improved the agreement with the principle of corresponding states.

The first four columns are the critical constants. The ratio $(pv/RT)^c$ is the same for each substance, within an experimental error of about 1 per cent in v^c.

The 5th column is the normal boiling point, which is not a corresponding state, as each substance has a different critical pressure. The 6th column is, therefore, the boiling point at a fixed fraction (1/50) of the critical pressure. The 7th and 8th columns show that the normal boiling point is not a constant fraction of the critical temperature but that the reduced boiling point at a given reduced pressure is sensibly constant.

The 9th column is the enthalpy of evaporation at the normal boiling point. The 10th column shows that the entropy of evaporation at this point

Table 7.1 The reduced properties of argon, krypton and xenon

	1 T^c/K	2 p^c/bar	3 $v^c/cm^3\,mol^{-1}$	4 $(pv/RT)^c$
Argon	150.9	49.0	74.6	**0.291**
Krypton	209.4	55.0	92.2	**0.291**
Xenon	289.8	58.8	118.8	**0.290**

	5 T^b/K	6 $T†/K^a$	7 T^b/T^c	8 $T†/T^c$
Argon	87.3	86.9	0.579	**0.577**
Krypton	119.9	121.0	0.573	**0.578**
Xenon	165.1	167.9	0.570	**0.580**

	9 Δh_e $J\,mol^{-1}$	10 $\Delta s_e(T^b)$ $J\,mol^{-1}\,K^{-1}$	11 $\Delta s_e(T†)$	12 $(c_p)^l(T^t)$ $J\,mol^{-1}\,K^{-1}$
Argon	6 520	74.7	**75.0**	**29.4**
Krypton	9 030	75.3	**74.6**	**32.0**
Xenon	12 640	76.6	**75.3**	**32.0**

	13 T^t/K	14 T^t/T^c	15 p^t/bar	16 p^t/p^c
Argon	83.8	**0.556**	0.689	**0.0141**
Krypton	116.0	**0.554**	0.730	**0.0133**
Xenon	161.3	**0.557**	0.815	**0.0139**

	17 $(v^l/v^s)^t$	18 Δh_f $J\,mol^{-1}$	19 Δs_f $J\,mol^{-1}\,K^{-1}$
Argon	**1.15**	11 76	**14.0**
Krypton	**1.15**	16 36	**14.1**
Xenon	**1.15**	22 98	**14.2**

a $T†$ is the temperature at which the vapour pressure is equal to $(p^c/50)$. Δh_e and Δh_f are the enthalpy of evaporation at the normal boiling point and the enthalpy of fusion.

(Trouton's constant) is not the same for each substance, but at a given reduced temperature (or pressure) it is a constant (11th column). The configurational heat capacity at constant pressure of the liquid at the triple point is about $30\,J\,mol^{-1}\,K^{-1}$ (12th column). This is one of the properties of the liquid state most sensitive to a lack of conformation in the potentials. The 14th and 16th columns indicate that the reduced triple-point temperatures and pressures are constant. The latter is again a very sensitive test of the principle, as vapour pressure changes rapidly with temperature. The last three columns show that the reduced volume and entropy changes on fusion are also constants.

These three liquids, and liquid argon in particular, may therefore be used as a standard of normal behaviour with which other liquids may be compared. *Table 7.2* is a summary of the reduced properties of liquid argon based on *Tables 2.3* and *2.11*.

This excellent conformation of the three noble gases to the principle of corresponding states implies that their potentials are of a common form, equation (7.106), although it does not exclude the presence of suitably conformal three-body potentials. It does not determine this pair form, but a Lennard-Jones (12; 6) potential, although not accurate, is the best simple effective potential for the liquid. However, the need to go from theory to experiments on, say, argon via such a simple model pair potential is now diminished since computer simulation of the LJ system provides surrogate experimental results.

Table 7.2 The reduced properties of liquid argon

T/T^c	p/p^c	v/v^c	$(\alpha_p T^c)$	$(\beta_T p^c)$	$(\gamma_v T^c/p^c)$
(0.55)	—	(0.377)	(0.66)	(0.0098)	(67)
0.555^t	0.0141	0.378	0.66	0.0100	66
0.60	0.0297	0.390	0.69	0.0118	59
0.65	0.0563	0.404	0.76	0.0155	51
0.70	0.101	0.420	0.86	0.019	45
0.75	0.166	0.439	0.95	0.024	40
0.80	0.258	0.462	1.22	0.036	34
0.85	0.381	0.490		—	28
0.90	0.539	0.532	—	—	21
0.95	0.742	0.600	—	—	15
1.00	1.000	1.000	∞	∞	5.5

T/T^c	$\Delta h_e/RT^c$	h'/RT^c	u'/RT^c	c_p'/R	c_v'/R
(0.55)	(5.22)	(−4.56)	(−4.56)	(3.52)	(0.81)
0.555^t	5.20	−4.54	−4.54	3.54	0.83
0.60	4.98	−4.40	−4.41	3.76	0.92
0.65	4.80	−4.20	−4.21	4.0	0.9
0.70	4.52	−4.04	−4.05	4.3	0.8
0.75	4.4	−3.87	−3.90	4.6	0.8
0.80	4.0	−3.52	−3.60	4.9	0.7
0.85	3.6	−3.28	−3.30	5.6	0.7
0.90	3.1	−2.8	−3.1	7.1	0.7
0.95	—	—	—	—	0.8
1.00	0.0	—	—	∞	∞

The reduced temperature of 0.55 is below the triple point. Extrapolated values are given for ease of interpolation.

t triple point

7.6 Perturbation theories: spherical molecules

We have seen that it is now possible to calculate accurately the properties of a system composed of hard spheres, and also that the pair distribution function, $g(r)$, of such a system is surprisingly similar at high fluid densities to that of model systems with more realistic potentials, such as the LJ(12, 6) potential. The conjunction of these two ideas suggests that a possible route to the calculation of the properties of systems composed of LJ and of more complicated molecules is to treat the difference of intermolecular potentials $(u_{\mathrm{LJ}} - u_0)$, where u_0 is the hard-sphere potential, as a perturbation applied to the hard-sphere system. If this difference (or a function of it) is, in some sense, a small quantity, then the standard techniques of a perturbation expansion can be expected to yield a highly convergent series for the properties of the more realistic system. This hope has been amply fulfilled in the last 15 years and is now the best way of calculating the properties of many realistic model systems. The methods have been reviewed in detail[43,90,91]; in this section we discuss those used for spherical molecules and in the next those for non-spherical molecules. The equally productive applications to mixtures are the subject of Sections 8.5 and 8.8.

A realistic spherical potential differs from a hard sphere in two ways: it has an attractive component ($u < 0$) at large separations, and it is softer than a hard sphere at short separations. The two differences pose different problems and are discussed separately. The first is the more straightforward. Consider, for example, the artificially simple case of hard spheres surrounded by a weak but long-ranged attractive potential,

$$u(r) = \infty \qquad (r < d)$$
$$u(r) = -\left(\frac{3a}{2\pi}\right)\gamma^3 \exp[\gamma^3(d^3 - r^3)] \qquad (r \geqslant d) \tag{7.116}$$

where a is a parameter which measures the strength of the attractive potential, and γ is a parameter which is small. In the limit $\gamma = 0$ the attractive potential is everywhere zero and its range, γ^{-1}, becomes infinite. The configurational energy U has then a finite non-zero limit, as can be seen by integrating equation (7.27). Since the range of $u(r)$ is infinite, it suffices to replace $g(r)$ by its asymptotic value of unity, and so $n^{(2)}(r)$ by n^2, equation (7.29), to give

$$U = -Nan \tag{7.117}$$

This is the form of U given by van der Waals's equation, and, indeed, his equation of state can be regarded as the first application of perturbation theory in statistical mechanics[92].

More generally, we can write

$$\mathscr{U} = \sum_{i<j}\sum u(r_{ij}) \qquad u(r_{ij}) = u_0(r_{ij}) + \lambda u_{\mathrm{a}}(r_{ij}) \tag{7.118}$$

where $u_0(r_{ij})$ is a hard-sphere potential $(r_{ij} < d)$, $u_a(r_{ij})$ is an attractive potential, and the parameter λ is a device for 'turning on' the attractive part. At $\lambda = 0$ we have a hard-sphere potential and at $\lambda = 1$ the full pair potential $u(r)$ which we wish to consider. The perturbation expansion can be written as a Taylor expansion in λ.

$$A = A_0 + \lambda(\partial A/\partial \lambda)_{N,V,T} + \tfrac{1}{2}\lambda^2(\partial^2 A/\partial \lambda^2)_{N,V,T} + \cdots \tag{7.119}$$

where the derivatives are to be evaluated at $\lambda = 0$. By differentiation of equation (7.13) and use of the expression for the canonical average, equations (7.19) and (7.25),

$$\left(\frac{\partial A}{\partial \lambda}\right) = -\frac{kT}{Q}\left(\frac{\partial Q}{\partial \lambda}\right) = \overline{\sum_{i<j}\sum u_a(r_{ij})} \tag{7.120}$$

$$= \tfrac{1}{2}\iint u_a(r_{12})n_0^{(2)}(r_{12})\,\mathrm{d}\mathbf{r}_1\,\mathrm{d}\mathbf{r}_2 \tag{7.121}$$

$$= \frac{N}{2}n\int u_a(r_{12})g_0(r_{12})\,\mathrm{d}\mathbf{r}_{12} \tag{7.122}$$

Thus, putting $\lambda = 1$ in equation (7.119), we see that to first order the effect of the attractive potential on A can be found by averaging $u_a(r)$ over the unchanged structure of the hard-sphere system, $g_0(r)$. The second-order term[93] is more complicated and requires a knowledge of how $g_0(r)$ is itself modified by the effect of the attractive potential, or, what is equivalent, a knowledge of the three- and four-body distribution functions of the unperturbed system. The first term in equation (7.122) is independent of temperature since $g_0(r)$ is a function only of density. The second term in equation (7.119) is proportional to $(1/kT)$, and higher terms to powers of the reciprocal temperature, so that the λ-expansion, as (7.119) is sometimes called, is an expansion in powers of $(1/kT)$. It has been tested for a square-well potential[94], for which the first-order term accounts for almost the whole of $(A - A_0)$ even at $\tau = kT/\varepsilon \approx 1$, provided that the fluid density is high. At lower densities, the higher terms are more important since the attractive forces have then a greater effect on the structure of the fluid. In the limit of the dilute gas, the attractive potential dominates the second virial coefficient at low temperatures. If, however, we are seeking a theory of dense liquids then these results are encouraging.

The best way of calculating the effects of the softening of the repulsive potential is not so obvious. Consider, for example, the set of LJ$(n,\tfrac{1}{2}n)$ potentials, of which the common choice is $n = 12$. A possible small expansion parameter is n^{-1}; that is, we write

$$A = A_0 + n^{-1}(\partial A/\partial n^{-1})_{N,V,T} + O(n^{-2}) \tag{7.123}$$

where the derivative is to be calculated at $n^{-1} = 0$. The differentiation is little more difficult than that which led to equation (7.122), and gives a result which can be expressed[95] (cf. equation (7.109))

$$A(V, T) = A_0(V/g^3, T) - 3NkT \ln g \qquad (7.124)$$

That is, the LJ$(n, \frac{1}{2}n)$ fluid behaves, to first order, as a hard-sphere fluid whose diameter (gd) differs from d, that of the LJ potential, where g is an explicit function of temperature which becomes[95]

$$g = (\varepsilon/kT)^{1/n}[1 + \gamma/n] \qquad (7.125)$$

in the limit of high temperatures, where γ is here Euler's constant.

This result is not directly useful since the convergence of equation (7.123) is slow; the first-order correction (the order to which equation (7.124) is correct) is adequate only for $\tau \gtrsim 10$. Barker and Henderson[96] generalized this method of expanding the free energy in powers of a parameter of inverse steepness, and obtained an effective diameter of the form

$$gd = \int_0^d [1 - \exp(-u_r(\mathbf{r})/kT)] \, dr \qquad (7.126)$$

where u_r is the repulsive part of the LJ (or similar) potential, that is, the part in the range $r < d$. The parameter g is again a function of temperature but not of density. There is, however, a still more effective solution to the problem. Mayer's f-function, equation (7.45), differs little for a soft repulsion and for a hard sphere of similar diameter. At small separations f is -1 for both and at large separations it is zero. There is a small range near $r = d$ where they differ; for $r < d$, f is greater for the soft potential, and for $r > d$ for the hard sphere, with a discontinuity of -1 in the difference at $r = d$. Let the range over which the difference is not negligible be of length ξd, and take ξ as the small perturbation parameter. This is equivalent to taking the difference $\Delta f(r)$ as a small function and making a functional Taylor expansion in this quantity, where

$$\Delta f(r) = f(r) - f_0(r) \qquad (7.127)$$

Here, $f(r)$ is the Mayer function of a smooth but wholly repulsive potential, and $f_0(r)$ is the hard-sphere function

$$f_0(r) = -1 \quad (r < d) \qquad f_0(r) = 0 \quad (r \geqslant d) \qquad (7.128)$$

The free energy difference is now[91,97]

$$A = A_0 + \tfrac{1}{2}n^2 \int \Delta f(r) y_0(r) \, d\mathbf{r} + \cdots \qquad (7.129)$$

where $y_0(r)$ is the hard-sphere function (7.46), which, by the argument below equation (7.55), is a smooth function of the separation. Hence the hard-sphere diameter d can be chosen to nullify the first-order term in equation (7.129) by putting the integral of the *blip-function*, $B(r)$, equal to zero.

$$B(r) = \Delta f(r) y_0(r) \qquad \int B(r) \, dr = 0 \qquad (7.130)$$

This choice of effective hard-sphere diameter is that of Andersen, Weeks and Chandler[91,97]; the diameter is a function of both temperature and density. In terms of ξ, the reduced range of the blip-function, we have

$$A = A_0[1 + O(\xi^4)]$$

$$g(r) = g_0(r)[1 + O(\xi^2)]$$

(7.131)

For a purely repulsive potential, e.g., $u(r) = \varepsilon(d/r)^{1/2}$, ξ is not particularly small, about 0.35 at $\tau \sim 0.8$, but for the repulsive part of a LJ(12, 6) potential it is only about 0.15. Hence this way of calculating the effect of the softening of the potential is an accurate way of obtaining the thermodynamic properties, and a useful but less accurate way of obtaining the structure.

It remains to combine the two treatments of the attractive and repulsive forces. This was first done successfully by Barker and Henderson[96] who divided the LJ potential into two parts at the collision diameter,

$$u_r(r) = u_{LJ}(r) \quad (r < d) \qquad u_a(r) = u_{LJ}(r) \quad (r \geq d)$$

(7.132)

where u_r and u_a are the soft repulsive and attractive parts, respectively. The first was handled by equation (7.126) and the second by equation (7.122) together with higher-order terms. For the first time, a theoretical calculation generated thermodynamic properties for a LJ fluid which were close to those of computer simulation. Their treatment was followed by that of Andersen, Weeks and Chandler, which was based on a different division of the LJ potential:

$$u_r(r) = u_{LJ}(r) + \varepsilon \quad (r < r_m)$$

$$u_a(r) = -\varepsilon \quad (r < r_m) \qquad u_a(r) = u_{LJ}(r) \quad (r \geq r_m)$$

(7.133)

where r_m is the separation at which u_{LJ} has its minimum, $u_{LJ}(r_m) = -\varepsilon$. The repulsive term was handled by the blip-function (7.130), and the first-order term of equations (7.119)–(7.122) then suffices for an adequate calculation of the thermodynamic properties, and of $g(r)$. Values of the energy and pressure for two typical liquid states are shown in *Table 7.3*. The energies are excellent, and even the pressures, which are extremely sensitive functions of density in a dense liquid, are satisfactory.

Table 7.3 A comparison[91] of calculated properties of a LJ(12, 6) fluid with those obtained by computer simulation (CS)

ρ	τ	$U/N\varepsilon$		$p/\rho kT$	
		calc	CS	calc	CS
0.75	0.84	−6.01	−6.04	0.38	0.37
0.85	0.76	−6.06	−6.07	0.74	0.82

A comparison of theory with reliable results from computer simulation is now acknowledged to be the best test of the accuracy of the statistical mechanical approximations. If we aspire further to a direct calculation of the properties of a real liquid then we must be sure that we know precisely the intermolecular forces. In particular, the assumption of pair-wise additivity is inadequate at liquid density even for the simplest molecules. The most important correction is the third-order dispersion energy—the Axilrod–Teller correction[98],

$$u(r_{12}, r_{13}, r_{23}) = v_3(1 + 3\cos\theta_{12}\cos\theta_{13}\cos\theta_{23})/r_{12}^3 r_{13}^3 r_{23}^3 \qquad (7.134)$$

where the three molecules are at the corners of a triangle of sides r_{12}, r_{13} and r_{23}, and internal angles θ_{12}, θ_{13} and θ_{23}. The coefficient v_3 can be calculated, at least approximately, from the coefficient v_2 of the two-body r^{-6} (London) energy[98]. For liquid argon the net effect of the three-body term is 'repulsive', since equation (7.134) is positive if all the internal angles of the triangle are acute, and negative only if one is sufficiently obtuse. In a dense liquid most close triplets of molecules form acute-angled triangles.

Since the three-body energy is small compared with the two-body, it is again appropriate to use perturbation theory to calculate its effects. In this way it has been shown[1,99] that the thermodynamic properties of liquid argon are well represented by adding the Axilrod–Teller correction to the properties appropriate to a fluid composed of molecules interacting with the accurately known two-body potential.

7.7 Perturbation theories: non-spherical molecules

The development in the preceding section suggests that perturbation theory is a natural way to go from the now well-understood properties of fluids with spherical molecules to the less-understood properties of those with non-spherical molecules. Thirty years ago it was realized that departures from the principle of corresponding states were neither large nor random, but could be characterized by one new small empirical parameter for each fluid[100–102], which is now called the *acentric factor*[102], ω (and is not to be confused with the same symbol used as an abbreviation for the set of angular variables). Thus if, for example, we compare the configurational free energy of ethane (1) and propane (2) with that of argon (0) by means of the principle of corresponding states, then the parameters f_{11}, g_{11} and f_{22}, g_{22} of equation (7.109) are not constants related to p_1^c, v_1^c, T_1^c and p_2^c, v_2^c, T_2^c by equation (7.113), but are slowly varying functions of reduced temperature and density. This slow variation is of the same functional form for both ethane and propane, but the function is everywhere larger for the latter in the ratio of 0.152 to 0.102, which are the values of the acentric factors. It is this common functional form of departure from the principle of corresponding states which suggests strongly that a first-order or linear perturbation theory is adequate for the calculation of the thermodynamic properties of at least the simpler non-spherical molecules. Such a theory was developed[100] and its conclusions matched many of the facts that can be deduced by plotting results, such as those in Chapter 2, in a corresponding states form. For example, the reduced vapour pressure curve becomes increasingly steep as the acentric factor increases, and, since this is an easy property to measure accurately, the slope at $(T/T^c) = 0.7$ is the property used empirically to determine ω. Typical values are Ar, Kr, Xe, 0.00; O_2, 0.021; N_2, 0.040; CH_4, 0.013; C_2H_6, 0.102, C_3H_8, 0.152; C_6H_6, 0.215; H_2O, 0.348. With water, ω is not sufficiently small for departures from corresponding states to be adequately represented by a linear theory. Theory also predicts, and experiments confirm, that the reduced density of the liquid, its reduced enthalpy of evaporation and configurational heat capacities at constant volume and pressure all increase with increasing values of ω. However, the specialized form of perturbation theory used in this work[100,103], has not been that on which most subsequent developments have been based (the work of Perram and White[104] is an exception), so we omit the details and turn instead to a more general treatment due to Pople[105].

As in equation (7.118), the pair potential is divided into two parts, a reference potential u_0 which is spherical and so a function only of r_{ij}, and a perturbation potential u_1, which is a function of the angular variables abbreviated by ω_i and ω_j and which is multiplied by a coupling parameter λ,

$$u(r_{ij}, \omega_i, \omega_j) = u_0(r_{ij}) + \lambda u_1(r_{ij}, \omega_i, \omega_j) \tag{7.135}$$

The division of u into u_0 and u_1 is made in such a way that the angle-average of the latter vanishes,

$$\left(\frac{1}{4\pi}\right)^2 \int u_1(r_{ij}, \omega_i, \omega_j) \, d\omega_i \, d\omega_j = 0$$

$$\left(\frac{1}{4\pi}\right)^2 \int u(r_{ij}, \omega_i, \omega_j) \, d\omega_i \, d\omega_j = u_0(r_{ij}) \tag{7.136}$$

For axially symmetric molecules, u can be expanded in spherical harmonics, equation (7.94), with coefficients $\chi_{l_1 l_2 m}(r_{ij})$. From equation (7.136),

$$u_0(r_{ij}) = \chi_{000}(r_{ij}) \tag{7.137}$$

A λ-expansion of A, equation (7.119), has no term of first-order because the angular average of u_1 is zero, equation (7.136), and so, after putting $\lambda = 1$, A can be written

$$A = A_0 + A_2 + A_3 + \cdots \tag{7.138}$$

where[105], with an obvious contraction of notation,

$$A_2 = -\frac{1}{4}\frac{n^2}{kT}\left(\frac{1}{4\pi}\right)^2 \int [u_1(12)]^2 g_0^{(2)}(12) \, d\omega_1 \, d\omega_2 \, dr_1 \, dr_2$$

$$-\frac{1}{12}\frac{n^3}{kT}\left(\frac{1}{4\pi}\right)^3 \int u_1(12)u_1(13)g_0^{(3)}(123) \, d\omega_1 \, d\omega_2 \, d\omega_3 \, dr_1 \, dr_2 \, dr_3 \tag{7.139}$$

The more complicated third-order term A_3 can be expressed[106,107] in terms of averages over distribution functions of the reference system up to the fourth, $g_0^{(4)}(1234)$. Both A_2 and A_3 are simplified if, as is true for the multipole expansion, there are no terms with $l_1 = l_2 = 0$ in the spherical harmonic expansion. If l is non-zero, then

$$\frac{1}{4\pi} \int u_1(12)\,d\omega_1 = \frac{1}{4\pi} \int u_1(12)\,d\omega_2 = 0 \tag{7.140}$$

and the three-body term in A_2 and the four-body terms in A_3 vanish. The latter is then[106,107]

$$A_3 = \frac{1}{12} \frac{n^2}{(kT)^2} \left(\frac{1}{4\pi}\right)^2 \int [u_1(12)]^3 g_0^{(2)}(12)\,d\omega_1\,d\omega_2\,d\mathbf{r}_1\,d\mathbf{r}_2$$

$$+ \frac{1}{6} \frac{n^3}{(kT)^2} \left(\frac{1}{4\pi}\right)^3 \int u_1(12)u_1(13)u_1(23)g_0^{(3)}(123)\,d\omega_1\,d\omega_2\,d\omega_3\,d\mathbf{r}_1\,d\mathbf{r}_2\,d\mathbf{r}_3 \tag{7.141}$$

If u_1 is a dipole–dipole term then the first term of equation (7.141) is also zero. The integrations over the angles can readily be made from the known properties of the spherical harmonic functions, and leads for A_2 to integrals of r^{-n} over $g_0^{(2)}$, which have been tabulated[108] for a LJ(12, 6) reference potential, using the results of computer simulation for $g_0(r)$. The calculation[106] of A_3 requires the superposition approximation for the term containing $g_0^{(3)}$.

The rate of convergence of equation (7.138) depends on the magnitude of u_1; if this arises from the interaction of dipoles, μ, or of quadrupoles, q, then the magnitude is determined by the reduced quantities μ^* and q^*;

$$(\mu^*)^2 = \mu^2/(4\pi\varepsilon_0)\varepsilon d^3 \qquad (q^*)^2 = q^2/(4\pi\varepsilon_0)\varepsilon d^5 \tag{7.142}$$

where ε_0 is the permittivity of a vacuum, ε is the (quite unrelated) energy and d the diameter of the reference potential, u_0. The free energy A_n of equation (7.138) is proportional to $(\mu^*)^{2n}$, or $(q^*)^{2n}$. Thus a small multipole in a small molecule produces as large a perturbation in the free energy as a large multipole in a large molecule, provided that the charge distribution of the latter is well-represented by a point multipole at its centre. This is a reasonable assumption for many small inorganic molecules but unrealistic for organic molecules such as CH_3OH or $(CH_3)_2CO$. Typical values[106] of μ^* and q^* range from about 0.5 (q^* for N_2), through unity (μ^* and q^* for HCl, μ^* for NH_3, q^* for CO_2), to 2 (μ^* for H_2O). When μ^* is above unity, the convergence of equation (7.138) is slow, but Stell, Rasaiah and Narang[107] suggest that the series be summed by approximating it by a geometrical progression:

$$A = A_0 + A_2[1 - (A_3/A_2)]^{-1} \tag{7.143}$$

This apparently arbitrary assumption works extremely well[107,108], and has the merit of behaving correctly at large μ^*. Rushbrooke, Stell and Høye[109] showed, from electrostatic arguments put forward by Onsager[109], that $(A - A_0) > -N(\mu^*)^2/(\pi\varepsilon_0)d^3$ for dipolar hard spheres; that is, A is bounded below by a term proportional to $(\mu^*)^2$, a feature which equation (7.143) and the mean-spherical approximation both preserve. A similar summation of the perturbation expansion of $g(r)$ about $g_0(r)$ is equally successful[110].

The thermodynamic properties of N_2, O_2, CO, CH_4 and CO_2, some of which are set out in the tables of Chapter 2, provide a test for this theory. The comparison has been made independently by two groups[108], using different versions of the theory and with different choices of intermolecular parameters; both, however, rely on equation (7.143). The agreement is excellent, but may owe something to the great flexibility of this treatment with a large set of parameters for the central and anisotropic parts of the pair potential, and for the three-body forces.

It is assumed above that multipoles are scalars, i.e., that each can be represented by a single numerical quantity. In general, this is not so, for the dipole is a vector and the quadrupole a second-order tensor, etc. If a molecule has charges e_i at positions $\mathbf{r}_i$, then

$$\boldsymbol{\mu} = \sum_i e_i \mathbf{r}_i \qquad \mathbf{q} = \sum_i e_i (\tfrac{3}{2}\mathbf{r}_i\mathbf{r}_i - \tfrac{1}{2}r_i^2 \mathbf{1}) \tag{7.144}$$

are the cartesian representations of $\boldsymbol{\mu}$ and $\mathbf{q}$. Since $\mathbf{q}$ is traceless,

$$q_{xx} + q_{yy} + q_{zz} = 0 \tag{7.145}$$

If $\mathbf{q}$ is zero then the axis of the dipole moment defines the only non-zero component of $\boldsymbol{\mu}$, say $\mu = \mu_z$, with $\mu_x = \mu_y = 0$. Further simplification depends on the molecular symmetry, for $\mathbf{q}$ has the same symmetry as the tensor of inertia if the molecule is rigid and isotopically pure. In a linear molecule (e.g., HCl), the z-axis is the molecular axis and μ_z and q_{zz} are the only non-vanishing components. The former is also zero in such molecules as N_2 and CO_2. In a spherical top (e.g., CH_4) all components of $\boldsymbol{\mu}$ and $\mathbf{q}$ are zero; the octopole is the first non-vanishing moment. In a symmetric top (e.g., NH_3, C_6H_6) the z-axis is the axis of symmetry and $\mu \equiv \mu_{zz}$ and $q \equiv q_{zz} = -2q_{xx} = -2q_{yy}$ are again the single quantities that enter into the dipolar and quadrupolar energies of interaction. The off-diagonal elements q_{xy}, etc., are zero. Asymmetric tops (e.g., H_2O, C_2H_4, CH_3OH) are more subtle. If the cartesian frame is that of the principal axes of inertia, then q_{xy}, etc. $= 0$, and so from equation (7.145) there are two independent elements in $\mathbf{q}$. If μ lies along the z-axis, as in H_2O, then $\mu = \mu_{zz}$ is the one non-zero component of $\boldsymbol{\mu}$. If this is not the case (e.g., CH_3OH) then μ_x, μ_y and μ_z are non-zero, and five quantities enter into the energy of interaction. Perturbation theory now becomes algebraically complicated[111]. It is sometimes useful (e.g., C_2H_4) to use only the scalar value of $\boldsymbol{\mu}$ (zero in this case) and a single effective quantity q defined by

$$q_{eff}^2 = \tfrac{2}{3}(q_{xx}^2 + q_{yy}^2 + q_{zz}^2) = q_{zz}^2 + \tfrac{1}{3}(q_{xx} - q_{yy})^2 \tag{7.146}$$

This approximation is seriously in error if the moments are large or the symmetry low[111]. Further errors are incurred by the neglect of higher moments.

The separation of the potential (7.135) is unsuitable if the repulsive core is non-spherical, for then (u_1/kT) is large and positive (infinite, if the core is hard) for some configurations, and so cannot be treated as a small perturbation. The Mayer function, $f(r, \omega)$, is always bounded, and generally better behaved than u itself, so that an alternative separation is

$$f(r, \omega) = (1 - \lambda)f_0(r) + \lambda f(r, \omega) \tag{7.147}$$

The right-hand side is $f_0(r)$ at $\lambda = 0$ and $f(r, \omega)$ at $\lambda = 1$. There are two ways of developing this expansion in λ, that of Perram and White[104], and the extension to non-spherical potentials[112] of the blip-function expansion, (7.127)–(7.131). The second has been more widely used for N_2, etc., but is inadequate[74] for molecules with cores as long as those of Cl_2 and Br_2. For these the only solution is to choose a reference potential that is itself non-spherical. The first attempts[113] to do this expressed the properties of this system in terms of centre-of-mass coordinates and Eulerian angles, and obtained these properties by means of a second expansion (e.g., of the kind developed by Bellemans[114], which is restricted to hard non-spherical bodies). We return to some of this work in Section 8.8 since it has interesting qualitative implications for the properties of mixtures; here we note only that such methods are difficult, and only partially successful in a quantitative sense for highly anisotropic potentials. However, now that there are reliable Monte Carlo results for hard-core interaction site model fluids[115] it is preferable to use these as the reference potential for the perturbation calculations. Such a theory has been put forward by Tildesley[116] and applied successfully to a calculation of the structure and properties of simulated chlorine[73,77]. A problem that is still to be solved is the combination of strongly asymmetric repulsive potentials with strong, possibly non-axial, multipoles. It appears that under these circumstances the multipolar forces are structure-determining and so not easily incorporated in a perturbation scheme. Unfortunately, such potentials are widespread among organic molecules.

The application of site-model and perturbation theories is still an active field of research, new developments are reported almost monthly, and until some consensus has been reached further discussion of the field must be left to more specialized reviews.

References

1 MAITLAND, G., RIGBY, M., SMITH, E. B., and WAKEHAM, W. A., *Intermolecular Forces*, Clarendon Press, Oxford (1981)
2 GRAY, C. G., and GUBBINS, K. E., *Theory of Molecular Fluids*, Vol. 1, Clarendon Press, Oxford (1982) [Vol. 2 to be published]
3 MÜNSTER, A., *Statistical Thermodynamics*, Vol. 1, 214 et seq., Springer, Berlin (1969)

4 Ref. 3, Vol. 1, Chap. 5
5 CHEN, S.-H., (Chap. 2) and BAXTER, R. J., (Chap. 4), in *Physical Chemistry: an Advanced Treatise*, Ed. Henderson, D., Vol. 8A, Academic Press, New York (1971)
6 HANSEN, J.-P., and McDONALD, I. R., *Theory of Simple Liquids*, Academic Press, London (1976)
7 SCOTT, R. L., ref. 5., Chap. 1
8 GREEN, H. S., *Proc. R. Soc.*, **A189**, 103 (1947)
9 ORNSTEIN, L. S., and ZERNIKE, F., *Proc. Sect. Sci. K. ned. Akad. Wet.*, **17**, 793 (1914)
10 FRISCH, H. L., and LEBOWITZ, J. L. (Eds.), *The Equilibrium Theory of Classical Fluids*, Benjamin, New York (1964)
11 Ref. 3, Vol. 1, Chap. 8–10
12 PINGS, C. J., in *The Physics of Simple Liquids*, Eds. Temperley, H. N. V., Rowlinson, J. S., and Rushbrooke, G. S., Chap. 10, North-Holland, Amsterdam (1968); YARNELL, J. L., KATZ, M. J., WENZEL, R. G., and KOENIG, S. H., *Phys. Rev.*, **A7**, 2130 (1973); KARNICKY, J. F., and PINGS, C. J., *Adv. chem. Phys.*, **34**, 157 (1976)
13 PAGE, D. I., in *Chemical Applications of Thermal Neutron Scattering*, Ed. Willis, B. T. M., Chap. 8, Clarendon Press, Oxford (1973); COPLEY, J. R. D., and LOVESEY, S. W., *Rep. Prog. Phys.*, **38**, 461 (1975); CHIEUX, P., in *Neutron Diffraction*, Ed. Dachs, H., Chap. 8, Springer, Berlin (1978)
14 BLUM, L., and NARTEN, A. H., *Adv. chem. Phys.*, **34**, 203 (1976); *Structure and Motion in Molecular Fluids, Chem. Soc. Faraday Discuss.*, **66** (1978)
15 MIKOLAJ, P. G., and PINGS, C. J., *J. chem. Phys.*, **46**, 1401, 1412 (1967); PINGS, C. J., *Discuss. Faraday Soc.*, **43**, 89 (1967)
16 Ref. 6, Chap. 3; VALLEAU, J. P., and WHITTINGTON, S. G., (Part A, Chap. 4); VALLEAU, J. P., and TORRIE, G. M., (Part A, Chap. 5); ERPENBECK, J. J., and WOOD, W. W., (Part B, Chap. 1); KUSHICK, J., and BERNE, B. J., (Part B, Chap. 2) in *Statistical Mechanics*, Ed. Berne, B. J., Plenum, New York (1977)
17 BARKER, J. A., and HENDERSON, D., *Molec. Phys.*, **21**, 187 (1971)
18 ALDER, B. J., and WAINWRIGHT, T. E., *J. chem. Phys.*, **33**, 1439 (1960); WOOD, W. W., and ERPENBECK, J. J., *A. Rev. phys. Chem.*, **27**, 319 (1976)
19 NICOLAS, J. J., GUBBINS, K. E., STREETT, W. B., and TILDESLEY, D. J., *Molec. Phys.*, **37**, 1429 (1979); HANSEN, J.-P., and VERLET, L., *Phys. Rev.*, **182**, 307 (1969)
20 WIDOM, B., *J. chem. Phys.*, **39**, 2808 (1963); ROMANO, S., and SINGER, K., *Molec. Phys.*, **37**, 1765 (1979); POWLES, J. G., *Molec. Phys.*, **41**, 715 (1980)
21 WATTS, R. O., in *Statistical Mechanics, Specialist Periodical Reports*, Ed. Singer, K., Vol. 1, Chap. 1, Chemical Society, London (1973)
22 ROWLINSON, J. S., and WIDOM, B., *Molecular Theory of Capillarity*, Chap. 4, Clarendon Press, Oxford (1982)
23 KIRKWOOD, J. G., *J. chem. Phys.*, **3**, 300 (1935); KIRKWOOD, J. G., and BOGGS, E. M., *J. chem. Phys.*, **10**, 394 (1942); KIRKWOOD, J. G., MAUN, E. K., and ALDER, B. J., *J. chem. Phys.*, **18**, 1040 (1950)
24 RAVECHÉ, H. J., and MOUNTAIN, R. D., *J. chem. Phys.*, **53**, 3101 (1970); WANG, S., and KRUMHANSL, J. A., *J. chem. Phys.*, **56**, 4287 (1972)
25 LEE, Y. T., REE, F. H., and REE, T., *J. chem. Phys.*, **48**, 3506 (1968); REE, F. H., LEE, Y. T., and REE, T., *J. chem. Phys.*, **55**, 234 (1971)
26 VAN LEEUWEN, J. M. J., GROENEVELD, J., and DE BOER, J., *Physica*, **25**, 792 (1959); RUSHBROOKE, G. S., *Physica*, **26**, 259 (1960); VERLET, L., and LEVESQUE, D., *Physica*, **28**, 1124 (1962); MEERON, E., *J. math. Phys.*, **1**, 192 (1960); MORITA, T., and HIROIKE, K., *Prog. theor. Phys.*, *Osaka*, **23**, 385 (1960); VERLET, L., *Nuovo Cim.*, **18**, 77 (1960)
27 ROWLINSON, J. S., *Molec. Phys.*, **9**, 217 (1965); **10**, 533 (1966); *Discuss. Faraday Soc.*, **43**, 243 (1967); LADO, F., *J. chem. Phys.*, **47**, 4828 (1967)
28 PERCUS, J. K., and YEVICK, G. J., *Phys. Rev.*, **110**, 1 (1958)
29 STELL, G., *Physica*, **29**, 517 (1963); *Equilibrium Theory of Classical Fluids*, Eds. Frisch, H. L., and Lebowitz, J. L., Chap. II–4, Benjamin, New York (1964)
30 THIELE, E., *J. chem. Phys.*, **39**, 474 (1963); WERTHEIM, M. S., *Phys. Rev., Lett.*, **10**, 321 (1963); *J. math. Phys.*, **5**, 643 (1964)
31 KLEIN, M., *J. chem. Phys.*, **39**, 1388 (1963); *Physics Fluids*, **7**, 391 (1964)
32 CARNAHAN, N. F., and STARLING, K. E., *J. chem. Phys.*, **51**, 635 (1969)
33 HUTCHINSON, P., *Molec. Phys.*, **13**, 495 (1967)

34 HURST, C., *Phys. Lett.*, **14**, 192 (1965); *Proc. phys. Soc.*, **88**, 533 (1966); HENDERSON, D., *Proc. phys. Soc.*, **87**, 592 (1966)

35 WERTHEIM, M. S., *J. math. Phys.*, **8**, 927 (1967)

36 PERCUS, J. K., *Phys. Rev. Lett.*, **8**, 462 (1962)

37 VERLET, L., *Physica*, **30**, 95 (1964); **31**, 959 (1965); *Phys. Rev.*, **159**, 98 (1967); LEVESQUE, D., *Physica*, **32**, 1985 (1966); VERLET, L., and LEVESQUE, D., *Physica*, **36**, 254 (1967)

38 GUERRERO, M. I., ROWLINSON, J. S., and SAWFORD, B. L., *Molec. Phys.*, **28**, 1603 (1974)

39 LEVESQUE, D., *Physica*, **32**, 1985 (1966)

40 WATTS, R. O., *Can. J. Phys.*, **47**, 2709 (1969)

41 MANDEL, F., BEARMAN, R. J., and BEARMAN, M. Y., *J. chem. Phys.*, **52**, 3315 (1970)

42 BARKER, J. A., HENDERSON, D., and WATTS, R. O., *Phys. Lett.*, **31A**, 48 (1970); HENDERSON, D., BARKER, J. A., and WATTS, R. O., *IBM. J. Res. Dev.*, **14**, 668 (1970)

43 BARKER, J. A., and HENDERSON, D., *Rev. mod. Phys.*, **48**, 587 (1976)

44 MADDEN, W. G., and FITTS, D. D., *J. chem. Phys.*, **61**, 5475 (1974)

45 MANDEL, F., and BEARMAN, R. J., *J. chem. Phys.*, **55**, 4762 (1971)

46 CHEN, M., HENDERSON, D., and BARKER, J. A., *Can. J. Phys.*, **47**, 2009 (1969)

47 WATTS, R. O., *J. chem. Phys.*, **48**, 50 (1968)

48 HENDERSON, D., and MURPHY, R. D., *Phys. Rev.*, **A6**, 1224 (1972)

49 GUERRERO, M., SAVILLE, G., and ROWLINSON, J. S., *Molec. Phys.*, **29**, 1941 (1975)

50 GREEN, M. S., *J. chem. Phys.*, **33**, 1403 (1960)

51 GREEN, K. A., LUKS, K. D., and KOZAK, J. J., *Phys. Rev. Lett.*, **42**, 985 (1979); GREEN, K. A., LUKS, K. D., LEE, E., and KOZAK, J. J., *Phys. Rev.*, **A21**, 356 (1980)

52 Ref. 2, Chap. 5; CHANDLER, D., *A. Rev. phys. Chem.*, **29**, 441 (1978)

53 LEBOWITZ, J. L., and PERCUS, J. K., *Phys. Rev.*, **144**, 251 (1966)

54 WERTHEIM, M. S., *J. chem. Phys.*, **55**, 4291 (1971)

55 NIENHUIS, G., and DEUTCH, J. M., *J. chem. Phys.*, **56**, 5511 (1971); RUSHBROOKE, G. S., STELL, G., and HØYE, J. S., *Molec. Phys.*, **26**, 1199 (1973)

56 WERTHEIM, M. S., *A. Rev. phys. Chem.*, **30**, 471 (1979); STELL, G., PATEY, G. N., and HØYE, J. S., *Adv. chem. Phys.*, **48**, 183 (1981)

57 PATEY, G. N., and VALLEAU, J. P., *J. chem. Phys.*, **61**, 534 (1974); VERLET, L., and WEIS, J. J., *Molec. Phys.*, **28**, 665 (1974); LEVESQUE, D., PATEY, G. N., and WEIS, J. J., *Molec. Phys.*, **34**, 1077 (1977)

58 HØYE, J. S., LEBOWITZ, J. L., and STELL, G., *J. chem. Phys.*, **61**, 3253 (1974)

59 HØYE, J. S., and STELL, G., *J. chem. Phys.*, **67**, 524 (1977)

60 ANDERSEN, H. C., and CHANDLER, D., *J. chem. Phys.*, **53**, 547 (1970); **57**, 1918 (1972)

61 PATEY, G. N., *Molec. Phys.*, **34**, 427 (1977)

62 WERTHEIM, M. S., *Molec. Phys.*, **25**, 211 (1973); **26**, 1425 (1973); **33**, 95 (1977); **34**, 1109 (1977); **36**, 1217 (1978); **37**, 83 (1979)

63 HENDERSON, R. C., and GRAY, C. G., *Can. J. Phys.*, **56**, 571 (1978); GRAY, C. G., and HENDERSON, R. C., *Can. J. Phys.*, **57**, 1605 (1979); SMITH, W. R., and HENDERSON, D., *J. chem. Phys.*, **15**, 319 (1979)

64 STELL, G., and WEIS, J. J., *Phys. Rev.*, **16A**, 757 (1977)

65 ROSE, M. E., *Elementary Theory of Angular Momentum*, Wiley, New York (1957)

66 STREETT, W. G., and GUBBINS, K. E., *A. Rev. phys. Chem.*, **28**, 373 (1977)

67 STEELE, W. A., *Chem. Soc. Faraday Discuss.*, **66**, 138 (1978)

68 ROWLINSON, J. S., and EVANS, M., *Rep. Prog. Chem.*, **72A**, 5 (1975)

69 FREASIER, B. C., JOLLY, D., and BEARMAN, R. J., *Molec. Phys.*, **31**, 255 (1976); BEARMAN, R. J., *Molec. Phys.*, **34**, 1687 (1977); NEZBEDA, I., *Molec. Phys.*, **33**, 1287 (1977); AVIRAM, I., TILDESLEY, D. J., and STREETT, W. B., *Molec. Phys.*, **34**, 881 (1977)

70 TWU, C. H., GRAY, C. G., and GUBBINS, K. E., *Molec. Phys.*, **27**, 1601 (1974); **30**, 1607 (1975); THOMPSON, S. M., TILDESLEY, D. J., and STREETT, W. B., *Molec. Phys.*, **32**, 711 (1976); NEZBEDA, I., *Molec. Phys.*, **33**, 1287 (1977); TILDESLEY, D. J., STREETT, W. B., and WILSON, D. S., *Chem. Phys.*, **36**, 63 (1979); TILDESLEY, D. J., STREETT, W. B., and STEELE, W. A., *Molec. Phys.*, **39**, 1169 (1980); NEZBEDA, I., and SMITH, W. R., *Chem. Phys. Lett.*, **64**, 146 (1979)

71 FURUMOTO, H. W., and SHAW, C. H., *Physics Fluids*, **7**, 1026 (1964)

72 DORE, J. C., WALFORD, G., and PAGE, D. I., *Molec. Phys.*, **29**, 565 (1975); CARNEIRO, K., and McTAGUE, J. P., *Phys. Rev.*, **A11**, 1744 (1975)

73 BAROJAS, J., LEVESQUE, D., and QUENTREC, B., *Phys. Rev.*, **A7**, 1092 (1973); CHEUNG, P. S. Y., and POWLES, J. G., *Molec. Phys.*, **30**, 921 (1975); **32**, 1383 (1976); EGELSTAFF, P. A., *Chem. Soc. Faraday Discuss.*, **66**, 7 (1978)

74 STREETT, W. B., and TILDESLEY, D. J., *Proc. R. Soc.*, **A348**, 485 (1976); **A355**, 239 (1977); *J. chem. Phys.*, **68**, 1275 (1978); WEIS, J. J., and LEVESQUE, D., *Phys. Rev.*, **A13**, 450 (1976)

75 GRUEBEL, R. W., and CLAYTON, G. T., *J. chem. Phys.*, **46**, 639 (1967); **47**, 175 (1967); NARTEN, A. H., AGRAWAL, R., and SANDLER, S. I., *Molec. Phys.*, **35**, 1077 (1978)

76 CLARKE, J. H., DORE, J. C., WALFORD, G., and SINCLAIR, R. N., *Molec. Phys.*, **31**, 883 (1976)

77 SINGER, K., TAYLOR, A. J., and SINGER, J. V. L., *Molec. Phys.*, **33**, 1757 (1977); **37**, 1239 (1979); AGRAWAL, R., SANDLER, S. I., and NARTEN, A. H., *Molec. Phys.*, **35**, 1087 (1978)

78 SANDLER, S. I., and NARTEN, A. H., *Molec. Phys.*, **32**, 1543 (1976)

79 SUZUKI, K., and EGELSTAFF, P. A., *Can. J. Phys.*, **52**, 241 (1974); GIBSON, I. P., and DORE, J. C., *Molec. Phys.*, in press

80 STREETT, W. B., and TILDESLEY, D. J., *Chem. Soc. Faraday Discuss.*, **66**, 26 (1978); TILDESLEY, D. J., and MADDEN, P. A., *Molec. Phys.*, **42**, 1137 (1981)

81 NARTEN, A. H., and LEVY, R., *J. chem. Phys.*, **55**, 2263 (1971); *Water: A Comprehensive Treatise*, Ed. Franks, F., Vol. 1, 311, Plenum, New York (1972); OLOVSSON, I., and JÖNSSON, P.-G., in *The Hydrogen Bond*, Eds. Schuster, P., Zundel, G., and Sandorfy, C., Vol. 2, Chap. 8, North-Holland, Amsterdam (1976); PALINKAS, G., KALMAN, E., and KOVACS, P., *Molec. Phys.*, **34**, 505, 525 (1977) [electron diffraction]

82 PAGE, D. I., and POWLES, J. G., *Molec. Phys.*, **21**, 901 (1971); BLUM, L., and NARTEN, A. H., *Adv. chem. Phys.*, **34**, 201 (1976)

83 RAHMAN, A., and STILLINGER, F. H., *J. chem. Phys.*, **55**, 3336 (1971); McDONALD, I. R., and KLEIN, M. C., *Chem. Soc. Faraday Discuss.*, **66**, 48, 80 (1978); PANGALI, C., RAO, M., and BERNE, B. J., *Molec. Phys.*, **40**, 661 (1980)

84 SCHUSTER, P., ZUNDEL, G., and SANDORFY, C. (Eds.), *The Hydrogen Bond*, Vol. 1: *Theory*; Vol. 2: *Structure and Spectroscopy*; Vol. 3: *Dynamics, Thermodynamics and Special Systems*, North-Holland, Amsterdam (1976)

85 CHEN, Y. D., and STEELE, W. A., *J. chem. Phys.*, **50**, 1428 (1969); **52**, 5284 (1970); MORRISON, P. F., and PINGS, C. J., *J. chem. Phys.*, **60**, 2323 (1974)

86 CHANDLER, D., and ANDERSEN, H. C., *J. chem. Phys.*, **57**, 1930 (1972); CHANDLER, D., *J. chem. Phys.*, **59**, 2749 (1973); LADANYI, B. M., and CHANDLER, D., *J. chem. Phys.*, **62**, 4308 (1975); CHANDLER, D., *Molec. Phys.*, **31**, 1213 (1976)

87 CHANDLER, D., HSU, H. S., and STREETT, W. B., *J. chem. Phys.*, **66**, 5231 (1977)

88 HSU, H. S., CHANDLER, D., and LOWDEN, L. J., *Chem. Phys.*, **14**, 213 (1976)

89 DE BOER, J., and MICHELS, A., *Physica*, **5**, 945 (1938); PITZER, K. S., *J. chem. Phys.*, **7**, 583 (1939); GUGGENHEIM, E. A., *J. chem. Phys.*, **13**, 253 (1945)

90 Ref. 2, Chap. 4; Ref. 6, Chap. 6; SMITH, W. R., *Statistical Mechanics, Specialist Periodical Reports*, Ed. Singer, K., Vol. 1, Chap. 2, Chemical Society, London (1973); BOUBLIK, T., NEZBEDA, I., and HLAVATY, K., *Statistical Thermodynamics of Simple Liquids and their Mixtures*, Chap. 6, Elsevier, Amsterdam (1980)

91 ANDERSEN, H. C., CHANDLER, D., and WEEKS, J. D., *Adv. chem. Phys.*, **34**, 105 (1976)

92 LONGUET-HIGGINS, H. C., and WIDOM, B., *Molec. Phys.*, **8**, 549 (1964); WIDOM, B., *Science*, **157**, 375 (1967); RIGBY, M., *Q. Rev. chem. Soc.*, **24**, 416 (1970); ROWLINSON, J. S., *Nature, Lond.*, **244**, 414 (1973); KLEIN, M. J., *Physica*, **73**, 28 (1974); LEVELT SENGERS, J. M. H., *Physica*, **73**, 73 (1974); **82A**, 319 (1976)

93 ZWANZIG, R., *J. chem. Phys.*, **22**, 1420 (1954); HENDERSON, D., and BARKER, J. A., in *Physical Chemistry*, Ed. Henderson, D., Vol. 8A, Chap. 6, Academic Press, New York (1971)

94 BARKER, J. A., and HENDERSON, D., *J. chem. Phys.*, **47**, 2856 (1967); ALDER, B. J., and HECHT, C. E., *J. chem. Phys.*, **50**, 3032 (1969); ALDER, B. J., YOUNG, D. A., and MARK, M. A., *J. chem. Phys.*, **56**, 3013 (1972)

95 ROWLINSON, J. S., *Molec. Phys.*, **8**, 107 (1964)

96 BARKER, J. A., and HENDERSON, D., *J. chem. Phys.*, **47**, 4714 (1967)

97 ANDERSEN, H. C., WEEKS, J. D., and CHANDLER, D., *Phys. Rev.*, **A4**, 1597 (1971)

98 AXILROD, B. M., and TELLER, E., *J. chem. Phys.*, **11**, 299 (1943); **19**, 724 (1951); BARKER, J. A., FOCK, W., and SMITH, F., *Physics Fluids*, **7**, 897 (1964)

99 BARKER, J. A., HENDERSON, D., and SMITH, W. R., *Molec. Phys.*, **17**, 579 (1969)
100 COOK, D., and ROWLINSON, J. S., *Proc. R. Soc.*, **A219**, 405 (1953); ROWLINSON, J. S., *Trans. Faraday Soc.*, **50**, 647 (1954); **51**, 131 (1955); *Molec. Phys.*, **1**, 414 (1958)
101 RIEDEL, L., *Chemie-Ingr-Tech.*, **26**, 83, 259, 679 (1954); **27**, 209, 475 (1955); **28**, 557 (1956)
102 PITZER, K. S., *J. Am. chem. Soc.*, **77**, 3427 (1955); PITZER, K. S., LIPPMAN, D. Z., CURL, R. F., HUGGINS, C. M., and PETERSEN, D. E., *ibid.*, 3433; DANON, F., and PITZER, K. S., *J. chem. Phys.*, **36**, 425 (1962)
103 BARKER, J. A., *J. chem. Phys.*, **19**, 1430 (1951)
104 PERRAM, J. W., and WHITE, L. R., *Molec. Phys.*, **24**, 1133 (1972); **28**, 527 (1974); QUIRKE, N., PERRAM, J. W., and JACUCCI, G., *Molec. Phys.*, **39**, 1311 (1980); SMITH, W. R., NEZBEDA, I., MELNYK, T. W., and FITTS, D. D., *Chem. Soc. Faraday Discuss.*, **66**, 130 (1978); MELNYK, T. W., and SMITH, W. R., *Molec. Phys.*, **40**, 317 (1980); CUMMINGS, P., NEZBEDA, I., and SMITH, W. R., *Molec. Phys.*, **43**, 1471 (1981)
105 POPLE, J. A., *Proc. R. Soc.*, **A221**, 498 (1954)
106 GUBBINS, K. E., and GRAY, C. G., *Molec. Phys.*, **23**, 187 (1972); FLYTZANI-STEPHANOPOULOS, M., GUBBINS, K. E., and GRAY, C. G., *Molec. Phys.*, **30**, 1649 (1975); GRAY, C. G., GUBBINS, K. E., and TWU, C. H., *J. chem. Phys.*, **69**, 182 (1978)
107 STELL, G., RASAIAH, J. C., and NARANG, H., *Molec. Phys.*, **23**, 393 (1972); **27**, 1393 (1974); McDONALD, I. R., *J. Phys.*, **C7**, 1225 (1974)
108 ANANTH, M. S., GUBBINS, K. E., and GRAY, C. G., *Molec. Phys.*, **28**, 1005 (1974); SHUKLA, K. P., SINGH, S., and SINGH, Y., *J. chem. Phys.*, **70**, 3086 (1979)
109 ONSAGER, L., *J. phys. Chem., Ithaca*, **43**, 189 (1939); RUSHBROOKE, G. S., STELL, G., and HØYE, J. S., *Molec. Phys.*, **26**, 1199 (1973)
110 MADDEN, W. G., and FITTS, D. D., *Molec. Phys.*, **31**, 1923 (1976); MADDEN, W. G., FITTS, D. D., and SMITH, W. R., *Molec. Phys.*, **35**, 1017 (1978)
111 GUBBINS, K. E., GRAY, C. G., and MACHADO, J. R. S., *Molec. Phys.*, **42**, 817 (1981); GRAY, C. G., and GUBBINS, K. E., *ibid.*, 843
112 SUNG, S., and CHANDLER, D., *J. chem. Phys.*, **56**, 4989 (1972); LADANYI, B. M., and CHANDLER, D., *J. chem. Phy.*, **62**, 4308 (1975)
113 STEELE, W. A., and SANDLER, S. I., *J. chem. Phys.*, **61**, 1315 (1974); SANDLER, S. I., *Molec. Phys.*, **28**, 1207 (1974); MO, K. C., and GUBBINS, K. E., *J. chem. Phys.*, **63**, 1490 (1975); BOUBLIK, T., *Molec. Phys.*, **32**, 1737 (1976); KOHLER, F., MARIUS, W., QUIRKE, N., PERRAM, J. W., HOHEISEL, C., and BREITENFELDER-MANSKE, H., *Molec. Phys.*, **38**, 2057 (1979)
114 BELLEMANS, A., *Phys. Rev. Lett.*, **21**, 527 (1968)
115 TILDESLEY, D. J., and STREETT, W. B., *Molec. Phys.*, **41**, 85 (1980)
116 TILDESLEY, D. J., *Molec. Phys.*, **41**, 341 (1980)

Chapter 8
The statistical thermodynamics of mixtures

8.1 Introduction

The extension of the statistical theory of fluids to fluid mixtures is not as straightforward as might be expected. It is often trivial to generalize the formal and exact equations of the last chapter to those for a multicomponent system, but it is the subsequent approximations that need care if useful theories are to be developed. The experimental results of Chapter 5 are expressed in the form of excess thermodynamic functions, and so it is often useful to cast the theories in a form which gives these functions directly. This can be done, for example, by an (approximate) extension of the principle of corresponding states to mixtures, or by means of a perturbation theory which takes the properties of one component as its reference system.

A successful theory can be expected to throw light on two problems central to our interest in the properties of mixtures. First, how can the excess functions of the binary systems of Chapter 5 be understood in terms of the balance of intermolecular potentials between like and unlike molecules, and, second, how can the great variety of phase behaviour revealed in Chapter 6 be interpreted in terms of these potentials. Sections 8.3–8.8 address the first of these problems and 8.9 the second.

Since less is known of the intermolecular potentials between unlike than between like molecules[1], the testing of theories is first done by comparison with the results of computer simulation rather than with those of real experiments. The early sections deal therefore with model systems, which have specified potentials, rather than with real systems which have imperfectly known potentials. We start with a note on the statistical basis of the laws of ideal mixtures, and proceed to random and related mixtures.

8.2 Ideal and random mixtures

A starting point of many theories of mixtures is the configuration integral, Q. This is given by an obvious generalization of equation (7.13) to an assembly of N_α molecules of species, α, N_β of species β, etc., all of which are—in this section—assumed to be spherically symmetrical

$$Q = \frac{1}{\prod_\alpha N_\alpha!} \int \cdots \int e^{-\mathcal{U}/kt}\, d\mathbf{r}_1 \ldots d\mathbf{r}_N \tag{8.1}$$

where

$$\sum_\alpha N_\alpha = N \tag{8.2}$$

The configuration energy, $\mathcal{U}$, now depends not only on the positions of the N molecules but also on what may be called their assignment by species to these positions. That is, the value of $\mathcal{U}$ is altered if the chemical species of a pair of molecules at two given positions $\mathbf{r}_i$ and $\mathbf{r}_j$ is changed from, say, α at i and β at j to, say, γ at i and β at j, or to γ at i and δ at j. In a pure liquid, $\mathcal{U}$ is a function only of the positions $\mathbf{r}_i$ and not of the assignment of the indistinguishable molecules to these N positions.

It was seen in the last chapter that one of the most fruitful manipulations of Q was that which led to the principle of corresponding states and to the calculation of small deviations from it. It is natural to seek first the analogous results for a mixture. However, the dimensional arguments of Section 7.5 cannot be used directly because of the dependence of $\mathcal{U}$ on the assignment of the molecules to the N positions of each configuration. The energy is independent of the assignment only if all types of molecule have the same potential, as, for example, in an isotopic mixture. The species are there distributed at random over all assignments and a comparison of the configuration integral of the mixture with that of one of the pure isotopic components shows that the mixture is ideal. The ideal free energy and entropy of mixing are obtained at once from

$$A(V,T) - A_0(V,T) = -kT \ln(Q/Q_0) = kT \ln\left(\prod_\alpha N_\alpha!/N! \right) \tag{8.3}$$

by using Stirling's approximation for the factorials. Here, A is the configurational free energy of the mixture and A_0 that of an equal amount of one of the pure components. This equation holds only for a mixture that obeys the laws of classical mechanics. Isotopic mixtures are not ideal at low temperatures, as was shown in Section 5.10.

The laws of the ideal mixture are obtained only if all the potentials are the same, but not necessarily spherical; that restriction was introduced into equation (8.1) for the convenient simplification of the following discussion. It is not sufficient for the unlike interactions to be an arithmetic or any other mean of the like interactions. For conformal potentials it is common to write

$$d_{12} = \tfrac{1}{2}d_{11} + \tfrac{1}{2}d_{22} \tag{8.4}$$

$$\varepsilon_{12} = \xi_{12}(\varepsilon_{11}\varepsilon_{22})^{1/2} \tag{8.5}$$

Equation (8.4) is exact for mixtures of hard spheres and the assumption that this rule applied to $b^{1/3}$, the parameter in van der Waals's equation equivalent to d, was made by Lorentz[2] in 1881. In 1898, Berthelot[3] suggested that $\xi_{12} = 1$ in equation (8.5), and so a conformal mixture to which these simplifications apply is now called a *Lorentz–Berthelot mixture*. The first rule is usually adopted for simple mixtures, but in general the second is not obeyed by real systems. We shall see below that ξ_{12} is significantly less than unity in most cases. Even if $\xi_{12} = 1$ (or if ε_{12} were the arithmetic mean of ε_{11} and ε_{22}) the mixture would not be ideal.

The result (8.3) suggests that a simplification of Q is most easily achieved for a non-ideal mixture by replacing $\mathscr{U}$ in equation (8.1) by its average over all assignments of molecules to the N positions of each configuration. This approximation defines a *random mixture*. The average over assignments is made separately for every possible configuration of the N molecules in the volume V—the configurations are not restricted, for example, to those of a lattice. The concept of a random mixture was important, since it is a system to which the principle of corresponding states may be extended and, as the approximation is mathematically well defined, it served as a starting point for more sophisticated attempts to evalute Q. Prigogine and his colleagues[4], Scott[5], Byers Brown[6] and Kirkwood and his colleagues[7] all formulated the theory of random mixtures and suggested methods of improvement. The treatments are essentially the same; this account follows the full exposition of Byers Brown[6].

The true value of $\mathscr{U}$ is

$$\mathscr{U} = \sum_{i>j} \sum u(r_{ij}) \tag{8.6}$$

where i and j denote a pair of molecules of any species. The average value of $\mathscr{U}$ over all assignments is

$$\langle \mathscr{U} \rangle = \sum_{i>j} \sum \langle u(r_{ij}) \rangle \tag{8.7}$$

where the average of u is found from the probability that position $\mathbf{r}_i$ is occupied by a molecule of species α and position $\mathbf{r}_j$ by one of species β, and a summation over all species. There are N positions and so the probability that position i is occupied by a molecule of species α is (N_α/N) or x_α, if the distribution is random. The probability of a molecule of species β at position j is similarly x_β. Hence

$$\langle u(r_{ij}) \rangle = \sum_\alpha \sum_\beta x_\alpha x_\beta u_{\alpha\beta}(r_{ij}) \qquad \langle \mathscr{U} \rangle = \sum_{i>j} \sum \sum_\alpha \sum_\beta x_\alpha x_\beta u_{\alpha\beta}(r_{ij}) \tag{8.8}$$

It is convenient to introduce a hypothetical pure substance whose intermolecular potential is the average over all assignments, $\langle u(r) \rangle$, and to call it the *equivalent substance* for composition x. The substitution of equation (8.8) into (8.1) shows that the random mixture is equivalent to an ideal isotopic mixture of molecules of the equivalent substance for that composition. That is,

$$A(V, T, x) = A_x(V, T) + NkT \sum_\alpha x_\alpha \ln x_\alpha \tag{8.9}$$

$$A_x(V, T) = -kT \ln Q_x$$

$$= -kT \ln \left[\frac{1}{N!} \int \cdots \int e^{-\langle \mathscr{U} \rangle / kT} \, \mathrm{d}\mathbf{r}_1 \ldots \mathrm{d}\mathbf{r}_N \right] \tag{8.10}$$

where A_x is the configurational free energy of an equal number of molecules of the equivalent substance. The latter changes with x and a further simplification of equations (8.9) and (8.10) can be made only after the introduction of a restriction on the form of the intermolecular potential. Without such a restriction, $\langle u(r) \rangle$ of equation (8.8) need not be conformal with the $u_{\alpha\beta}(r)$ of which it is composed, and so the thermodynamic properties of the equivalent substance could not be calculated from equation (8.10) by the principle of corresponding states. It has been proved rigorously that this extension of the principle of corresponding states to mixtures can be made only for Lennard-Jones potentials, with no restriction on the indices n and m except that they should each be the same for all interactions[6]. This condition is more restrictive than the corresponding requirements for pure substances, equation (7.106), although the Lennard-Jones potential is, of course, a particular case of that equation. The averaging of $u(r)$ over all assignments for a Lennard-Jones potential gives

$$\langle u(r) \rangle = \langle \lambda \rangle r^{-n} - \langle v \rangle r^{-m} \tag{8.11}$$

$$\langle \lambda \rangle = \sum_\alpha \sum_\beta x_\alpha x_\beta \lambda_{\alpha\beta} \qquad \langle v \rangle = \sum_\alpha \sum_\beta x_\alpha x_\beta v_{\alpha\beta} \tag{8.12}$$

The equivalent result for the distribution function is[8]

$$g_{\alpha\beta}(r) = g_x(r) \qquad (\text{all } \alpha, \beta) \tag{8.13}$$

Now the Lennard-Jones potential is conformal with that of a reference species and related to it by

$$u_{\alpha\beta}(r) = f_{\alpha\beta} u_{00}(r/g_{\alpha\beta}) \qquad (\text{cf. } 7.106) \tag{8.14}$$

where f and g have the meanings given in Section 7.5. In terms of the parameters λ and v,

$$f_{\alpha\beta} = \left(\frac{v_{\alpha\beta}}{v_{00}} \right)^{\frac{n}{n-m}} \left(\frac{\lambda_{\alpha\beta}}{\lambda_{00}} \right)^{\frac{-m}{n-m}} \tag{8.15}$$

$$g_{\alpha\beta} = \left(\frac{v_{\alpha\beta}}{v_{00}} \right)^{\frac{-1}{n-m}} \left(\frac{\lambda_{\alpha\beta}}{\lambda_{00}} \right)^{\frac{1}{n-m}} \tag{8.16}$$

and so f_x and g_x, the ratios for the equivalent substances, are given by

$$f_x = \left(\sum_\alpha \sum_\beta x_\alpha x_\beta f_{\alpha\beta} g_{\alpha\beta}^m \right)^{\frac{n}{n-m}} \left(\sum_\alpha \sum_\beta x_\alpha x_\beta f_{\alpha\beta} g_{\alpha\beta}^n \right)^{\frac{-m}{n-m}} \qquad (8.17)$$

$$g_x = \left(\sum_\alpha \sum_\beta x_\alpha x_\beta f_{\alpha\beta} g_{\alpha\beta}^m \right)^{\frac{-1}{n-m}} \left(\sum_\alpha \sum_\beta x_\alpha x_\beta f_{\alpha\beta} g_{\alpha\beta}^n \right)^{\frac{1}{n-m}} \qquad (8.18)$$

It follows at once from equation (7.109) and the corresponding equation for G that

$$A_x(V, T) = f_x A_0(V/g_x^3, T/f_x) - 3NkT \ln g_x \qquad (8.19)$$

$$G_x(p, T) = f_x G_0(pg_x^3/f_x, T/f_x) - 3NkT \ln g_x \qquad (8.20)$$

Differentiation of equation (8.19) with respect to volume gives for a system of one phase

$$p(V, T, x) = p_x(V, T) = (f_x/g_x^3)p_0(V/g_x^3, T/f_x) \qquad (8.21)$$

Subtraction of the Gibbs function of the components from equation (8.20) gives

$$G^E(p, T, x) = f_x G_0(pg_x^3/f_x, T/f_x)$$
$$- \sum_\alpha x_\alpha f_{\alpha\alpha} G_0(pg_{\alpha\alpha}^3/f_{\alpha\alpha}, T/f_{\alpha\alpha}) - 3NkT \sum_\alpha x_\alpha \ln(g_x/g_{\alpha\alpha}) \qquad (8.22)$$

Other excess functions are obtained by differentiation.

We do not explore this theory any further, although it played an important role in the development of theories of mixtures, since it has been known for more than a decade that it rests on an unrealistic physical basis, unless all the molecules happen to be of the same size. Simple liquids are geometrically fairly close-packed; the same is true for simple mixtures, and so it is unrealistic to suppose that their molecules can be exchanged at random if they are of different sizes; a small molecule cannot be exchanged for a large one in a densely packed fluid. The mathematical origin of the trouble can be seen in the singular nature of the basic equations for f_x and g_x, equations (8.17) and (8.18), when the repulsive index of the pair potential, n, approaches infinity. Consider, for example, equation (8.18). If n becomes infinite whilst m remains finite, then g_x becomes equal to the largest coefficient of the set $g_{\alpha\beta}$ and is independent of composition. That is, we have the absurd result that the single substance chosen to represent the mixture is composed of molecules whose size is that of the largest molecule present, even if the mole fraction of that species is vanishingly small. If the theory were applied to a mixture of hard spheres of different sizes, then G^E would be positive infinite. Indeed, it is clear that this must be so, since it is the essence of random mixing that the chance of finding a large molecule in a given molecular environment is the same as that of finding a small one. The replacement of a small molecule by a large one in an already close-packed local configuration must lead to the overlapping of the hard spheres and so to a positive infinite value of $\mathcal{U}$ for that configuration.

Thus the theory of random mixing must be abandoned as a starting point for the discussion of mixtures of hard spheres, and hence also for that of mixtures of real molecules, since these have steep repulsive potentials, that is, large values of n. Clearly, the problem of the structure of a dense fluid of hard spheres of different sizes must be solved before we can hope to put the theory of mixtures with more realistic potentials on a sound basis.

8.3 Mixtures of spheres

The structure of a fluid composed of uniform hard spheres is governed by the geometrical problem of their packing; the same principle holds also for mixtures of hard spheres. Thus, if the size ratio of the spheres in a binary mixture is large, then the small spheres can be accommodated advantageously within the holes of the array formed by the mutually touching large spheres, rather as sand could be packed into the interstices of a coarse gravel. Such a geometry is the antithesis of a random mixture, and leads to a smaller overall volume than can be achieved even with a system of uniform size. This contraction is smaller, but still present, in mixtures of less extreme ratios. Smith and Lea[9] (ratio of diameters, 5/3) and Rotenberg[10] (ratio of diameters, 11/10) have determined the pressure, and the former the excess thermodynamic properties also, from Monte Carlo simulation. Alder[11] has made similar molecular dynamic calculations for a size ratio of 3/1. These mixtures are close to ideal, but where v^E can be determined it is small and negative. Since mixtures of hard spheres have no configurational energy, it follows that g^E is also negative:

$$g^E(p^*, T^*) = \int_0^{p^*} v^E(p^*, T^*)\,dp \tag{8.23}$$

It has been tacitly assumed so far that, as for mixtures of real hard spheres, the Lorentz equation (8.4) determines the cross-diameter. Since, however, the systems are only models, it is interesting to depart from this rule. If d_{12} exceeds the mean of d_{11} and d_{22}, then the kind of packing discussed above becomes unfavourable, and v^E and g^E are positive. A sufficiently large cross-diameter drives the system into two fluid phase at high densities[12]. Such unrealistic systems are of interest for two reasons: first, they may be more useful reference systems for perturbation calculations than the more realistic models[8], and secondly, they are easier to use than more realistic systems for the study of general problems of phase separation, and even of the properties of the surfaces between phases. This second point has been demonstrated by the rich harvest of results for the extreme model system in which $d_{11} = d_{22} = 0$, with d_{12} non-zero, the primitive version of the penetrable sphere model[13].

The computer studies show also that v^E of mixtures that conform to the Lorentz rule becomes large and positive as soon as the mixture enters the solid phase. Here the geometry of the crystal constrains the spheres of different size into a common structure and so makes the system approach more nearly that of their random assignment to the sites of the lattice. These considerations are, however, irrelevant in the fluid phase.

It was shown in Section 7.4 that the Percus–Yevick (PY) theory led to an accurate representation of the fluid branch of the isotherms of a one-component system of hard spheres. The PY equation for a binary mixture was solved by Lebowitz[14], and it has been shown[15] that this solution leads to the same results as those found by computer, namely, that v^E, and hence g^E, are small and negative. Unfortunately, the excess functions cannot be obtained explicitly beyond second-order differences in the sizes of the spheres, but the numerical solutions agree well with the computer results. Neither PY theory nor the computer experiments leads to a separation of fluid phases at any density, composition or size ratio. In this they differ radically from the implications of the random-mixture theory.

Since the structure of real fluids at high densities is, to a large degree, determined by the repulsive forces, the PY result for mixtures of hard spheres serves as a useful guide for the development of a theory of mixtures of real molecules. The PY results can be brought a little nearer reality by extending them to a mixture of soft spheres by means of the perturbation expansion, (7.123)–(7.125). We have in the soft-sphere potential a common ground on which several theories can be compared, since the equations of the random mixture, (8.17)–(8.20) can also be applied to such an assembly. Furthermore, it is a potential in which each molecular interaction is characterized by one parameter only (conveniently a volume, $h_{\alpha\beta} = g_{\alpha\beta}^3$) and which is, therefore, ideally suited for the study of the specific effects of differences of size upon the excess functions.

Let us consider the excess Helmholtz free energy, $A_{T,\rho}^E$, which is related to a process of mixing in which the temperature and reduced number density, $\rho = Nd^3/V$, are the same for each component before mixing, and also in the resulting mixture. Let the Lorentz rule (8.4) be assumed to apply to each cross-interaction, and let the equation of state of the reference substance be the PY compressibility equation (7.89), where the reduced number density, $\eta = \pi\rho/6$, is now that for an assembly of soft spheres. All first-order contributions to $A_{T,\rho}^E$ are eliminated by the Lorentz rule, and so, to the second order, we can write

$$A_{T,\rho}^E = J_n(\eta)NkT \sum_\alpha \sum_\beta x_\alpha x_\beta (h_{\alpha\alpha} - h_{\beta\beta})^2 \tag{8.24}$$

where the coefficient J is given by the following theories[16]:

Random mixture

$$J_n(\eta) = \left(\frac{n+7}{12}\right)\frac{1+\eta+\eta^2}{(1-\eta)^3} - \left(\frac{n+1}{12}\right)\frac{1}{2}\frac{(1+2\eta)^2}{(1-\eta)^4} \tag{8.25}$$

Percus–Yevick

$$J_n(\eta) = \frac{5}{6} \frac{1 + \eta + \frac{8}{5}\eta^2}{(1 - \eta)^3} - \frac{1}{3} - \frac{1}{2} \frac{(1 + 2\eta)^2}{(1 - \eta)^4} \tag{8.26}$$

The coefficient J_n in equation (8.25) becomes positive infinite if n is infinite. In contrast, the PY expression for J is negative, independent of n, and is always substantially below the other result if $n \sim 12$ and $\eta < 0.6$, the approximate density at which the mixture starts to solidify.

8.4 The van der Waals approximation

The PY result cannot be applied to mixtures of Lennard-Jones molecules, or to other sets of conformal potentials, without the numerical solution of the integral equation for each particular mixture. This has been done and the results are good[17], but the solutions are difficult to obtain and it is not a practicable theory of mixtures. However, there is a one-fluid model which leads quite generally to a form of J that resembles closely the PY results, and this is the approximation introduced by van der Waals[18]. In his theory of mixtures, he assumed that the parameters a_x and b_x of his equation of state were quadratic sums of the parameters $a_{\alpha\beta}$ and $b_{\alpha\beta}$. This is the only combination that leads to the correct second virial coefficient for the mixture and, as Reid and Leland[19] showed, it can be justified by appeal to the more generalized van der Waals model (*see* Section 7.6) in which the attractive potential is weak compared with kT and of long range. Since a is proportional to $(T^c V^c)$ and b to V^c, the assumption of van der Waals is, in more modern notation[16],

$$f_x = \left(\sum_\alpha \sum_\beta x_\alpha x_\beta f_{\alpha\beta} h_{\alpha\beta} \right) \left(\sum_\alpha \sum_\beta x_\alpha x_\beta h_{\alpha\beta} \right)^{-1} \tag{8.27}$$

$$h_x = \sum_\alpha \sum_\beta x_\alpha x_\beta h_{\alpha\beta} \qquad \text{(where } h = g^3 \text{)} \tag{8.28}$$

The complicated average of equation (8.18) is now replaced by a simple quadratic sum of volume parameters $h_{\alpha\beta}$. These equations give A and G for the mixture when substituted in equations (8.19) and (8.20). The value of $J_n(\eta)$ in (8.24) for the van der Waals approximation is remarkably close to that of the PY result (8.26) and hence to the computer experiments, namely[16],

van der Waals approximation

$$J_n(\eta) = \frac{5}{6} \frac{1 + \eta + \eta^2}{(1 - \eta)^3} - \frac{1}{3} - \frac{1}{2} \frac{(1 + 2\eta)^2}{(1 - \eta)^4} \tag{8.29}$$

If the PY result is a good discriminant, then it follows that the van der

Waals approximation (8.27)–(8.28) is a better model than the random-mixture approximation for mixtures of molecules of different size. It is readily seen that both yield the same simple result if all the sizes are equal (i.e., all $h_{\alpha\beta} = 1$):

$$f_x = \sum_\alpha \sum_\beta x_\alpha x_\beta f_{\alpha\beta} \qquad h_x = 1 \tag{8.30}$$

The random mixing and the van der Waals approximations are examples of what is generally called a *one-fluid approximation*[5]; that is, they assume that the configurational properties of the mixture can be approximated by those of a hypothetical fluid (subscript x) whose properties are determined by potential parameters that are quadratic averages of those of real mixture, equations (8.17)–(8.18) or (8.27)–(8.28). An apparently less drastic form of averaging is to replace a mixture of n components not by a single hypothetical substance but by an ideal mixture of n hypothetical substances. Since binary mixtures are the most commonly discussed, this kind of theory is called a *two-fluid approximation*[5]. Thus, for the two-fluid version of the random mixing approximation, we assume that each component has a mean energy of interaction with all other molecules given by

$$u_\alpha(r) = \sum_\beta x_\beta u_{\alpha\beta}(r) = f_\alpha u_{00}(r/g_\alpha) \tag{8.31}$$

where f_α and g_α are functions of the composition given by

$$f_\alpha = \left(\sum_\beta x_\beta f_{\alpha\beta} g_{\alpha\beta}^m\right)^{\frac{n}{n-m}} \left(\sum_\beta x_\beta f_{\alpha\beta} g_{\alpha\beta}^n\right)^{\frac{-m}{n-m}} \tag{8.32}$$

$$g_\alpha = \left(\sum_\beta x_\beta f_{\alpha\beta} g_{\alpha\beta}^m\right)^{\frac{-1}{n-m}} \left(\sum_\beta x_\beta f_{\alpha\beta} g_{\alpha\beta}^n\right)^{\frac{1}{n-m}} \tag{8.33}$$

(compare equations (8.17) and (8.18) for the one-fluid version).

The thermodynamic properties are now obtained by substituting these parameters in either an expression for the Helmholtz[4,6,20] or Gibbs[4,5] free energies of a mixture of these hypothetical components. (The numerical consequences depend on which of these is chosen.)

$$A(V, T, x) = NkT \sum_\alpha x_\alpha \ln x_\alpha$$

$$+ \sum_\alpha x_\alpha [f_\alpha A_0(V/g_\alpha^3, T/f_\alpha) - 3NkT \ln g_\alpha] \tag{8.34}$$

$$G(p, T, x) = NkT \sum_\alpha x_\alpha \ln x_\alpha + \sum_\alpha x_\alpha [f_\alpha G_0(pg_\alpha^3/f_\alpha, T/f_\alpha)$$

$$- 3NkT \ln g_\alpha] \tag{8.35}$$

This approximation is a slight improvement over the one-fluid version, but is still unsatisfactory for mixtures of molecules of different sizes. Its analogue for the van der Waals approximation is[21]

$$f_\alpha = \left(\sum_\beta x_\beta f_{\alpha\beta} h_{\alpha\beta}\right)\left(\sum_\beta x_\beta h_{\alpha\beta}\right)^{-1} \tag{8.36}$$

$$h_\alpha = \sum_\beta x_\beta h_{\alpha\beta} \qquad \text{(where } h = g^3\text{)} \tag{8.37}$$

This approach can be carried a stage further and a *three-fluid model* generated in which the preliminary averaging of equations (8.27)–(8.28) or (8.36)–(8.37) is avoided and A or G is obtained directly from the set of coefficients $f_{\alpha\beta}$ and $g_{\alpha\beta}$. Thus for G we can write

$$G(p, T, x) = NkT \sum_\alpha x_\alpha \ln x_\alpha$$

$$+ \sum_\alpha \sum_\beta x_\alpha x_\beta [f_{\alpha\beta} G_0(pg_{\alpha\beta}^3/f_{\alpha\beta}, T/f_{\alpha\beta}) - 3NkT \ln g_{\alpha\beta}] \tag{8.38}$$

For small differences of size, it is found that the one-fluid van der Waals approximation is better than the two-fluid, but that the latter is better if the ratio is two or more[22]. However, for LJ molecules that differ also in energy this advantage is lost, particularly in the calculation of excess volumes[23]. Moreover, the two-fluid version is more difficult to use in practice and suffers from internal inconsistencies in the critical region[24], so we do not consider it further. The three-fluid approximation[5] imputes an unrealistically high degree of ordering to a dense fluid system; it is exact in the limit of the slightly imperfect gas but inappropriate for liquids.

These preliminary considerations show that only one of these theories need be carried forward as a serious contender for the representation of the excess properties of mixtures of simple molecules of arbitrary size, namely, the one-fluid van der Waals approximation. Before comparing it with computer simulations or real results, we develop first other theoretical methods.

8.5 Perturbation theories of mixtures of spherical molecules

The power of perturbation theories of pure liquids was seen in the last chapter. Many of the treatments discussed there have been extended to mixtures with, on the whole, equally good results. Thus Section 7.6 opened with a theoretical derivation of a primitive perturbation theory, namely that which leads to van der Waals's equation of state

$$p/nkT = f(n) - (an/kT) \tag{8.39}$$

where $f(n)$ is the compression factor for hard spheres which van der Waals represented

$$f(n) = (1 - bn)^{-1} \qquad b = \tfrac{2}{3}\pi d^3 \tag{8.40}$$

As was shown in the last section, the extension of this result to mixtures[18], by letting parameters a and b be equivalent to quadratic functions of composition, provided a model for the more general approximation that can be abbreviated vdW1; McGlashan[25] has shown that the original version is worth consideration in its own right. An improvement is to replace $f(n)$ by a more realistic compression factor for hard spheres, e.g., that given by computer simulation, or the PY approximation to this function[26], or Guggenheim's empirical representation of it, and so forth[25,27]. To these must be added a semi-theoretical representation of $f(n)$ by Flory[28] which contains a third (empirical) parameter, denoted c, to represent the mean number of external degrees of freedom per segment of a polyatomic molecule. The discussion of this treatment properly belongs later (*see* Section 8.7) but it is mentioned here for completeness since it belongs to the same general van der Waals family, equation (8.39), and since it has been widely used for fitting the properties of non-polar mixtures (Sections 8.6 and 8.7).

Such treatments are surprisingly successful in calculating excess functions at low pressures and temperatures, but they have no advantage over the one-fluid van der Waals approximation in accuracy or convenience for the simplest systems. They have, however, proved of value in interpreting the phase behaviour of mixtures at high pressures, and we return to them in Section 8.9.

Another form of perturbation theory is historically important, for in 1951 it was the first serious attack on the excessive use of lattice models for the discussion of liquid mixtures. This is the treatment called by Longuet-Higgins[29] the *theory of conformal solutions*, although it must be noted that his definition of *conformal* is a little more restrictive than that now generally used. His theory is an expansion of the properties of a mixture of molecules with conformal potentials (in the sense of equation (7.106)) about those of a reference substance, subscript zero, which may or may not be one of the components. The expansion parameters are the differences $(f_{\alpha\beta} - f_{00})$ and $(g_{\alpha\beta} - g_{00})$. The excess functions of the mixture take a simple form at zero pressure for a system which conforms to the Lorentz rule (8.4)

$$f^{E} = \sum_{\alpha > \beta}\sum x_{\alpha}x_{\beta}(2f_{\alpha\beta} - f_{\alpha\alpha} - f_{\beta\beta}) \tag{8.41}$$

$$G^{E} = (U)_{0}f^{E} \tag{8.42}$$

$$H^{E} = (U - TC_{p})_{0}f^{E} \tag{8.43}$$

$$TS^{E} = (-TC_{p})_{0}f^{E} \tag{8.44}$$

$$V^{E} = (-VT\alpha_{p})_{0}f^{E} \tag{8.45}$$

Thus, to the first order of the energy differences, G^{E}, H^{E}, TS^{E} and V^{E} are all proportional to each other and, because of the uniformly negative coefficients of f^{E}, are either all negative or all positive. Both random mixtures and the vdW1 approximation reduce to equations (8.41)–(8.45) if only first-order differences are retained, i.e., they are exact to this order. However, first-order differences do not suffice for the interpretation of the properties of simple mixtures, and this result now serves more as a criterion of consistency to which more ambitious theories should conform.

The first successful perturbation theory of pure liquids was that of Barker and Henderson (Section 7.6), and this was also the first of this type to be extended to mixtures. There are several ways of doing this for LJ molecules. First, the reference system can be chosen to be a single-component system of hard spheres, when the free energy A is expanded about A_0 in powers of one parameter of inverse steepness, and one of the strength of the attractive well. The diameter of the hard spheres can be chosen to annul the first-order term in the inverse-steepness parameter. This method was used, in a degenerate form, by Henderson and Barker, for a calculation of the properties of a mixture of hard spheres from those of a single hard-sphere fluid[30]. For a mixture of LJ molecules[31], it is better to expand about a reference system that is a mixture of hard-spheres, although, if these have additive (Lorentz) diameters, it is not now possible to choose diameters to annul entirely the first-order term in the inverse-steepness parameter. This second version requires an accurate representation of the properties of the hard-sphere mixture, and, as for a pure substance, the best[32] is that obtained[33] by adding two-thirds of the PY-compressibility result for the pressure to one-third of the PY-virial result. Both versions are generally used only to first-order in the perturbation parameters. The second-order terms have been calculated but need great accuracy in the integrations over $g_0(r)$; they yield but a modest improvement[34]. It is shown below that this perturbation theory yields results that agree well both with the results of computer simulation and with the experimental results for real simple systems.

The perturbation theory of Weeks, Chandler and Andersen (Section 7.6) is more accurate at first order for pure LJ liquids than that of Barker and Henderson, but this superiority does not extend to mixtures. This extension[35] can be made by dividing each pair potential as in equation (7.133), and then taking the mixture with the repulsive potentials $(u_r)_{\alpha\beta}$ as the reference system, whose properties are obtained from those of a mixture of hard spheres of 'equivalent' diameters by using the blip-function. The properties of the hard-sphere fluid are, in turn, obtained from the mean PY calculation of the pressure[33] and a generalization to mixtures of the recipe of Verlet and Weis[36] for improving the PY expression for $g_{HS}(r)$. Each step in this chain is doubtless quite accurate and leads to an acceptable calculation of the configurational properties of both pure liquids and the mixture (~ 1 per cent accuracy), but on subtraction the excess functions are generally worse than those obtained from the theory of Barker and Henderson.

Variational methods are akin to perturbation methods but have usually proved to be disappointing in classical statistical mechanics[37]. The

Gibbs–Bogoliubov inequality provides an upper bound to the Helmholtz free energy[37,38].

$$A \leqslant A_0 + \overline{(\mathscr{U} - \mathscr{U}_0)_0} \tag{8.46}$$

where $\mathscr{U}$ and $\mathscr{U}_0$ are the configurational energy of the system and of some reference system whose free energy, A_0, is known. The canonical average is taken over the distribution functions of the reference system. For pure fluids this upper limit is not useful since A is rarely more than about 95 per cent of the right-hand side. If, however, the upper bound is used as an estimate both for A of a mixture and of the unmixed components then the excess thermodynamic properties may be given more accurately, although the sign of the error in the difference is unknown. This route to the excess properties of simple mixtures has been followed by Mansoori and Leland[39] who used the properties of a mixture of hard spheres as the reference system. The excess functions for mixtures of spherical molecules are generally as good as those for the best perturbation theory, but unfortunately the treatment has not been used at high pressures nor extended to non-spherical molecules.

8.6 Excess functions of simple mixtures

The accurate simulation of the thermodynamic properties of LJ fluids has transformed the task of testing statistical theories of simple mixtures. Until such results were available, the comparison of theory and experiment was bedevilled by our ignorance of the form and magnitudes of the intermolecular potentials. There are now reliable sets of excess functions at essentially zero pressure for mixtures of LJ(12,6) molecules that conform to the Lorentz–Berthelot rules, equations (8.4)–(8.5) with $\xi = 1$. They have all been obtained by Monte Carlo simulation since it is essential to work in an ensemble in which the temperature is an independent variable, and so known precisely. In principle, the isobaric ensemble (N, p, T) has an advantage over the isochoric (N, V, T), since excess functions are to be calculated at fixed, and essentially zero pressures. In practice, the computer calculations are more difficult in this ensemble and there is little to choose between the two routes.

McDonald[40] has used the isobaric and Singer and Singer[41] have used the isochoric. It is not easy to compare their results since they chose different values of the parameters. Both take ε_{12} and d_{12} as their standards and vary ε_{11} and d_{11}, whence ε_{22} and d_{22} follow from equations (8.4)–(8.5). Singer and Singer chose $\tau_{12} = 0.727$ and 0.876, and McDonald $\tau_{12} = 0.819$, where $\tau = kT/\varepsilon$. Let the 1–2 interaction define the reference substance, so that, cf. equation (7.107),

$$f_{11} = \varepsilon_{11}/\varepsilon_{12} \qquad f_{12} = 1 \qquad f_{22} = \varepsilon_{22}/\varepsilon_{12} \tag{8.47}$$

$$g_{11} = d_{11}/d_{12} \qquad g_{12} = 1 \qquad g_{22} = d_{22}/d_{12} \qquad h_{ij} = g_{ij}^3$$

Singer and Singer choose values of $f_{11} = 0.81$, 0.9, $(0.9)^{-1} = 1.111$, and $(0.81)^{-1} = 1.235$, with values of g_{11} from 1 to 1.12, that is, up a volume ratio h_{11}/h_{22} from 1 to 3.01. McDonald's choice, after changing his sign convention to match that of Singer and Singer, is a wider range of energies, f_{11} from 0.663 to 1.337, but a smaller range of sizes, g_{11} from 1 to 1.0628, that is, a volume ratio h_{11}/h_{22} from 1 to 1.46. He has also used pairs of parameters chosen to match as well as possible the properties of particular real systems. Singer and Singer obtain g^E, h^E, v^E and μ^E for the equimolar mixtures, while McDonald has g^E, h^E and v^E at several compositions. We restrict the discussion to the equimolar mixtures since few excess functions for simple systems show any marked asymmetry. There are no relevant calculations for mixtures that depart from the Berthelot rule, i.e., $\xi \neq 1$, but this lack is probably not serious since all theories agree with the conformal solution theory, i.e., are exact to the first order in the difference $(\varepsilon_{12} - \frac{1}{2}\varepsilon_{11} - \frac{1}{2}\varepsilon_{22})$. Torrie and Valleau[42] studied the system with $d_{11} = d_{12} = d_{22}$ and $\varepsilon_{12} = \frac{3}{4}\varepsilon_{11} = \frac{3}{4}\varepsilon_{22}$ in order to investigate a technique of importance sampling in Monte Carlo simulation. The system separates into two phases but the parameters are not intended to be used for testing theories that are to be applied to real systems. Finally, we note, but do not discuss, the comparison with theory of distribution functions of LJ mixtures obtained by molecular dynamic simulation[43].

Figures 8.1 and *8.2* show G^E/NkT, H^E/NkT and V^E/Nd^3 for a Lorentz–Berthelot mixture of LJ(12,6) molecules at $\tau_{12} = 0.727$. The Monte Carlo results[41] are from the (N, V, T) ensemble and so the calculation of the excess functions at fixed pressure ($p \approx 0$) is lengthy. It is presumably such manipulation that is responsible for the small discontinuities in the curves for H^E and V^E at $g_{11} = 1$, which may give some measure of the accuracy of the final results.

Molecules which differ only in size ($f_{11} = f_{22} = 1$) mix with small negative values of G^E and V^E, as do hard spheres, and with small positive values of H^E, which is zero for hard spheres. The van der Waals approximation and the perturbation theory of Barker and Henderson lead to the same signs but the latter is the more accurate[23,32]. Molecules which differ only in energy ($g_{11} = g_{22} = 1$; $f_{11} = 0.81$, $f_{22} = 1.235$) have positive G^E and H^E, and negative S^E and V^E. If the larger molecule has the larger energy, as is the case in most real mixtures, then all four functions may be negative. Both theories conform to these patterns. For $f_{11} < 1$ and $g_{11} > 1$, the van der Waals approximation is the better for G^E and H^E but worse for V^E, while in the more important region $f_{11} < 1$, $g_{11} < 1$, the perturbation theory is clearly superior for all four functions. Indeed, the van der Waals approximation becomes unacceptable for $g_{11} < 0.9$, but that is a volume ratio h_{22}/h_{11} of nearly 2 and there are many simple mixtures of interest with less extreme ratios. Ewing and Marsh[44] set higher standards and suggest that the van der Waals approximation be used for G^E and H^E only if the volume ratio is less than 1.4, and for V^E if it is less than 1.2.

The variational treatment of Mansoori and Leland[39] gives results similar to those of the perturbation theory[23], being a little more accurate for G^E, about as accurate for H^E, and less accurate for V^E. This pattern changes little with temperature for all three theories, at least up to[23,32] $\tau_{12} = 0.876$, which corresponds to T/T^c of about 0.65 for a pure LJ liquid. At this reduced temperature, argon has a vapour pressure of more than 3 bar, a pressure above which it is not easy or useful to use excess functions (Sections 4.3–4.4).

Before comparing theory with experimental results on simple systems, we have to face the problems of deciding, first, if the LJ(12,6) or any other set of conformal pair potentials is an adequate representation of the interactions in the system, and, secondly, what parameters $\varepsilon_{\alpha\beta}$ and $d_{\alpha\beta}$ are to be assigned to each pair potential. The conventional answer to the first question is that it is known that simple molecules, such as Ar, Kr, N_2, CH_4, etc., do not have spherical pair potentials of the LJ(12,6) form, that there are three-body forces that cannot be neglected in the liquid, but that nevertheless the spherical LJ(12,6) potential is a reasonable effective pair potential for the calculation of the properties of the liquid state, and that, unless we are unlucky, some of the errors involved in its use will disappear on making the subtractions needed in the calculation of the excess functions. This conventional view can be tested by seeing if there is a set of LJ(12,6) parameters for which the Monte Carlo simulations match the experimental results. The use of conformal potentials implies the validity of the principle of corresponding states (Section 7.5) so that the like parameters $f_{\alpha\alpha}$ and $g_{\alpha\alpha}$ can be chosen (with respect to, say, argon as the reference liquid) from the critical properties[21] or from the properties of the liquids[23]; the results are similar. For the cross-parameters it is usual to retain the Lorentz rule and so some imperfections, particularly in v^E, may be expected. It is clear, however, that the Berthelot rule cannot be retained. We consider nine of the simple systems of Chapter 5 for which there are good results for g^E and at least one other excess function. These are listed in *Table 8.1*. Lorentz–Berthelot mixtures in which the larger molecule has the larger energy have small or even negative values of g^E and h^E (*Figure 8.1*), while the experimental values in *Table 8.1* are all positive. The remaining parameter in a binary mixture, ε_{12} or f_{12}, or, what is equivalent, ξ_{12} of equation (8.5), is therefore chosen to fit one of the three excess functions, in practice g^E of the equimolar mixture. The test of the potentials and of the theories is therefore the accuracy with which they reproduce h^E and v^E. Since all the systems are close to ideal, we cannot expect great relative accuracy. The absolute magnitudes of h^E and v^E are to be measured against scales on which the configurational energies of these liquids are about $-6000 \, \text{J mol}^{-1}$ and their volumes about $30 \, \text{cm}^3 \, \text{mol}^{-1}$.

The Monte Carlo simulations (*Table 8.1*) yield tolerable values of h^E and less satisfactory values of v^E; in both cases the results are too low. These discrepancies are a measure of the failure of the LJ(12,6) parameters and the

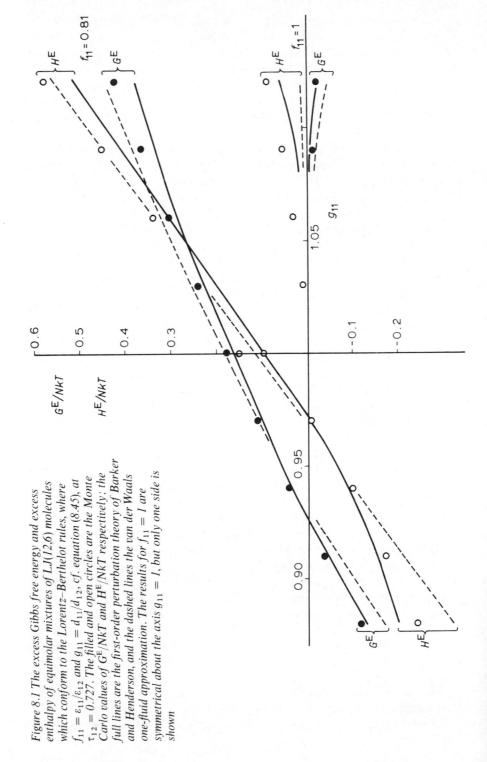

Figure 8.1 The excess Gibbs free energy and excess enthalpy of equimolar mixtures of LJ(12,6) molecules which conform to the Lorentz–Berthelot rules, where $f_{11} = \varepsilon_{11}/\varepsilon_{12}$ and $g_{11} = d_{11}/d_{12}$, cf. equation (8.45), at $\tau_{12} = 0.727$. The filled and open circles are the Monte Carlo values of G^E/NkT and H^E/NkT respectively; the full lines are the first-order perturbation theory of Barker and Henderson, and the dashed lines the van der Waals one-fluid approximation. The results for $f_{11} = 1$ are symmetrical about the axis $g_{11} = 1$, but only one side is shown

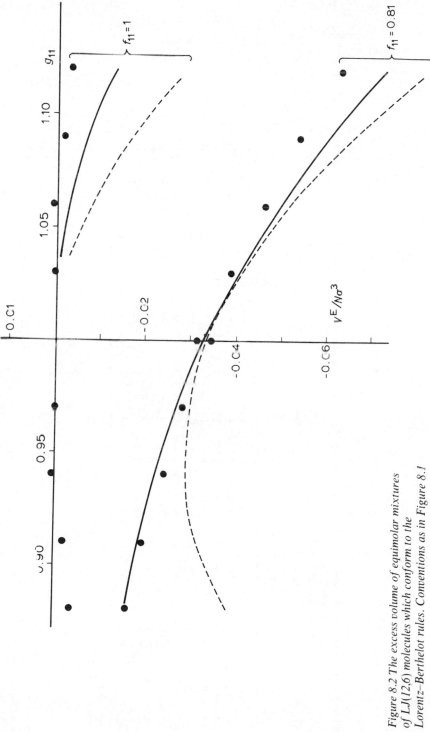

Figure 8.2 The excess volume of equimolar mixtures of LJ(12,6) molecules which conform to the Lorentz–Berthelot rules. Conventions as in Figure 8.1

Table 8.1 Comparison of theory and experiment for nine simple systems. The equimolar excess functions, g^E and h^E in J mol^{-1} and v^E in cm^3 mol^{-1}, are taken from Sections 5.4 and 5.7

System	T/K	Experiment			Monte Carlo[23]			Perturbation[32]			vdW 1[8]			S–H[23]		
		g^E	h^E	v^E	ξ_{12}	h^E	v^E	ξ_{12}	h^E	v^E	ξ_{12}	h^E	v^E	ξ_{12}	h^E	v^E
Ar + Kr	116	84	43	−0.52	0.989	36	−0.53	0.983	45	−0.47	0.996	24	−0.66	0.991	7	−0.97
Ar + N$_2$	84	34	52	−0.18	1.000	40	−0.23	1.001	35	−0.27	1.002	33	−0.34	1.003	38	−0.37
Ar + O$_2$	87	36	61	+0.14	0.988	52	+0.06	0.987	51	+0.06	0.988	51	+0.05	0.987	59	+0.09
Ar + CO	84	57	—	+0.09	0.989	79	−0.11	0.986	79	−0.07	0.988	78	−0.13	0.989	92	−0.07
Ar + CH$_4$	91	75	102	+0.18	0.975	83	−0.01	0.972	91	+0.03	0.972	78	−0.18	0.976	94	+0.02
N$_2$ + O$_2$	78	40	61	−0.21	0.999	42	−0.25	1.002	41	−0.26	1.003	39	−0.34	1.006	39	−0.47
N$_2$ + CO	84	23	—	+0.13	—	—	—	0.990	34	+0.07	0.991	35	+0.06	0.990	40	+0.11
CO + CH$_4$	91	110	105	−0.32	0.988	67	−0.61	0.983	92	−0.48	0.987	77	−0.77	0.992	87	−0.81
CH$_4$ + CF$_4$	{98 110	360	540	+0.88	0.900	—	+0.80	(0.909)	580	+1.76	(0.909)	482	0.90	—	—	—

Lorentz rule to describe the real mixtures. One obvious fault is the use of spherical potentials to describe N_2, O_2 and CO (*see* Section 8.8). As expected, ξ_{12} has generally to be less than unity to reproduce the observed values of g^E.

Three theories are chosen for discussion, the first order perturbation theory of Barker and Henderson[32], the one-fluid van der Waals approximation[8,16], and the treatment of Snider and Herrington[23,26], which can be taken to be representative of all treatments based on explicit equations of state of the van der Waals family[23,25]. The results from all three theories resemble each other and those of the Monte Carlo simulations more closely than they agree with experiment. This is seen most clearly with h^E for $Ar + N_2$, $Ar + CH_4$, $N_2 + O_2$, and $CO + CH_4$, and with v^E for $Ar + O_2$, $Ar + CO$, $Ar + CH_4$, and $CO + CH_4$. This agreement suggests that there is no point in trying to improve the statistical theories of simple mixtures until we know more of the intermolecular potentials. On the whole, the perturbation theory does best in the sense that it is never far from the Monte Carlo or experimental results. The van der Waals approximation gives results that are lower and generally less accurate. The treatment of Snider and Herrington is erratic, with some success, such as $Ar + O_2$, and some failure, such as $Ar + Kr$ and v^E for $N_2 + O_2$.

The system $CH_4 + CF_4$ is of interest since there are here large departures from ideality and, moreover, the parameter ξ_{12} can be determined from the second virial coefficients of the gas[45]. Its value of 0.909 is confirmed by that needed to obtain agreement for g^E in the Monte Carlo simulations. Since h^E almost certainly decreases with temperature, the agreement with experiment is here better for the van der Waals approximation than for perturbation theory.

From all these results it can be concluded that the perturbation theory is marginally the best for these mixtures. It yields also good values for the properties of the pure fluids. It is, however, much the most difficult to use. The van der Waals approximation is less accurate, easier to use, and reduces to the principle of corresponding states for the pure fluids. The explicit equations of state are, on the whole, of good accuracy, but with the occasional lapse. They are easy to use, but generally poor for the calculation of the properties of the pure components.

Cycloalkanes, highly branched alkanes and the large pseudo-spherical molecules known by the abbreviation OMCTS and TKEBS (*see* Section 5.5) are a set of liquids whose molecules are non-polar and 'round', if not truly spherical. Marsh and his colleagues[44,46] have applied many of the theories discussed above to these systems. They find that the van der Waals approximation is generally worse than the calculations made with explicit equations of state. Of these, that of Flory[28] performs well, as does an extension by Gibbons[47] of the treatment of Snider and Herrington in which the pressure of a mixture of hard spheres is replaced by that of suitable hard non-spherical convex bodies. The poor performance of the van der Waals approximation for large volume ratios was shown in *Figure 8.2*, but many of their other conclusions are more an assessment of the inadequacies of the supposed intermolecular potentials than of the statistical mechanics of the theories of mixtures.

8.7 Mixtures of alkanes

Normal alkanes form a family in which the physical and chemical properties change regularly with n, the number of carbon atoms. As is to be expected, their mixtures also show strong regularities, some of which were summarized in Section 5.5. The molecules are non-polar (or of negligible polarity) and far from spherical. Beyond the first few members, however, the cores are not rigid and so the departures from spherical shape are not to be treated by theories of the kind discussed for pure liquids in Section 7.7, and extended to mixtures in Section 8.8. Special methods of treatment have been devised for these molecules with flexible cores, many of which have their historical origin in efforts to understand the thermodynamic properties of polymer solutions. They are based on extensions of the principle of corresponding states.

At an early stage of Chapter 7 it was necessary to factorize the total distribution function for a group of h molecules, $f^{(h)}$, into the product of the internal molecular partition functions and a positional or configurational function $n^{(h)}$. This factorization is justified for simple molecules and is used in the derivation of the usual principle of corresponding states in Section 7.5. However, in chain molecules this division is, strictly, impossible since a flexible chain can rotate either as a whole or in groups of segments. The latter motions cannot be classified as wholly internal or wholly external, since they clearly affect the energy of interaction of the molecule with its surroundings. Any division of the degrees of freedom into molecular and configurational is necessarily empirical and to be judged by the success of its macroscopic predictions.

The extension of the principle of corresponding states to chain molecules was made first by Prigogine and his colleagues[48] on the basis of a cell model. The extension is developed by regarding a chain molecule as a series of quasi-spherical segments whose interaction with the segments of neighbouring chains is characterized in the usual way by an energy, ε, and a distance, d. (The segments may be of more than one kind, e.g., CH_3— and —CH_2—, each with its appropriate ε and d for like and unlike interactions.) The chain is specified, first, by the number of segments, conventionally denoted r, secondly, by the number of external contacts that these make with other segments, qz (where z is a coordination number of about 10) and, thirdly, by that of external degrees of freedom, $3c$. Clearly, q and c increase with r, and it is often assumed that $q = r$, so that the ratio (c/r) or (c/q) becomes a measure of the flexibility of a chain. The ratio is unity for a monomer and often falls to about 0.3 in the infinite chain.

It is reasonable, and can be supported by plausible but not rigorous arguments, that r, q and c should serve as reduction parameters for obtaining the reduced volume, energy and entropy, respectively, of a chain molecule[49–51]. Thus a reduced volume $\tilde{V}$ at a reduced temperature of

$$\tilde{T} = (c/r)(kT/\varepsilon) \tag{8.48}$$

is defined by

$$\tilde{V}(\tilde{T}) = V_n(T)/rN\sigma^3 \tag{8.49}$$

where $V_n(T)$ is the orthobaric liquid volume of the alkane C_nH_{2n+2}. The energy U_n and the entropy S_n are reduced similarly

$$\tilde{U}(\tilde{T}) = U_n(T)/qN\varepsilon \qquad \tilde{S}(\tilde{T}) = S_n(T)/cNk \tag{8.50}$$

The extension of the principle of corresponding states is, therefore, the assumption that there is for all alkanes a single reduced equation of state which, for example, gives $\tilde{U}$ as a function of $\tilde{T}$ (or of $\tilde{V}$ and $\tilde{T}$ off the orthobaric line).

In this treatment, the chain molecule is replaced by a set of segments. The predicted properties of the fluid do not change if some of these segments are imagined to be moved from one molecule to another, since such a move changes neither the total number of segments nor that of the molecules. Hence the properties of a mixture of alkanes should depend only on the mean chain length and not upon the individual lengths, and so this principle of corresponding states formally includes the older, and more empirical, *principle of congruence* put forward by Brønsted and Koefoed[52] in 1946. This states that the thermal properties of a multicomponent system of alkanes are determined not by the individual chain lengths but by the mean $n = \sum_\alpha x_\alpha n_\alpha$, where n_α is here the number of carbon atoms in the normal alkane C_nH_{2n+2}.

Thus the properties of *congruent mixtures* (mixtures with the same value of n) differ only in the ideal entropy of mixing. This principle and its verification have been the motive behind much of the work on alkane mixtures summarized in Section 5.5. Some of the features of these systems, such as the negative values of V^E, are consistent with the principle of corresponding states of Prigogine. (The negative volumes are a simple consequence of the fact that the curve that represents $\tilde{V}$ as a function of $\tilde{T}$ is concave upwards.) Other features, notably the change of H^E from positive to negative as the temperature rises, are not derivable from Prigogine's cell model and are only qualitatively in accord with the wider principle of congruence[49,51].

Clearly, there is no need to be committed to a particular equation of state of the reference substance in this field, since a theory of mixtures is often held to have fulfilled its function if it describes accurately the properties of the mixture in terms of the properties of the pure substances and a set of intermolecular parameters. However, many recent efforts have been attempts to set up model configurational integrals which describe the reference substance. Amongst the most successful of these is that of Flory and his colleagues[28], mentioned briefly in the last section. It is a synthesis of the ideas behind Prigogine's extension of the principle of corresponding states with theories based on a generalized van der Waals equation. Delmas and Patterson[50] compare critically this and other theories that can be loosely described as belonging to the extended principle of corresponding states; they conclude that Flory's version is the most successful and use it in subsequent papers on alkane mixtures[49]. The details are too complex to admit of a brief summary.

Mixtures containing branched alkanes are still outside the scope of any theory with a claim to a foundation in statistical mechanics. Some of the ideas on what is physically important in determining their properties were given in Section 5.5.

8.8 Perturbation theories of mixtures of non-spherical molecules

The methods of Section 7.7 can be extended to mixtures and used to interpret the properties of small rigid molecules which have non-spherical cores and non-zero multipole moments, that is, to mixtures in which the components are, for example, N_2, CO, CO_2, N_2O, C_2H_2, C_2H_4, C_2H_6, HCl, and HBr. Mixtures of these with spherical molecules, such as Xe, or almost spherical molecules, such as CH_4, have also been studied. Some mixtures of this kind (e.g., $Ar + N_2$, $CO + CH_4$) have been considered in Section 8.5 by treating their molecules as spherical. This is a reasonable approximation for N_2, CO and CH_4 but it is not useful for the longer linear molecules, such as CO_2 or C_2H_2 which also have large quadrupole moments, nor for polar molecules such as HCl and HBr.

In Sections 7.6, 7.7 and 8.6, the primary test of a statistical theory was its ability to reproduce the results of computer simulations. This can rarely be done for non-spherical mixtures since the few simulations are all for hard-core bodies, such as mixtures of diatomic molecules composed of fused hard spheres[53], or mixtures of hard spheres with hard sphero-cylinders[54,55]. The excess volumes of the dense systems are extremely small. A perturbation theory[56] and an extension of the blip-function expansion[53] have been proposed but neither is successful at high densities[53,54]. Empirical equations of state[57] in which the parameters are mean values of the molecular volumes, surface areas and mean curvatures, and combinations of these, reproduce well the computer simulations[54,55], and may later serve as a reference state for other perturbation theories.

Polar molecules (a category which is defined here to include molecules with multipoles of any order) were discussed in Section 7.7, where two perturbation schemes were set out. The first, which had its origins in a treatment of deviations from the principle of corresponding states, has been extended to mixtures via conformal solution theory[58]. The empirical analogue of this treatment is the use of acentric factors to describe the deviations; this too has been extended to mixtures and used extensively in engineering calculations[59].

One result of this simple form of perturbation theory is that a lack of balance of the polar interactions, as, for example, in a mixture of polar and non-polar molecules, gives large positive contributions to g^E, h^E, s^E and v^E. Since s^E is increased it follows that the ratio (h^E/g^E) is a useful discriminant between systems in which the excess functions arise from a lack of balance of the central forces and those in which there is a lack of balance of the non-central forces[58]. The ratio is 0.51 for Ar + Kr, for which perturbation theory (*Table 8.1*) gives 0.54. For non-spherical molecules, the observed ratios are noticeably higher than the calculated: 1.53 as against 1.03 for Ar + N_2; 1.69 as against 1.42 for Ar + O_2; 1.53 as against 1.02 for $N_2 + O_2$; and 0.95 as against 0.84 for CO + CH_4. Less simple systems can have even higher ratios, e.g., $N_2O + C_2H_4$, 1.60; $C_2H_4 + C_2H_6$, 1.95; Xe + N_2O, 2.01; C_6H_6 + c-C_6H_{12}, 2.67. In all these, it is probable that strong quadrupolar or other non-central forces are active in the mixtures. Systems that contain C_2H_4 have higher ratios than those with C_2H_6, e.g., Kr + C_2H_4, 1.31 as against Kr + C_2H_6, 0.64. The ratio for $C_2H_4 + C_2H_6$ (1.95) has the size of a typical polar + non-polar system. This interpretation is in accord with the sizes of the quadrupole moments[60], which are probably about -0.6×10^{-26} esu cm^2 for C_2H_6 and $+1.5 \times 10^{-26}$ esu cm^2 for the mean moment of C_2H_4.

More sophisticated calculations have been based on the extension to mixtures of the λ-expansion of the free energy, equation (7.138), with the summation of the higher terms implied by equation (7.143). The extension requires A_0 for a reference mixture, and some means of estimating the distribution functions $(g_{\alpha\beta}^{(2)})_0$ and $(g_{\alpha\beta\gamma}^{(3)})_0$ for the calculation of the complicated integrals that comprise A_2 and A_3. The first problem is usually solved by calculating A_0 for a mixture of LJ(12,6) molecules in the van der Waals one-fluid approximation. The second problem has been tackled either by using again this approximation, or a similar one, or by computer simulation of a reference fluid.

The first work of this kind[61] was for a mixture of hard spheres, some with and some without dipoles, and led to qualitatively the same kind of results as those described above, namely, positive g^E, h^E and v^E, with a large ratio (h^E/g^E). Chambers and McDonald[62] introduced the more realistic Stockmayer potential (i.e., LJ(12,6) with dipoles), and first used the summed expansion (7.143). Most of the later developments have been made by Gubbins and Gray[63-68] who have explored several ways[64] of calculating the integrals in A_2 and A_3. Their non-central forces have included dipoles, quadrupoles, octopoles, and anisotropic dispersion forces and cores. The last are a r^{-12} repulsion with the shape of a prolate or oblate ellipsoid, and although this is not the kind of perturbation which is best handled by a λ-expansion (*see* Section 7.7) it is apparently quite effective[66] for cores as elongated as those of N_2O and CO_2. Their procedures have changed a little with the years but a good guide for those wanting to use their methods is in the first of two papers addressed to engineers[67]. Others have added the effects of polarizability[69] and three-body forces[70].

Comparison of this theory with experiment is difficult because of the wealth of parameters in each pair potential. It is customary to take both dipole and quadrupole moments as known from other experiments, and to adjust the parameters of the central forces to reproduce the properties of the pure liquid. The cross-energy parameter ξ is adjusted to fit the equimolar value of g^E at one temperature. A cross-volume parameter is sometimes used to fit v^E also. The test of theory is then its ability to reproduce the correct shapes of the g^E–x and v^E–x curves, and, more importantly, the value of h^E. The theory passes the first tests[68] for Xe + HCl and Xe + HBr, and all tests for Xe + N_2O, but predicts much too large a value of h^E for Xe + HCl, a failure which is probably due to imperfections in the potential rather than failure of the statistical mechanics, although the treatment of overlap forces by the λ-expansion is an obvious weakness. With more difficult mixtures, such as $CH_3OH + C_2H_6$, the agreement with experiment is worse[71]. The use of all the components of the quadrupole, rather than an effective or axial average moment (*see* Section 7.7) is the next logical step, and preliminary calculations for simple systems such as Xe + C_2H_4, and $C_2H_4 + C_2H_6$ show that this change has an appreciable effect on the thermodynamic properties[72].

Attempts to use non-spherical reference potentials in perturbation theory were mentioned briefly at the end of the last chapter. They have so far been little used either for pure liquids or for mixtures, but one qualitative consequence for the latter is worth noting. If a multipole expansion is made about a spherical reference potential then the leading terms are of second order, i.e., proportional to the squares of the moments, $(\mu_\alpha^2 \mu_\beta^2)$, $(q_\alpha^2 q_\beta^2)$, etc., and so are independent of the signs of the moments. Thus CO_2, in which the outer ends of the molecule are negative ($q_1 < 0$), and C_2H_2, in which they are positive ($q_2 > 0$), have similar properties, and the cross interaction in a mixture would, to leading order $(q_1^2 q_2^2)$, be expected to be similar to that in the pure fluids, q_1^4 and q_2^4. If, however, the expansion is made about a non-spherical reference potential[73], or if the reference potential is spherical but the shape of the core can interact with q in the perturbation expansion[67], then the leading terms are first-order in q_α and q_β, whose signs are then relevant. In practice, this is found to be the case, for $CO_2 + C_2H_2$ forms a negative azeotrope at low temperatures, which is a clear indication of the strong mutual attraction expected from quadrupoles of opposite sign.

No doubt there will be further theoretical refinements but it is hard to resist the conclusion that real progress is, for the moment, becoming very slow in this part of the field. A full description of the like and unlike intermolecular forces in, say, a mixture of two polar organic molecules is bound to be extremely complicated, and it is difficult to generalize from one system to another. Thermodynamic properties are measures of averages of these forces from which it is impossible to infer their details. Until there are other, non-thermodynamic ways of determining the forces in detail, then it is hard to see much further improvement in comparing theory with experiment. Both for spherical molecules, in the last section, and for non-spherical molecules, in this section, we have seen that the comparisons tell us quite a lot about the virtues of the theories and about the strengths of the more simple components of the intermolecular potentials, but that the bar to further quantitative progress is our ignorance of the precise forms of the potentials.

8.9 Phase equilibria at high pressures

The theoretical interpretation of the material of Chapter 6 is a formidable task with nearly a century of history behind it. In 1889, van der Waals presented to the Academy of Sciences in Amsterdam the first molecular theory of phase equilibrium in binary mixtures[18]. This was the simple generalization of his equation of state, (8.39)–(8.40), to a binary mixture, but the physics implied by this generalization is so rich that it is only within the last few years that it has been thoroughly worked out; even today there are still some points unresolved. Van der Waals, van Laar and Kamerlingh Onnes published about 50 papers between 1890 and 1914 in which they tried to work out the kinds of phase equilibria and critical lines to which this equation could give rise, but, although much was achieved, there were many omissions and errors that arose from trying to solve numerically equations of high order in more than one variable without the aid of modern computers. The first reasonably comprehensive and accurate sol-utions were obtained in 1968 by Scott and van Konynenburg, and have only recently been published in full[74]. Some minor errors and omissions in their work were rectified by Furman and Griffiths[75].

Van der Waals's generalization of his equation to a mixture was the assumption (cf. Section 8.4) that the parameters a and b are quadratic functions of composition

$$a_x = \sum_\alpha \sum_\beta x_\alpha x_\beta a_{\alpha\beta} \qquad b_x = \sum_\alpha \sum_\beta x_\alpha x_\beta b_{\alpha\beta} \qquad (8.51)$$

where, cf. equations (8.27)–(8.28), the parameters $a_{\alpha\beta}$ are proportional to the intermolecular parameters in the product $(f_{\alpha\beta} g_{\alpha\beta}^3)$, and $b_{\alpha\beta}$ proportional to $g_{\alpha\beta}^3$. In a binary mixture, there are six independent parameters, but it is usual to eliminate b_{12} by putting it equal to the arithmetic mean of b_{11} and b_{22}. The Lorentz rule (8.4) might be more accurate but the difference is generally small. In fact, almost all work has been done on mixtures of molecules of equal size ($b_{11} = b_{12} = b_{22}$), to which the discussion below is therefore restricted. The pattern of phase behaviour shown by any system is determined only by the ratio of parameters and is therefore determined by the values of a_{12}/a_{11} and a_{22}/a_{11}, or, following Scott and van Konynenburg, by the values of the two independent variables ζ and Λ,

$$\zeta = (a_{22} - a_{11})/(a_{11} + a_{22}) \qquad \Lambda = (a_{11} - 2a_{12} + a_{22})/(a_{11} + a_{22}) \quad (8.52)$$

which are constrained to lie in the ranges $(-1 \leqslant \zeta \leqslant 1)$ and $(\Lambda \leqslant 1)$. Since the labelling of the species is irrelevant, it follows that the system (ζ, Λ) has the same behaviour as $(-\zeta, \Lambda)$, and so the range of ζ can be further restricted to $(0 \leqslant \zeta \leqslant 1)$.

Scott and van Konynenburg calculated the phase diagrams for many binary van der Waals mixtures, by solving the equations of Section 6.2 to locate the phase boundaries and critical lines. They found that these mixtures could adopt any one of the first five types shown in *Figure 6.1*, according to the values chosen for ζ and Λ. The full lines in *Figure 8.3* are the boundaries that separate these types. The parameter ζ is a measure of the difference of T^c, and so also of p^c, between the two pure components,

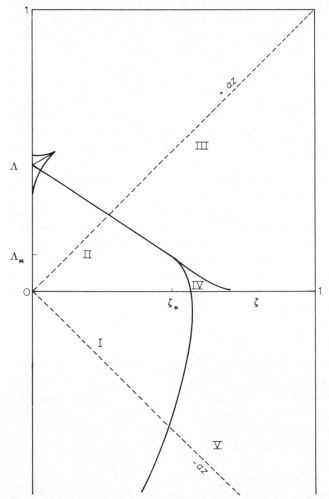

Figure 8.3 The boundaries between the types I–V of the phase diagrams of Figure 6.1, as calculated from the van der Waals equation, in terms of the intermolecular energy ratios ζ and Λ of equation (8.52). The dashed lines are the boundaries outside which azeotropes occur

whilst Λ is a measure of the lack of balance of the 1–2 energy with respect to the arithmetic mean of the 1–1 and 2–2 energies. If the molecules are similar ($\zeta \sim 0$) then the system is of type I if $\Lambda \leqslant 0$ and of type II if $\Lambda > 0$; that is, there is a continuous GL critical line which is supplemented by a LL critical line at low temperatures if a_{12} is less than the mean of a_{11} and a_{22}. However, the van der Waals equation yields no solid phases and the predicted UCST retreats to $T = 0$ as Λ approaches 0^+. Hence for real systems, no UCST is observed above the quadruple point (SSLG) unless the weakness of the 1–2 energy is pronounced. If ζ is small but Λ large, then the system is of type III, that is the UCST has moved to a temperature above that of T_1^c. The change from type II to type III for $|\zeta| < 0.08$ is complicated (and not labelled on the diagram), with a series of minor variants of types II and III interpolated between the two main types. It is the region which Griffiths[75,76] has called the 'shield region'.

If the molecules are sufficiently different, then types IV and V appear. (The minimum value of ζ for type IV is $\zeta_* = (10\sqrt{6} - 6\sqrt{2})/33 = 0.4851$.) These are the diagrams with a short range of liquid–liquid immiscibility near the critical temperature of the more volatile component, and, in agreement with the statement above about the minimum value of ζ, occur in mixtures of similar chemical type but different size, e.g., $CH_4 + C_6$ alkanes, and almost any alkane + a hydrocarbon polymer, cf. Section 6.6. The distinction between types IV and V, like that between I and II, rests on the sign of Λ. The LL critical line will again be lost below the quadruple point of a real system unless the 1–2 energy is sufficiently weak, but if it is too weak then the system will belong to type III. The maximum value of Λ on the III–IV border is $1 - \sqrt{(1 - \zeta_*^2)} = 0.1256$. It follows that few real systems can be expected to belong to type IV, a result which is in accord with the facts recorded in Chapter 6.

The boundary between II and IV marks the point at which the three-phase line LLG has shrunk to zero length, and is therefore a line of tricritical points, cf. Sections 4.10 and 6.6. The phase rule forbids such points in binary systems unless there are 'accidental' symmetries or, as here, particular sets of intermolecular parameter pairs, (ζ, Λ). In practice, we cannot choose real binary systems with specified ratios of their intermolecular energies and so tricritical points are not found. However, as described in Section 6.6, it is possible to choose a pseudo-binary system of the kind $(CH_4 + C_6' + C_6'')$, where C_6' and C_6'' are two isomeric hexanes, in order to bring the system exactly on to the II–IV border.

The lines that form the II–IV and III–IV boundaries are not tangential at (ζ_*, Λ_*) but cross there to produce an extremely narrow zone (type IV*, *Figure 8.4*) which separates II from III. This zone is so narrow (typically 10^{-6} in ζ and Λ) that no examples of it are known. The changes from II to IV* and from IV* to III are marked by rearrangements of the way the critical points and critical end points are connected. There is again a line of tricritical points on the IV*–III boundary as the three-phase line shrinks to a point, but no examples of this are known.

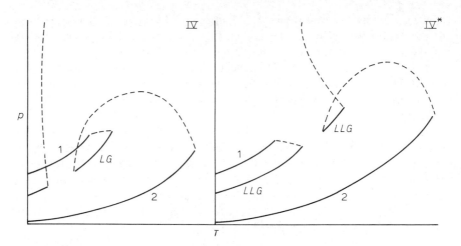

Figure 8.4 The predicted type IV*, compared with type IV of Figure 6.1

The two dashed lines in *Figure 8.3* are the boundaries of the azeotropic systems. The condition for azeotropy is $0 < x_2^{az} < 1$, where, for molecules of equal size with $a_{22} > a_{11}$,

$$x_2^{az} = \frac{a_{11} - a_{12}}{a_{11} - 2a_{12} + a_{22}} = \frac{\Lambda - \zeta}{2\Lambda} \tag{8.53}$$

Thus azeotropy occurs at $x_2 = 0$ when $a_{12} = a_{11}$, and disappears at $x_2 = 1$ when $a_{12} = a_{22}$; that is, the limits are $\zeta = \pm \Lambda$. Systems with $0 < \zeta < \Lambda$ form positive azeotropes, and those with $0 < \zeta < -\Lambda$ negative azeotropes. The azeotropy may be homogeneous or heterogeneous; if the latter then the three-phase line LLG lies either above both or below both pure vapour pressure curves in the (p, T) projections in *Figure 6.1*.

Type VI of *Figure 6.1* is not found for van der Waals systems; it is confined to highly non-spherical molecules. With this exception (to which we return later) we see that this remarkably simple equation of state can reproduce all the classes of phase behaviour discussed in Chapter 6, and indeed some new types still to be found in practice. There is a good qualitative connection between the values of ζ and Λ in each area of *Figure 8.3* and what is known or estimated of the intermolecular energies of real systems that exhibit the behaviour of types I–V. It is therefore to be expected that, with a more realistic equation of state and with more powerful theoretical methods, it should be possible to obtain quantitative fits to the experimental curves or, conversely, to make more precise deductions about intermolecular energies from the shapes of critical lines.

The more realistic equations used have included the replacement[77] of $f(n)$ of equation (8.40) by one of the more accurate representations of the equation of state of hard spheres, the Redlich–Kwong equation and others of the same family[78,79], and a multi-parameter equation[80] due to Vennix and Kobayashi[81] which represents tolerably well the equation of state of methane. For none of these has so exhaustive an exploration been made as for the van der Waals equation, but it is clear that they provide quantitatively better fits, for example, to the critical and azeotropic lines of mixtures of alkanes (including those of types IV and V), and even for those systems of type III that show gas–gas immiscibility. The numerical solution of these equations to give all the stable phase, critical and azeotropic lines is not an easy task; the most systematic of the various algorithms that have been tried is that of Hicks and Young[82].

The most accurate information on intermolecular energies can be derived from T^c as a function of composition, for systems of molecules of similar sizes and energies, which, therefore, belong to type I. Here it is possible to use analytic methods of solution. If the van der Waals or other one-fluid approximation is used, then the calculation of f_{12} of equation (8.14), and so of ξ_{12} of equation (8.5), from T^c, is made in two steps. The first is the calculation of f_x, the ratio of the critical temperature of the equivalent substance, T_x^c, to that of the reference substance; $T_x^c = f_x T_0^c$. The second is the calculation of the difference between T_x^c and T^c, the actual critical temperature of the mixture.

The first problem is readily solved for the van der Waals one-fluid approximation since f_x and h_x $(= g_x^3)$ are given by equations (8.27) and (8.28). It is convenient (and usual) to eliminate h_{12} by the Lorentz assumption (8.4). For a binary mixture

$$f_x = x_1^2 f_{11} + 2x_1 x_2 f_{12}[1 + \tfrac{1}{4}(f_{11} - f_{22})(h_{11} - h_{22})] + x_2^2 f_{22} \qquad (8.54)$$

Thus, to leading order, f_x is a quadratic function of composition but the coefficient of the cross-term $(2x_1 x_2)$ is only f_{12} to first order; it is modified by a small second-order term that is absent from the terms in x_1^2 and x_2^2.

The critical point of the equivalent substance is not, however, the same as that of the mixture, since the former is determined by the conditions of mechanical and the latter by those of material stability. The second are the more restrictive conditions, as was shown in Chapter 6. The difference between T^c and T_x^c must be calculated before any use can be made of equation (8.54). Bellemans, Mathot and Zuckerbrodt[83] have shown that this difference is zero to the first order in those of f and h parameters. However, it is not zero for many important mixtures for which second-order differences in these parameters cannot be neglected. For these, the difference has been calculated by Bellemans and Zuckerbrodt and by Byers Brown[6], whose treatment is followed here.

The two equations whose solution gives the classical critical point of the mixture are

$$\left(\frac{\partial^2 G}{\partial x^2}\right)_{p,\,T} = \frac{nRT}{x(1-x)} + \left(\frac{\partial^2 G_x}{\partial x^2}\right)_{p,\,T} = 0 \qquad (8.55)$$

$$\left(\frac{\partial^3 G}{\partial x^3}\right)_{p,\,T} = \frac{nRT(2x-1)}{x^2(1-x)^2} + \left(\frac{\partial^3 G_x}{\partial x^3}\right)_{p,\,T} = 0 \qquad (8.56)$$

where there are n moles in the system and x is the mole fraction of the second component. These equations are more useful if written in terms of $A_x(V, T)$ than of $G_x(p, T)$, since the derivatives of the latter have singularities at the critical point of the equivalent substance. The second and third derivatives of G_x can be converted to derivatives of A_x by using equations (6.4)–(6.16). The two equations above then reduce to

$$\left(\frac{\partial p_x}{\partial V}\right)^{\mathrm{c}}_{x, T} + \frac{x(1-x)}{RT}\left[\left(\frac{\partial p_x}{\partial x}\right)^2_{V, T}\right]^{\mathrm{c}} = 0 \tag{8.57}$$

$$3\left(\frac{\partial p_x}{\partial V}\right)^{\mathrm{c}}_{x, T}\left(\frac{\partial^2 p_x}{\partial V \partial x}\right)^{\mathrm{c}}_{T} - \left(\frac{\partial p_x}{\partial x}\right)^{\mathrm{c}}_{V, T}\left(\frac{\partial^2 p_x}{\partial V^2}\right)^{\mathrm{c}}_{x, T} = 0 \tag{8.58}$$

when the molecular energies and sizes are reasonably close. The differences between T^{c} and T^{c}_x and V^{c} and V^{c}_x are small, and so the derivatives of p_x with respect to volume at the critical point of the mixture are given by the Taylor expansions

$$\left(\frac{\partial p_x}{\partial V}\right)^{\mathrm{c}}_{x, T} = (T^{\mathrm{c}} - T^{\mathrm{c}}_x)\left(\frac{\partial^2 p_x}{\partial V \partial T}\right)^{\mathrm{c}}_{x} \tag{8.59}$$

$$\left(\frac{\partial^2 p_x}{\partial V^2}\right)^{\mathrm{c}}_{x, T} = (T^{\mathrm{c}} - T^{\mathrm{c}}_x)\left(\frac{\partial^3 p_x}{\partial V^2 \partial T}\right)^{\mathrm{c}}_{x} + (V^{\mathrm{c}} - V^{\mathrm{c}}_x)\left(\frac{\partial^3 p_x}{\partial V^3}\right)^{\mathrm{c}}_{x, T} \tag{8.60}$$

Hence, from equations (8.57) and (8.58),

$$T^{\mathrm{c}} - T^{\mathrm{c}}_x = -\frac{x(1-x)}{RT}\left[\frac{(\partial p_x/\partial x)^2_{V, T}}{(\partial^2 p_x/\partial V \partial T)_x}\right]^{\mathrm{c}} \tag{8.61}$$

$$V^{\mathrm{c}} - V^{\mathrm{c}}_x = (T^{\mathrm{c}} - T^{\mathrm{c}}_x)\left[\frac{3\left(\frac{\partial^2 p_x}{\partial V \partial T}\right)_x\left(\frac{\partial^2 p_x}{\partial V \partial x}\right)_T - \left(\frac{\partial p_x}{\partial x}\right)_{V, T}\left(\frac{\partial^3 p_x}{\partial V^2 \partial T}\right)_x}{\left(\frac{\partial p_x}{\partial x}\right)_{V, T}\left(\frac{\partial^3 p_x}{\partial V^3}\right)_{x, T}}\right]^{\mathrm{c}} \tag{8.62}$$

The derivative of p_x with respect to x may be eliminated by differentiating equation (8.21)

$$\left(\frac{\partial p_x}{\partial x}\right)^{\mathrm{c}}_{V, T} = \frac{f'_x}{f_x}\left[p_x - T\left(\frac{\partial p_x}{\partial T}\right)_{V, x}\right]^{\mathrm{c}} - \frac{h'_x}{h_x}p^{\mathrm{c}}_x \tag{8.63}$$

whence

$$\left(\frac{\partial^2 p_x}{\partial V \partial x}\right)^{\mathrm{c}}_{T} = -\frac{f'_x}{f_x}\left[T\left(\frac{\partial^2 p_x}{\partial V \partial T}\right)_x\right]^{\mathrm{c}} \tag{8.64}$$

where f'_x is the derivative of f_x with respect to x.

These equations may be substituted in equations (8.61) and (8.62), which then give $(T^{\mathrm{c}} - T^{\mathrm{c}}_x)$ and $(V^{\mathrm{c}} - V^{\mathrm{c}}_x)$ in terms of accessible functions of the equivalent substance. Numerical values of these differences can be calculated only by assuming an explicit equation of state for the reference, and so for the equivalent substance. The simplest choice is van der Waals's equation which leads to

$$\frac{T^c - T_x^c}{T_x^c} = x(1 - x)\left[\frac{3f_x'}{4f_x} - \frac{h_x'}{4h_x}\right]^2 \tag{8.65}$$

$$\frac{V^c - V_x^c}{V_x^c} = -2x(1 - x)\left[\left(\frac{3f_x'}{4f_x}\right)^2 - \left(\frac{h_x'}{4h_x}\right)^2\right] \tag{8.66}$$

$$\frac{p^c - p_x^c}{p_x^c} = \frac{T^c - T_x^c}{T_x^c}\left(\frac{\partial \ln p_x^c}{\partial \ln T_x^c}\right)_{V,x}^c = 4\left(\frac{T^c - T_x^c}{T_x^c}\right) \tag{8.67}^a$$

It is seen from these equations that T^c is always greater than T_x^c and p^c is greater than p_x^c. The relative difference is largest for the critical pressure and, indeed, it is a well-known fact that the critical pressures of apparently normal non-polar mixtures show large maxima (*see Figure 6.10*(a)). These are presumably the consequence of this difference between p^c and p_x^c. It is seen that the critical ratio $(pv/RT)^c$ is not the same for the mixture as for the equivalent substance and the pure components. The derivatives f_x' and h_x' ($= (g_x^3)'$) are expressed most readily in terms of the starred parameters of Byers Brown[6]. They are given very simply for a Lorentz–Berthelot mixture of molecules of equal size by

$$f_x' = -\frac{T_1^c - T_2^c}{T_0^c} \qquad h_x' = 0 \tag{8.68}$$

whence

$$\frac{T^c - T_x^c}{T_x^c} = \tfrac{9}{16}x(1 - x)\left(\frac{T_1^c - T_2^c}{T_x^c}\right)^2 \tag{8.69}$$

This difference is often small. For example, for toluene ($T^c = 594\,\mathrm{K}$) + cyclohexane ($T^c = 553\,\mathrm{K}$), whose molecules are almost the same size, the maximum value of $(T^c - T_x^c)$ is $0.4\,\mathrm{K}$. For a Lorentz–Berthelot mixture of molecules of equal energy,

$$f_x' = 0 \qquad h_x' = -\frac{V_1^c - V_2^c}{V_0^c} \tag{8.70}$$

whence

$$\frac{T^c - T_x^c}{T_x^c} = \tfrac{1}{16}x(1 - x)\left(\frac{V_1^c - V_2^c}{V_x^c}\right)^2 \tag{8.71}$$

Thus for benzene ($v^c = 260\,\mathrm{cm^3\,mol^{-1}}$) + cyclohexane ($v^c = 308\,\mathrm{cm^3\,mol^{-1}}$) the maximum value of $(T^c - T_x^c)$ is $0.2\,\mathrm{K}$. Neither of these figures is

---

[a] The logarithmic slope of the critical isochore is about 6 for the noble gases, a value appreciably larger than given by van der Waals's equation.

appreciably affected by the departures of these two mixtures from the Lorentz–Berthelot equations.

In simple mixtures, experiment shows that T^c is close to a quadratic function of x that can be represented[79,84,85]

$$T^c - x_1 T_1^c - x_2 T_2^c = x_1 x_2 t_{12}^E T_0^c \tag{8.72}$$

where T_0^c is the critical temperature of a reference substance, taken here to be the arithmetic mean of T_1^c and T_2^c. From equations (8.54) and (8.65), the parameter t_{12}^E can be related to f_{12} by

$$f_{11} - 2f_{12} - f_{22}$$
$$= -t_{12}^E + \tfrac{1}{2}(f_{11} - f_{22})(h_{11} - h_{22}) + [\tfrac{3}{4}(f_{11} - f_{22}) + \tfrac{1}{4}(h_{11} - h_{22})]^2 \tag{8.73}$$

The second and third terms on the right-hand side are of second order in the differences of energies and sizes but are rarely negligible compared with the first-order difference on the left. The second term arises from the one-fluid van der Waals approximation for f_x (8.54), and the third from the difference between T^c of a mixture and that of a hypothetical pure substance equivalent to it. The coefficients of this last term are probably inaccurate, since they are based on van der Waals's equation of state for the reference fluid—a much more drastic approximation than the use of equation (8.54). Hence measurements of critical temperatures, that is, of t_{12}^E, can be used to obtain values of f_{12} only when the second and third terms of equation (8.73) are, at worst, no larger than the first.

The critical temperature is known as a function of composition for all ten binary mixtures formed from argon, nitrogen, oxygen, carbon monoxide and methane[84], but only four of these have sufficiently small values of $(f_{11} - f_{22})$ and $(h_{11} - h_{22})$ to satisfy the criterion of the last paragraph. For these we can derive values of f_{12} which can be compared with those obtained above from g^E, etc., of the liquid mixtures at low temperatures. This comparison is made in terms of ξ_{12} in *Table 8.2*. The agreement is good for the first and third systems, for which the second-order terms in equation (8.73) are negligible, but not so good for the second, nor would it be expected to be good for the fourth, for which the terms are larger.

Table 8.2 A comparison of the energy differences ξ_{12} obtained from the critical temperatures with those from the excess free energies for four simple systems

System	Critical temperature t_{12}^E	$f_{11} - 2f_{12} + f_{22}$	ξ_{12}	Free energy ξ_{12}
$Ar + O_2$	−0.018	0.018	0.991	0.988
$Ar + CO$	−0.028	0.020	0.994	0.989
$N_2 + CO$	−0.012	0.014	0.994	0.991
$O_2 + CO$	−0.023	0.010	1.000	—

Table 8.3 A comparison of the energy differences ξ_{12} obtained from the critical temperatures with those from the excess free energies for some hydrocarbon and fluorocarbon systems[85]

System	Critical temperature ξ_{12}	Free energy ξ_{12}
$C_6H_{12} + C_6H_6$	0.987	0.979
$C_6H_{12} + C_6F_6$	0.950	0.947
$C_6H_{12} + C_6F_{12}$	0.898	0.902

A similar analysis[85] (*Table 8.3*) of the gas–liquid critical temperatures of hydrocarbon and hydrocarbon + fluorocarbon mixtures shows that ξ_{12} is significantly less than unity for mixtures of aromatic + aliphatic hydrocarbons, and even lower for mixtures of aliphatic fluorocarbons with aliphatic or aromatic hydrocarbons. These results are in qualitative accord with the evidence from g^E, h^E, etc., set out in Sections 5.5 and 5.7.

The perturbation theory of Gubbins and his colleagues has been applied also to the prediction of phase, critical and azeotropic lines in polar mixtures at high pressures[65-68]. By varying the strength of dipoles, quadrupoles and octopoles on two otherwise spherical non-polar molecules chosen to represent Ar and Kr, they were able to generate the first five types of critical behaviour of *Figure 6.1*. Little of this work was designed to represent the behaviour of real systems—it was mainly topological exploration, similar to that carried out by Scott and van Konynenburg, but with a more sophisticated theory. However, they were able to show reasonable quantitative fits to Xe + HCl and $CH_4 + CF_4$, both type II systems. A more ambitious but only qualitatively successful attempt was made by Gibbs to fit the critical lines of mixtures of ethane and methanol, and similar systems[71].

This work shows that polarity alone, whether dipolar, quadrupolar or octopolar, never leads to type VI behaviour. Calculations based on lattice models[86], and the fact that this behaviour is found in practice only in strongly hydrogen-bonded systems, suggested that unusually strong or specific orientational forces are needed to turn type II into type VI; that is, to induce mixing to occur at the lowest temperature. A model with these properties has been found by Clancy and Gubbins[87] by combining polar forces with strong shape dependence of the repulsive core.

The final point to be made in this discussion of critical lines is that all the theories described above lead to classical behaviour, and so fail to yield the correct set of critical exponents described in Section 4.9. Most of the theories are based on an explicitly classical equation of state (e.g., van der Waals), but even those that use for their reference substance the true equation of a real one-component fluid, with the correct singularities, still lead to classical critical points for the mixture. The reason for this is the distinction between the critical point of the equivalent substance, at (V_x^c, T_x^c), and that of the actual mixture at (V^c, T^c). The singularities of the reference equation are transferred correctly to singularities of the one-component reference system at (V_x^c, T_x^c). However, as *Figure 6.2* makes clear, this point lies in an unstable region of the phase diagram of the real mixture, and so no transfer of the correct singularities to (V^c, T^c) is achieved. To do this we must, as Leung and Griffiths[88] first showed, re-formulate the equation of state of the reference substance in terms of field variables (*see* Section 4.7), using, for example, $\lambda_1/(\lambda_1 + \lambda_2)$, as the composition variable to replace the mole fraction, which is a density. The transcription from a one-component system to a two-component system in a set of field variables preserves the correct singularities and leads to a proper description of the critical point of the binary mixture. They applied this technique first to $^3He + {}^4He$, and, more recently, Moldover and Gallagher[89] have used it to make a particularly impressive fit to the phase, critical and azeotropic lines of the system[90] $SF_6 + C_3H_8$. This technique could, with advantage, be applied to other systems.

8.10 Conclusions

The theory of the equilibrium properties of liquid mixtures, like that of pure liquids, is now an essentially solved problem. By this assertion we mean that for any arbitrarily chosen set of intermolecular potentials it is now possible to calculate the thermodynamic properties, with adequate accuracy, by using one or more of the statistical theories described above. Of these, perturbation theory has proved to be the best, although its use requires a lot of work for each separate application. Naturally, like all generalizations, the assertion is not proof against minor criticisms in particular instances. Even for the simplest systems, the accuracy of the calculation of excess functions could be improved; highly polar systems raise questions about the adequacy of the convergence of the multipolar expansion; and, perhaps most seriously, flexible chain molecules present problems which are not yet satisfactorily solved, even in principle. *A fortiori*, the equilibrium theory of polymer solutions, which is outside the scope of this book, is still without a rigorous basis in statistical mechanics. Nevertheless, we believe that the advances in statistical theory needed for mixtures of small molecules are mainly matters of detail.

We do not assert, however, that it follows that these theories can yet be applied satisfactorily to interpret the experimental properties of more than a small number of the systems discussed in this book. The missing step is an accurate knowledge of intermolecular potentials, both two-body and multibody. Until we know much more about these, particularly for polar and polarizable molecules, then the recent advances in statistical theory cannot be used to their full potential. Moreover, the trials and errors of the last 20 years have shown that the study of liquids has contributed little precise information to our knowledge of intermolecular energies. For that we have had to look to the properties of gases (and to a lesser extent solids) and to more recent spectroscopic and beam-scattering techniques. This trend is likely to continue. It could be argued that the study of the thermodynamic properties of liquids and liquid mixtures has been too successful. Both experiment and statistical theory have reached so high a degree of accuracy and sophistication that they have put too great a strain on the weak link between them—our knowledge of intermolecular energies. It is improvement in this area that is needed next.

References

1 MAITLAND, G., RIGBY, M., SMITH, E. B., and WAKEHAM, W. A., *Intermolecular Forces*, Chap. 8 and 9, Clarendon Press, Oxford (1981)
2 LORENTZ, H. A., *Annln Phys.*, **12**, 127 (1881)
3 BERTHELOT, D., *C. r. hebd. Séanc. Acad. Sci. Paris*, **126**, 1703, 1857 (1898)
4 PRIGOGINE, I., BELLEMANS, A., and ENGLERT-CHWOLES, A., *J. chem. Phys.*, **24**, 518 (1956); PRIGOGINE, I., *The Molecular Theory of Solutions*, North-Holland, Amsterdam (1957)
5 SCOTT, R. L., *J. chem. Phys.*, **25**, 193 (1956)
6 BYERS BROWN, W., *Phil. Trans.*, **A250**, 175, 221 (1957); CHAUNDY, T. W., and McLEOD, J. B., *Q. Jl Math.*, **9**, 202 (1958)
7 SALSBURG, Z. W., WOJTOWICZ, P. J., and KIRKWOOD, J. G., *J. chem. Phys.*, **26**, 1533 (1957); **27**, 505 (1957)
8 HENDERSON, D., and LEONARD, P. J., in *Physical Chemistry*, Ed. Henderson, D., Vol. 8B, Chap. 7, Academic Press, New York (1971)
9 SMITH, E. B., and LEA, K. R., *Trans. Faraday Soc.*, **59**, 1535 (1963)
10 ROTENBERG, A., *J. chem. Phys.*, **43**, 4377 (1965)
11 ALDER, B. J., *J. chem. Phys.*, **40**, 2724 (1964)
12 MELNYK, T. W., and SAWFORD, B. L., *Molec. Phys.*, **29**, 891 (1975)
13 WIDOM, B., and ROWLINSON, J. S., *J. chem. Phys.*, **52**, 1670 (1970); ROWLINSON, J. S., *Adv. chem. Phys.*, **41**, 1 (1980)
14 LEBOWITZ, J. L., *Phys. Rev.*, **133**, A895 (1964); BAXTER, R. J., *J. chem. Phys.*, **52**, 4559 (1970); LEONARD, P. J., HENDERSON, D., and BARKER, J. A., *Molec. Phys.*, **21**, 107 (1971)
15 LEBOWITZ, J. L., and ROWLINSON, J. S., *J. chem. Phys.*, **41**, 133 (1964)
16 LELAND, T. W., ROWLINSON, J. S., and SATHER, G. A., *Trans. Faraday Soc.*, **64**, 1447 (1968)
17 THROOP, G. J., and BEARMAN, R. J., *J. chem. Phys.*, **44**, 1423 (1966); **47**, 3036 (1967); GRUNDKE, E. W., HENDERSON, D., and MURPHY, R. D., *Can. J. Phys.*, **49**, 1593 (1971); **51**, 1216 (1973)
18 VAN DER WAALS, J. D., *Z. phys. Chem.*, **5**, 133 (1890); *Die Kontinuität des gasförmigen und flüssigen Zustandes*, Vol. 2, Chap. 1, Barth, Leipzig (1900)
19 REID, R. C., and LELAND, T. W., *A.I.Ch.E.Jl.*, **11**, 228 (1965); **12**, 1277 (1966); LEACH, J. W., CHAPPELEAR, P. S., and LELAND, T. W., *A.I.Ch.E.Jl.*, **14**, 568 (1968)
20 RICE, S. A., *J. chem. Phys.*, **24**, 357 (1956); **29**, 141 (1958)
21 LELAND, T. W., ROWLINSON, J. S., SATHER, G. A., and WATSON, I. D., *Trans. Faraday Soc.*, **65**, 2034 (1969)
22 ROWLINSON, J. S., *Discuss. Faraday Soc.*, **49**, 30 (1970)
23 McDONALD, I. R., in *Statistical Mechanics, Specialist Periodical Reports*, Ed. Singer, K., Vol. 1, Chap. 3, Chemical Society, London (1973)
24 HICKS, C. P., *J. chem. Soc., Faraday Trans. 2*, **72**, 423 (1976)
25 McGLASHAN, M. L., *Trans. Faraday Soc.*, **66**, 18 (1970); MARSH, K. N., McGLASHAN, M. L., and WARR, C., *ibid.*, 2453
26 SNIDER, N. S., and HERRINGTON, T. M., *J. chem. Phys.*, **47**, 2248 (1967)
27 GUGGENHEIM, E. A., *Molec. Phys.*, **9**, 199 (1965)
28 FLORY, P. J., *J. Am. chem. Soc.*, **87**, 1833 (1965); ABE, A., and FLORY, P. J., *ibid.*, 1838;

HÖCKER, H., and FLORY, P. J., *Trans. Faraday Soc.*, **64**, 1188 (1968); FLORY, P. J., *Discuss. Faraday Soc.*, **49**, 7 (1970)

29 LONGUET-HIGGINS, H. C., *Proc. R. Soc.*, **A205**, 247 (1951); BYERS BROWN, W., and LONGUET-HIGGINS, H. C., *Proc. R. Soc.*, **A209**, 416 (1951); BYERS BROWN, W., *Proc. R. Soc.*, **A240**, 561 (1957); BARKER, J. A., *Proc. R. Soc.*, **A241**, 547 (1957); BUFF, F. P., and SCHINDLER, F. M., *J. chem. Phys.*, **29**, 1075 (1958); SMITH, W. R., *Can. J. chem. Engng*, **50**, 271 (1972); MANSOORI, G. A., and LELAND, T. W., *J. chem. Soc., Faraday Trans. 2*, **68**, 320 (1972)

30 HENDERSON, D., and BARKER, J. A., *J. chem. Phys.*, **49**, 3377 (1968); SMITH, W. R., *Molec Phys.*, **21**, 105 (1971); SMITH, W. R., and HENDERSON, D., *Molec. Phys.*, **24**, 773 (1972)

31 LEONARD, P. J., HENDERSON, D., and BARKER, J. A., *Trans. Faraday Soc.*, **66**, 2439 (1970)

32 GRUNDKE, E. W., HENDERSON, D., BARKER, J. A., and LEONARD, P. J., *Molec. Phys.*, **25**, 883 (1973)

33 MANSOORI, G. A., CARNAHAN, N. F., STARLING, K. E., and LELAND, T. W., *J. chem. Phys.*, **54**, 1523 (1971)

34 ROGERS, B. L., and PRAUSNITZ, J. M., *Trans. Faraday Soc.*, **67**, 3474 (1971); GIBBONS, R. M., *J. chem. Soc., Faraday Trans. 2*, **71**, 346, 353 (1975); PROCHAZKA, K., *Coll. Czech. chem. Commun.*, **41**, 1273 (1976)

35 LEE, L. L., and LEVESQUE, D., *Molec. Phys.*, **26**, 1351 (1973); BOUBLIK, T., *Coll. Czech. chem. Commun.*, **38**, 3694 (1973)

36 VERLET, L., and WEIS, J.-J., *Phys. Rev.*, **A5**, 939 (1972)

37 GIRARDEAU, M. D., and MAZO, R. M., *Adv. chem. Phys.*, **24**, 187 (1973)

38 MÜNSTER, A., *Statistical Thermodynamics*, Vol. 2, 739, Springer, Berlin (1974); HANSEN, J. P., and McDONALD, I. R., *Theory of Simple Liquids*, 148, Academic Press, London (1976)

39 MANSOORI, G. A., and LELAND, T. W., *J. chem. Phys.*, **53**, 1931 (1970); MANSOORI, G. A., *J. chem. Phys.*, **56**, 5335 (1972)

40 McDONALD, I. R., *Molec. Phys.*, **23**, 41 (1972); **24**, 391 (1972); FIORESE, G., and PITTION-ROSSILLON, G., *Chem. Phys. Lett.*, **77**, 562 (1981)

41 SINGER, J. V. L., and SINGER, K., *Molec. Phys.*, **24**, 357 (1972)

42 TORRIE, G. M., and VALLEAU, J. C., *J. chem. Phys.*, **66**, 1402 (1977)

43 MO, K. C., GUBBINS, K. E., JACUCCI, G., and McDONALD, I. R., *Molec. Phys.*, **27**, 1173 (1974); **29**, 324 (1975)

44 EWING, M. B., and MARSH, K. N., *J. chem. Thermodyn.*, **9**, 863 (1977)

45 DOUSLIN, D. R., HARRISON, R. H., and MOORE, R. T., *J. phys. Chem.*, **71**, 3477 (1967)

46 MARSH, K. N., *J. chem. Thermodyn.*, **3**, 355 (1971); EWING, M. B., and MARSH, K. N., *J. chem. Thermodyn.*, **9**, 357 (1977); **10**, 267 (1978); TOMLINS, R. P., and MARSH, K. N., *J. chem. Thermodyn.*, **9**, 651 (1977)

47 GIBBONS, R. M., *Molec. Phys.*, **17**, 81 (1969); **18**, 809 (1970)

48 PRIGOGINE, I., TRAPPENIERS, N., and MATHOT, V., *Discuss. Faraday Soc.*, **15**, 93 (1953); PRIGOGINE, I., *The Molecular Theory of Solutions*, North-Holland, Amsterdam (1957)

49 PATTERSON, D., BHATTACHARYYA, S. N., and PICKER, P., *Trans. Faraday Soc.*, **64**, 648 (1968); PATTERSON, D., TEWARI, Y. B., and SCHREIBER, H. P., *J. chem. Soc., Faraday Trans. 2*, **68**, 885 (1972); LAM, V. T., PICKER, P., PATTERSON, D., and TANCRÈDE, P., *J. chem. Soc., Faraday Trans. 2*, **70**, 1465 (1974); CROUCHER, M. D., and PATTERSON, D., *ibid.*, 1479

50 DELMAS, G., and PATTERSON, D., *Discuss. Faraday Soc.*, **49**, 98, 173 (1970)

51 HIJMANS, J., *Physica*, **27**, 433 (1961); HOLLEMAN, TH., and HIJMANS, J., *Physica*, **28**, 604 (1962); **31**, 64 (1965); HOLLEMAN, TH., *Physica*, **29**, 585 (1963); **31**, 49 (1965)

52 BRØNSTED, J. N., and KOEFOED, J., *K. danske Vidensk. Selsk. (Mat. Fys. Skr.)*, **22**, No. 17 (1946); LONGUET-HIGGINS, H. C., *Discuss. Faraday Soc.*, **15**, 73 (1953)

53 AVIRAM, I., and TILDESLEY, D. J., *Molec. Phys.*, **35**, 365 (1978)

54 MONSON, P. A., and RIGBY, M., *Molec. Phys.*, **39**, 977 (1980)

55 BOUBLIK, T., and NEZBEDA, I., *Czech. J. Phys.*, **B30**, 121, 953 (1980)

56 NEZBEDA, I., and LELAND, T. W., *J. chem. Soc., Faraday Trans. 2*, **75**, 193 (1979)

57 PAVLICEK, J., NEZBEDA, I., and BOUBLIK, T., *Czech. J. Phys.*, **B29**, 1061 (1979); BOUBLIK, T., *Molec. Phys.*, **42**, 209 (1981)

58 BARKER, J. A., *J. chem. Phys.*, **19**, 1430 (1951); POPLE, J. A., *Discuss. Faraday Soc.*, **15**, 35 (1953); ROWLINSON, J. S., and SUTTON, J. R., *Proc. R. Soc.*, **A229**, 271, 396 (1955)

59 REID, R. C., and LELAND, T. W., *A.I.Ch.E.Jl.*, **11**, 228 (1965); **12**, 1277 (1966); ROWLINSON, J. S., and WATSON, I. D., *Chem. Engng Sci.*, **24**, 1565, 1575 (1969); MOLLERUP, J., and FREDENSLUND, A., *Chem. Engng Sci.*, **28**, 1295 (1973); MOLLERUP, J., and ROWLINSON, J. S., *Chem. Engng Sci.*, **29**, 1373 (1974)
60 STOGRYN, D. E., and STOGRYN, A. P., *Molec. Phys.*, **11**, 371 (1966)
61 MELNYK, T. W., and SMITH, W. R., *Chem. Phys. Lett.*, **28**, 213 (1974)
62 CHAMBERS, M. V., and McDONALD, I. R., *Molec. Phys.*, **29**, 1053 (1975)
63 TWU, C. H., GUBBINS, K. E., and GRAY, C. G., *Molec. Phys.*, **29**, 713 (1975)
64 FLYTZANI-STEPHANOLOULOS, M., GUBBINS, K. E., and GRAY, C. G., *Molec. Phys.*, **30**, 1649 (1975)
65 TWU, C. H., GUBBINS, K. E., and GRAY, C. G., *J. chem. Phys.*, **64**, 5186 (1976)
66 GRAY, C. G., GUBBINS, K. E., and TWU, C. H., *J. chem. Phys.*, **69**, 182 (1978)
67 GUBBINS, K. E., and TWU, C. H., *Chem. Engng Sci.*, **33**, 863, 879 (1978); NICOLAS, J. J., GUBBINS, K. E., STREETT, W. B., and TILDESLEY, D. J., *Molec. Phys.*, **37**, 1429 (1979)
68 CLANCY, P., GUBBINS, K. E., and GRAY, C. G., *Chem. Soc., Faraday Discuss.*, **66**, 116 (1978); CALADO, J. C. G., GRAY, C. G., GUBBINS, K. E., PALAVRA, A. M. F., SOARES, V. A. M., STAVELEY, L. A. K., and TWU, C. H., *J. chem. Soc., Faraday Trans. 1*, **74**, 893 (1978); LOBO, L. Q., STAVELEY, L. A. K., CLANCY, P., and GUBBINS, K. E., *J. chem. Soc., Faraday Trans. 1*, **76**, 174 (1980); MACHADO, J. R. S., GUBBINS, K. E., LOBO, L. Q., and STAVELEY, L. A. K., *ibid.*, 2496; LOBO, L. Q., McCLURE, D. W., STAVELEY, L. A. K., CLANCY, P., and GUBBINS, K. E., *J. chem. Soc., Faraday Trans. 2*, **77**, 425 (1981)
69 SEVCIK, M., and BOUBLIK, T., *Coll. Czech. chem. Commun.*, **44**, 3541 (1979)
70 SHUKLA, K. P., and SINGH, Y., *J. chem. Phys.*, **72**, 2719 (1980)
71 GIBBS, G. M., *Chemical Thermodynamics Data on Fluids and Fluid Mixtures*, NPL Conference, 149, IPC Scientific and Technical Press, Guildford (1979); *Thesis*, Univ. Oxford (1979)
72 GUBBINS, K. E., GRAY, C. G., and MACHADO, J. R. S., *Molec. Phys.*, **42**, 817, 843 (1981)
73 SANDLER, S. I., *Molec. Phys.*, **28**, 1207 (1974); VALDERRAMA, J. O., SANDLER, S. I., and FLIGNER, M., *Molec. Phys.*, **42**, 1041 (1981)
74 SCOTT, R. L., and VAN KONYNENBURG, P. H., *Discuss. Faraday Soc.*, **49**, 87 (1970); VAN KONYNENBURG, P. H., and SCOTT, R. L., *Phil. Trans*, **A298**, 495 (1980)
75 FURMAN, D., and GRIFFITHS, R. B., *Phys. Rev.*, **A17**, 1139 (1978)
76 FURMAN, D., DATTAGUPTA, S., and GRIFFITHS, R. B., *Phys. Rev.*, **B15**, 441 (1977)
77 HURLE, R. L., JONES, F., and YOUNG, C. L., *J. chem. Soc., Faraday Trans. 2*, **73**, 613 (1977); HURLE, R. I., TOCZYLKIN, L., and YOUNG, C. L., *ibid.*, 618; McGLASHAN, M. L., STEAD, K., and WARR, C., *ibid.*, 1889
78 SPEAR, R. R., ROBINSON, R. L., and CHAO, K. C., *Ind. & Eng. Chem. Fundam.*, **8**, 2 (1969); HISSONG, D. W., and KAY, W. B., *A.I.Ch.E.Jl.*, **16**, 580 (1970); PAK, S. C., and KAY, W. B., *Ind. & Eng. Chem. Fundam.*, **11**, 255 (1972); **13**, 298 (1974)
79 HICKS, C. P., and YOUNG, C. L., *J. chem. Soc., Faraday Trans. 1*, **72**, 122 (1976)
80 TEJA, A. S., and ROWLINSON, J. S., *Chem. Engng Sci.*, **28**, 529 (1973); TEJA, A. S., and KROPHOLLER, H. W., *Chem. Engng Sci.*, **30**, 435 (1975); TEJA, A. S., *Chem. Engng Sci.*, **33**, 623 (1978)
81 VENNIX, A. J., and KOBAYASHI, R., *A.I.Ch.E.Jl.*, **15**, 930 (1969)
82 HICKS, C. P., and YOUNG, C. L., *J. chem. Soc., Faraday Trans. 2*, **73**, 597 (1977)
83 BELLEMANS, A., MATHOT, V., and ZUCKERBRODT, P., *Bull. Acad. r. Belg. Cl. Sci.*, **42**, 631, 643 (1956)
84 PARTINGTON, E. J., ROWLINSON, J. S., and WESTON, J. F., *Trans. Faraday Soc.*, **56**, 479 (1960); JONES, I. W., and ROWLINSON, J. S., *Trans. Faraday Soc.*, **59**, 1702 (1963)
85 HICKS, C. P., and YOUNG, C. L., *Trans. Faraday Soc.*, **67**, 1598, 1605 (1971); *Chem. Rev.*, **75**, 119 (1975)
86 BARKER, J. A., and FOCK, W., *Discuss. Faraday Soc.*, **15**, 188 (1953); WHEELER, J. C., and ANDERSEN, G. R., *J. chem. Phys.*, **73**, 5778 (1980)
87 CLANCY, P., and GUBBINS, K. E., *Molec. Phys.*, **44**, 581 (1981)
88 LEUNG, S. S., and GRIFFITHS, R. B., *Phys. Rev.*, **A8**, 1670 (1973)
89 MOLDOVER, M. R., and GALLAGHER, J. S., in *Phase Equilibrium and Fluid Properties in the Chemical Industry*, Eds. Storvick, T. S., and Sandler, S. I., 498, American Chemical Society, Washington DC (1977)
90 CLEGG, H. P., and ROWLINSON, J. S., *Trans. Faraday Soc.*, **51**, 1333 (1955)

General index

Index of systems

Systems are of one, two, three or four components, and are in the alphabetical order of the group of chemical elements that denote the first component. There are no cross-references. Components are named if they have isomers. Unnamed paraffins and perfluoroparaffins are the normal isomers.

The alphabetical order is subject to three conventions:

(a) Inorganic components precede organic.
(b) Carbon is placed first in organic components and is followed by hydrogen (if present).
(c) Substances containing one carbon atom, etc., precede those with two, etc..